A Competitive Book of
AGRICULTURE

For

ARS/SRF/JRF, UPTGT, UPPGT, AGTA, Pre PG & Ph.D. Entrance, UPSC, PSCs Examinations and Interviews of all Agriculture Services

Dr. Nem Raj Sunda,
Ph. D (Entomology)

Published By

SURAHEE PUBLICATIONS

JAIPUR, Rajasthan

First Edition :2011
Sixteenth Edition:March, 2026

Publication & Distributers

Surahee Publications, Bobas,
Jaipur-303338 (Rajasthan)

Contacts for supply:

Dr. P D Chaudhary +91-9784243509
Mr R N Rundla +91-9782087701
Plot no 132, Shyampuri, Near Paniki Tanki, Kanta Chouraha, Jhotwara, Jaipur, Rajasthan-302012
Email.-prabhu.surahee@gmail.com

ISBN 978-81-937885-6-1

Price: ₹ **395/- (Three Hundred and Ninety Five only)**

Preface

Agriculture is a broad subject. It covers all the basic subjects of science. Therefore students have to read all the agriculture related subjects with elementary courses during under graduation. After passing graduation, higher degrees are done in specialized field of agriculture so there is no need to read all agriculture subjects in M. Sc. and Ph.D. But for the preparation of various agriculture competition examinations students have to read all the basic books of agriculture to cover syllabus. That time most of students don't have all the necessary books and too much time to read them.

Therefore to overcome these problems the author tried his best to write this book through reading various books and other sources of agriculture to covers brief and best information of all subjects in one book.

To compiling most of subjects of agriculture in one book is very hard task so some mistake may be left, therefore your suggestions are invited to further improvement of this book. But author would like to say that students have to treat this book as a competitive book for various competitions, not a text book for degree courses.

I heartily thankful to, Dr. R. P. Ghasolia, Associate Prof., Pl. Pathology, Dr. O. P. Garhwal, Associate Prof., Horticulture and Mr. C. R. Hakla, Agriculture Officer, DOA, Rajasthan for their valuable suggestions and correction made by them in their respective subjects.

I hope this book would be very useful for students who appear in various agricultural competition examinations.

Nem Raj Sunda

CONTENTS

Unit- 1.
HISTORICAL MOVEMENT OF AGRICULTURE

Important events in ancient history of agriculture

Years ago	Event
40 million	Evolution of monkey and apes
1.5 million	First true man, ***Homo erectus*** emerged in Indonesia and Africa
0.5 million	*Homo erectus pekinensis* learnt use and control of fire
0.4 million	Man-made shelters out of tree branches
32,000	Cro-Magnon man, ***Homo sapiens sapiens*** emerged in Europe
10,000 B.C.	First pelt animal of human was **Dog** for Hunting in Iran & Israel
8700 B.C.	Domestication of **sheep**
7700 B.C.	Domestication of **goat**
7500 B.C.	Cultivation of **Wheat and barley**
6000 B.C.	Domestication of **cattle and pigs**
4400 B.C.	Cultivation of **maize**
4000 B.C.	Cultivation of **rice**
3500 B.C.	Cultivation of **potato**
3400 B.C.	**Wheel** was invented
3000 B.C.	**Bronze** was used to make tools
2900 B.C.	**Plough** was invented & **Irrigated farming** was started.
2700 B.C.	Domestication of **silk moth in China.**
2300 B.C.	Cultivation of **chick pea, mustard and cotton**
2200 B.C.	Domestication of **Fowl, buffalo and elephant**
1800 B.C.	Cultivation of **finger millet** (*Ragi)*
1725 B.C.	Cultivation of **sorghum**
1500 B.C.	Cultivation of **sugarcane** & Irrigation through **wells** was started
1400 B.C.	Use of **Iron** in Middle East
1000-600 B.C.	Age of iron, **iron-plough share** and **axe** invented, crop cultivation in Punjab, Uttar Pradesh and Bihar
543-491 B.C.	Reference to farming operations in **Kullavagga** and **Mahavagga** and **Budhist literature in Pali**
760-800 A.D.	**Persian wheel** used Araghatta (India)
900-1000	Construction of anicuts and tanks, ***Krishi Parashara*** and

A.D.	*Vrikshayurveda*, manuals on agricultureand botany written, A text book on Agriculture was prepared by Kashyapa by the name, ***Krishi Sukti***
15 Century A.D.	Cultivation of **Sweet orange**, sour orange, **Pomegranate**
16 Century A.D.	Introduction of several crops in India by Portuguese. They are **Potato, Sweet Potato, Arrow Root, Cassava, Tomato, Chillies, Pumpkin, Papaya, Pineapple, Guava, Custardapple, Groundnut, Cashewnut, Tobacco, American cotton, rubber**
1550 A.D.	Portugese introduced **grafting technique** in horticulture
1795 A.D.	Flora India, published**, Lieutenant William Frazer** started study farm at PUSA (Bihar) to breed horses
1820 A.D.	G.L. Blane constructed Western **Jamuna Canal**
1824 A.D.	Journal of Agriculture and Horticulture were published
1845 A.D.	Occurrence of **Iris Famine** due to (late blight of potato)
1865 A.D.	Famine in Bengal and Orissa (15 lakh people died)

Historical events of Modern Agriculture	
1871	Joint Department of Revenue, Agriculture and Commerce (DRAC) was established.
1874	Occurrence of severe famine in Bihar
1875	India Meteorological Department (**IMD**) was established with its HQ at **Culcutta.** The head quarter was shifted to Shimla in 1905, to Pune in 1928 and later to **Delhi in 1944**.
1876	Famine in Madras and Bombay presidencies
1877	Famine in Punjab, Central province (10 lacs people died)
1880	**Famine commission** was appointed.
1882	Veterinary college established at Lahore (now in Pakistan)
1891	Dr. J.A. Volcker's report on improvement of Indian Agriculture
1899-1900	Severe Famine occured (due to Vikram samvat 1956 this was known as **Chhapaniya Kal**)
1906	Established Forest Research Institute (FRI) at Dehradun
1901-1905	Agricultural Colleges were established at Pune, Kanpur, Sabour, Nagpur, Lyallapur and Coimbatore
1901	Appointed First Irrigation Commission.
1901	First entomologist (Nionel de Nicevelli) was appointed
1903	Model Village (Rural Reconstruction) Scheme was launched by Daniel Hamilton

1903	H. M. Lefroy was appointed as second Entomologist
1905	Established the Imperial (now Indian) Agricultural Research Institute (IARI) at Pusa, (Bihar), but after severe earth quake and flood (in 1934) it was shifted to New Delhi in 1936.
1919	Constitutional Reforms made Agriculture as a state subject
1920	**GURGAON** Project by F L Brayne
1921	**SHRINIKETAN** Project by R.N:Tagore
1921	**MARTHANDAM** Project by Spencer Hatch
1921	**SEVAGRAM** Project by Mahatma Gandhi. Adarsh Seva Sangh, Pohri (Gwalior) by Col. Shitole
1921	Constituted, Indian Central Cotton Committee (ICCC).
1926	Appointment of Royal Commission on Agriculture (RCA) headed by Lord Linlithgow
1928	RCA submitted its report
1929	Establishment of Imperial (now Indian) Council of Agricultural Research (ICAR), New Delhi
1931	Indian Central Lac Committee
1936	Indian Central Jute committee.
1940	Monthly magazine "**Indian Farming**" was started by Imperial (Indian) Council of Agricultural Research (ICAR), New Delhi.
1942	Department of Food was created.
1942	Grow more food campaign (Failed and again started in 1947)
1943	Occurred **Great Bengal Famine** (Caused by ***Helminthosporium oryzae***)
1944	Indian Central Sugarcane Committee.
1945	Indian Central Coconut Committee.
1945	Indian Central Tobacco Committee.
1945	Indian Village Service Established by A.T. Mosher, NewYork and B.N. Gupta.
1946	Directorate of plant protection, quarantine & storage (DPPQ&S), Faridabad (HR)
1946	Central Rice Research Institute (CRRI), Cuttack (Orissa).
1946- 1947	Firka Development of Madras State started by T.Prakashan and D. Tharugad
1947	Indian Central Oil Seeds Committee.
Jan.1948	GATT (General Agreement on Tariffs &Trade)
1948	Etawah Pilot Project by Albert Mayer
1949	Arecanut Committee

2Oct.1952	Community Development Programme (**CDP**)
2Oct.1953	Started National Extension Services (**NES**)
1956	Project for intensification of regional research on cotton, oil, seeds, millets (**PIRRCOM**)
1957	All India Coordinated Maize Improvement Project (**AICMIP**)
1958	Status of Deemed University accorded to IARI under UGC Act, 1956.
Oct. 1958	**NAFED** (National Agricultural Cooperative Marketing Federation)
2nd Oct.1959	**Panchayati Raj** (Democratic decentralization), 3-tier system on the recommendation of Balwant Roy Mehta, inaugurated by J.L.Nehru on 2nd Oct.1959 in Nagaur District of Rajasthan.
1960	**First agricultural University** in India, G.B Pant University of Agriculture & Technology at Pantnagar **(GBUAT)**, on the pattern of Land-Grant system of USA.
1960-61	Intensive Agriculture District Programme (IADP). It is known as **'Package Programme'**.
1963	National Seed Corporation (NSC).
1964	Intensive Cattle Development Project (ICDP)
1965	National Demonstration Programme (NDP)
1965	Nilokheri project (Mazdoor Manzil) by Shri S.K. Dey
1965	**First DG** or Vice President of ICAR, Dr. B.P. Pal
1966	Initiated 'High Yielding Varieties (**HYV's**) Programme'
1966	Established **'Directorate of Extension'**
1966-67	This year regarded as 'Green Revolution year'
1967-68	Farmers Training and Education Centre (**FTEC**)
1969	Second Irrigation Commission.
1970	National Commission on Agriculture.
1971	Minikit Programme (MKP)
1971	All India coordinated research project for dry land agriculture.
1971	Small Farmers Development Agency (SFDA)
1971	Marginal Farmers and Agricultural Labor Project (MFALP)
1973	Creation of Department of Agricultural Research and Education (DARE) in the Ministry of Agriculture
1973	Started **'Agricultural Research Service'** (ARS)
1st Nov, 1973	Establishment of Agricultural Scientists Recruitment Board (ASRB) based on recommendation of G. Gadkar report.
1973	**'Project Tiger'** started

1973	Started **'Minikit trials Programme'**
1973-74	Drought Prone Area Programme (**DPAP**)
1974	Training & Visit system (**T&V**) formulated by Daniel Benor and Baxter of Israel. First started in Canal Area of Rajasthan and Chambal area in Madhya Pradesh.
1974	Command Area Development Authority (**CADA**)
1974	Minimum Needs Programme (**MNP**)
1974	Krishi Vigyan Kendra (**KVK**). Mohan Singh Mehta committee recommended (First KVK at Pondicherry, under TNAU, Coimbatore).
1975	Started **'20 Points Programme'**
1976	**Started, AntyodyaYojana**
1976	National Academy of Agricultural Research and Management (**NAARM**), Hyderabad
1976	**Ashok Mehta Committee** proposed two tiers Panchayat Raj System
1977-78	Desert Development Programme (**DDP**)
April, 1978	Integrated Rural Development Programme (IRDP) launched in financial year 1978 and extended throughout India by 1980. It is known as **Poverty Eradication Programme.**
1979	Training for Rural Youth for Self Employment (**TRYSEM**)
Jan. 1979	National Agricultural Research Project (**NARP**)
1st June, 1979	Started 'Lab to Land Programme' (**LLP**), on the occasion of 'golden jubilee year of ICAR'
1980	National Rural Employment Programme (**NREP**)
1982	Development of Women and Children in Rural Areas (**DWCRA**).
1982	Council for Advancement of Peoples Action and Rural Technology (**CAPART**).
12th July, 1982	National Bank for Agriculture & Rural development (**NABARD**)
1983	National Agricultural Extension Project (**NAEP**)
1983	Rural Landless Employment Guarantee Programme (**RLEGP**)
1985	Crop Insurance Scheme (**CIS**) now National Agricultural Insurance Scheme (**NAIS**)
1985	Indira Awaas Yojana (**IAY**) launched.
1986	Technology Mission on Oilseeds (**TMO**)
1986	National Watershed Development Programme for Rainfed

	Areas
1986	National Agricultural Research Project (Phase-II)
1989	Integrated Watershed Development Programme (**IWDP**)
1994	Mahila Samrudhi Yojana
1995	ICAR's Institutional Village Linkage Programme (**IVLP**)
1996	Establishment of National **Gene Bank** at New Delhi
1998	ICAR's National Agricultural Technology Project (**NATP**), funded by World Bank.
1998	Kisan Credit Card Scheme started
1999	Jawahar Gram Samridhi Yojna (**JGSY**)
1999	Swarana Jayanti Gram Swarozgar Yojana (SGSY) (IRDP, TRYSEM, DWACRA, SITRA, GKY& MWS- merged in to SGSY). It is renamed as **'National Rural Livehood Mission'** (NRLM), 2011
Rabi 1999	National Agricultural Insurance Scheme (**NAIS**)
Dec.2000	Antyodaya Anna Yojana (**AAY**)
Dec. 2000	PradhanMantri Gram SadakYojana (**PGSY**)
April, 2004	Integrated Scheme of Oilseeds, Pulses, Oil palm and Maize (ISOPOM) being implemented from 1.4. 2004. Later (2014-15) **ISOPOM** merged into new scheme National Mission on Oilseeds and Oil Palm (**NMOOP**). Now **NMEO-OS**
2004	National Food for Work Programme (**NFWP**)
2004	Inauguration of National Agricultural Science Complex (**NASC**) at Now Delhi.
2004	Marine Fishing Policy
2004	**National Commission on Farmers** constituted on Nov., 2004 under the chairmanship of Professor M.S. Swaminathan.
May, 2005	Started **'National Horticulture Mission'** (NHM)
2005-06	**Bharat Nirman** is an Indian business plan for creating and augmenting basic rural infrastructure including irrigation, projects on irrigation roads (pradhan mantri gram sadak yojana), housing (indira awaas yojana), water supply and electricity supply.
2006-07	National Bamboo Mission (**NBM**) started
2006	ICAR Launched National Agricultural Innovation Project (**NAIP**) with Financial support of 'World Bank'.
2006	Mahatma Gandhi National Rural Employment Guarantee Act. (MGNREGA).Now replaced by the **Viksit Bharat-Guarantee**

	for Rozgar and Ajeevika Mission (**VB-G RAM G**), Act 2025.
2006	National Rainfed Area Authority (NRAA) on 3.11.2006
August, 07	National Food Security Mission (NFSM)
August, 07	Rashtriya KrishiVikashYojana (RKVY)
April,2014	NHM merged into a new Programme Mission For Integrated Development of Horticulture (MIDH)

INDIAN COUNCIL OF AGRICULTRAL RESEARCH (ICAR)

1926	Established **Royal Commission on Agriculture**, headed by **Lord Linlithgow,** and recommended setting up Imperial Council of Agricultural Research to promote, guide and coordinate agricultural research throughout the India.
1928	Royal Commission on Agriculture, **submitted its report**
16th July, 1929	**ICAR was set up on 16th July, 1929,** as registered society under the Societies Registration Act 1860. On the recommendation of Royal Commission on Agriculture.
March,1946	Under the president ship of Jogendra Singh it was decided to change the '**Imperial**' in to "**Indian**" and now it is called **Indian Council of Agricultural Research**.
1965	1st DG of ICAR was **Dr. B.P. Pal,** he Initiated All India Co-ordinate Research Projects (**AICRPs**)
1966	ICAR was made a **fully autonomous organization** on the recommendation of Agricultural Review Team.
June,1972	**Gajendra gadkar Committee** was constituted to review the recruitment and personal policies of ICAR and its institutes, submitted its report in 1973.
1973	Department of Agricultural Research and Education (**DARE**) were created in Ministry of food and Agriculture.
1973	The Agricultural Research Services (**ARS**) was started.
1973	The Agricultural Scientists Recruitment Board (**ASRB**) was established on 1 November 1973 as recommendations of the **Gajendra gadkar Committee**.

President of ICAR**:** ***Union Minister of Agriculture***
Present president of ICAR is*:* ***Shivraj Singh Chouhan***
First president of ICAR was**:** ***Khan Bahadur Sir MD. Habibullah***

DG of ICAR is also the **Secretary** of **DARE**: (Department of Agricultural Research and Education) which established in 1973.
Present DG and secretary of ICAR: ***Dr. M. L. Jat (2025)***
First DG of ICAR was: ***Dr. B.P. Pal***

INDIAN AGRICULTURAL RESEARCH INSTITUTE (IARI)	
1905	Agricultural Research Institute was established at Pusa, Bihar under the viceroyalty of Lord Curzon. Earlier it was in West Bengal. Help of sum of rupees 9 lakh was donated by an American citizen **Hennry Phipps**, hence the main building known as **Phipps laboratory or naulakha.**
1911	Renamed as Imperial Agricultural Research Institute
1923	Institute started offering Diploma of Associate ship.
1934	**Major Earth quake** and flood damaged the building at Pusa, Bihar.
1936	IARI shifted to Pusa Road, New Delhi
1936	**B. Vishwanath**, was **first Indian Director of IARI**
1947	Name has been changed from Imperial Agricultural Research Institute to Indian Agricultural Research Institute.
1958	Recognized as **"Deemed University"** under UGC Act 1956. PG school was established.

ICAR Institutions

Deemed Universities - 4

1. ICAR-Indian Agricultural Research Institute (IARI), New Delhi
2. ICAR-National Dairy Research Institute (NDRI), Karnal
3. ICAR-Indian Veterinary Research Institute (IVRI), Izatnagar
4. ICAR-Central Institute on Fisheries Education (CIFE), Mumbai

Central Agricultural Universities-4

1. Banaras Hindu University (BHU), Varanasi – U.P.
2. Central Agricultural University (CAU), Imphal – Manipur
3. Dr Rajendra Prasad Central Agricultural University (RPCAU), Pusa, Bihar
4. Rani Lakshmi Bai Central Agricultural University (RLBCAU), Jhansi – U.P.

ICAR- National Bureaux - 6

1. National Bureau of Plant Genetics Resources (NBPGR), New Delhi
2. National Bureau of Agriculturally Important Micro-organisms (NBAIM), Mau, UP
3. National Bureau of Agricultural Insect Resources (NBAIR), Bengaluru
4. National Bureau of Soil Survey and Land Use Planning (NBSSLUP), Nagpur
5. National Bureau of Animal Genetic Resources (NBAGR), Karnal
6. National Bureau of Fish Genetic Resources (NBFGR), Lucknow

ICAR-Directorates/Project Directorates - 13

1. Directorate of Groundnut Research (DGR), Junagarh, Gujarat
2. Directorate of Soybean Research (DSR), Indore, MP
3. Directorate of Rapeseed & Mustard Research, Bharatpur (DRMR), Rajasthan
4. Directorate of Mushroom Research (DMR), Solan, HP
5. Directorate on Onion and Garlic Research (DOGR), Pune, MH
6. Directorate of Cashew Research, Puttur (DCR), Karnataka
7. Directorate of Medicinal and Aromatic Plants Research (DMAPR), Anand, GJ
8. Directorate of Floricultural Research (DFR), Pune, Maharashtra
9. Directorate of Weed Research (DWR), Jabalpur, MP
10. Project Directorate on Foot & Mouth Disease (PD--FMD), Mukteshwar, Uttarakhand
11. Directorate of Poultry Research (DPR), Hyderabad, Telangana
12. Directorate of Knowledge Management in Agriculture (DKMA), New Delhi

ICAR-National Research Centres - 14

1. National Research Centre for Banana (NRCB), Trichi
2. National Research Centre for Grapes (NRCG), Pune
3. National Research Centre for Litchi (NRCL), Muzaffarpur
4. National Research Centre for Pomegranate (NRCP), Solapur
5. National Research Centre on Camel (NRCC), Bikaner
6. National Research Centre on Equines (NRCE), Hisar

8.. National Research Centre on Mithun (NRCM), Medziphema, Nagaland

9. National Research Centre on Orchids (NRCO), Pakyong, Sikkim
10. National Research Centre on Pig (NRCP), Guwahati
11. National Research Centre on Plant Biotechnology (NRCPBT), New Delhi
12. National Research Centre on Seed Spices (NRCSS), Ajmer
13. National Research Centre on Yak (NRCY), West Kemang
14. Mahatma Gandhi Integrated Farming Research Institute (MGIFRI), Motihari

ICAR- Institutions - 71

1. Central Island Agricultural Research Institute (CIARI), Port Blair, A&N Islands
2. Central Arid Zone Research Institute (CAZRI), Jodhpur, Rajasthan, 1959
3. Central Avian Research Institute (CARI), Izatnagar, UP, 1979
4. Central Inland Fisheries Research Institute (CIFRI), Barrackpore, WB, 1947
5. Central Institute Brackishwater Aquaculture (CIBA), Chennai, Tamil Nadu, 1987
6. Central Institute for Research on Buffaloes (CIRB), Hissar, HR, 1985
7. Central Institute for Research on Goats (CIRG), Makhdoom, UP, 1979
8. Central Institute of Agricultural Engineering (CIAE), Bhopal, MP, 1976
9. Central Institute for Arid Horticulture (CIAH), Bikaner, Rajasthan, 1993
10. Central Institute of Cotton Research (CICR), Nagpur, MH, 1976
11. Central Institute of Fisheries Technology (CIFT), Kochi, Kerala , 1957
12. Central Institute of Freshwater Aquaculture (CIFA), Bhubneshwar, Odisha, 1987
13. Central Institute of Research on Cotton Technology (CIRCOT), Mumbai, MH
14. Central Institute of Sub Tropical Horticulture (CISH), Lucknow, UP, 1972
15. Central Institute of Temperate Horticulture (CITH), Srinagar, J&k, 1994
16. Central Institute on Post harvest Engineering &Technology (CIPHET), Ludhiana
17. Central Marine Fisheries Research Institute (CMFRI), Kochi, Kerala, 1947
18. Central Plantation Crops Research Institute (CPCRI), Kasargod, Kerala, 1970
19. Central Potato Research Institute (CPRI), Shimla, HP, 1949
20. Central Research Institute for Jute and Allied Fibres (CRIJAF), Barrackpore, WB
21. Central Research Institute of Dryland Agriculture (CRIDA), Hyderabad, Telangana
22. National Rice Research Institute (NRRI), Cuttack, 1946
23. Central Sheep and Wool Research Institute (CSWRI), Avikanagar, Rajasthan, 1962

ICAR- Institutions - 71

24. Indian Institute of Soil and Water Conservation (IISWC), Dehradun, Uttrakhand
25. Central Soil Salinity Research Institute (CSSRI), Karnal, HR, 1969
26. National Institute for Research on Commercial Agriculture (NIRCA), Rajahmundry, AP
27. Central Tuber Crops Research Institute (CTCRI), Trivandrum, Kerala, 1963
28. ICAR-Research Complex for Eastern Region (ICAR-RCER), Patna, Bihar
29. ICAR-Research Complex for NEH Region (ICAR-RC-NEHR), Barapani, Meghalya
30. Central Coastal Agricultural Research Institute (CCARI), Ela, Goa, 1979
31. Indian Agricultural Statistics Research Institute (IASRI), New Delhi, 1970
32. Indian Grassland and Fodder Research Institute (IGFRI), Jhansi, UP, 1962
33. Indian Institute of Agricultural Biotechnology (IIAB), Ranchi, Jharkhand,
34. Indian Institute of Horticultural Research (IIHR), Bengaluru, Karnataka, 1967
35. National Institute of Secondary Agriculture (NISA), Ranchi, Jharkhand, 1924
36. Indian Institute of Pulses Research (IIPR), Kanpur, UP, 1993
37. Indian Institute of Soil Sciences (IISS), Bhopal, MP, 1988
38. Indian Institute of Spices Research (IISR), Calicut, Kerala, 1975
39. Indian Institute of Sugarcane Research (IISR), Lucknow, UP, 1952
40. Indian Institute of Vegetable Research (IIVR), Varanasi, UP, 19999
41. National Academy of Agricultural Research & Management (NAARM), Hyderabad, 1976
42. National Institute of Biotic Stresses Management (NIBSM), Raipur, 2012
43. National Institue of Abiotic Stress Management (NIASM), Malegaon, MH, 2009
44. National Institute of Animal Nutrition and Physiology (NIANP), Bengaluru, 1995
45. National Institute of Natural Fibre Engineering and Technology (NINFET), Kolkata
46. National Inst. of Veterinary Epidemiology & Disease Informatics (NIVEDI), Bengaluru
47. Sugarcane Breeding Institute (SBI), Coimbatore, Tamil Nadu, 1912
48. Vivekananda Parvatiya Krishi Anusandhan Sansthan (VPKAS), Almora, UK.
49. Central Institute for Research on Cattle (CIRC), Meerut, UP., 1987
50. National Institute of High Security Animal Diseases (NIHSAD), Bhopal, MP, 2014
51. Indian Institute of Maize Research (IIMR), Ludhiana, PB, 2015
52. Central Agroforestry Research Institute (CAFRI), Jhansi, UP, 1988
53. National Institute of Agricultural Economics and Policy Research, New Delhi
54. Indian Institute of Wheat and Barley Research (IIWBR), Karnal, HR, 2014
55. Indian Institute of Farming Systems Research (IIFSR), Modipuram, UP, 1989
56. Indian Institute of Millets Research (IIMR), Hyderabad, Telangana, 1958

ICAR- Institutions - 71

57. Indian Institute of Oilseeds Research (IIOR), Hyderabad, Telangana, 1977
58. Indian Institute of Oil Palm Research (IIOPR), Pedavegi, Andhara Pradesh, 1995
59. Indian Institute of Water Management (IIWM), Bhubaneshwar, Odisha, 1988
60. Indian Institute of Rice Research (IIRR), Hyderabad, Telangana,1965
61. Central Institute for Women in Agriculture (CIWA), Bhubaneshwar, Odisha, 2014
62. Central Citrus Research Institute (CCRI), Nagpur, MH, 1985
63. National Institute of Seed Science & Technology (NISST), Mau, UP,
64. Indian Agricultural Research Institute (IARI), Hazaribag, Jharkhand
65. National Meat Research Institute (NMRI), Hyderabad, Telangana
66. National Research Institute for Integrated Pest Management (NRIIPM), New Delhi
67. National Institute of Plant Biotechnology (NIPBT), New Delhi
68. Central Institute of Coldwater Fisheries Research (CICFR), Bhimtal, UK (2025)
69. National Institute on Foot and Mouth Disease (NIFMD), Bhubaneswar, Odissa (2023)
70. Indian Institute of Soyabean Research (IISR), Indore, MP, 2016
71. Indian Institute of Rapeseed- Mustard Research (IIRMR), Bharatpur, Raj. (2025)

International institutes:

1.	IRRI:	International Rice Research Institute at **Manila** (Philippines) **1960**.
2.	CIMMYT:	It is a **Mexican** word which denotes International Center for **Maize** and **Wheat** improvement at **Mexico** in 1966 ***(RAS Pre.-09)***
3.	ICARDA:	International Center for Agricultural Research in Dry land Area, **Aleppo** (Syria), 1977
4.	ICRISAT:	International Crop Research Institute for Semi-Arid Tropics, **Hyderabad** (India), 1972
5.	ILRAD:	International Laboratory for Research on Animal Diseases, Nairobi, Kenya, 1973
6.	ICGEB:	International Center for Genetic Engineering & Biotechnology, Triesta, Italy & New Delhi, India
7.	IBPGR :	International Board of Plant Genetic Resources, Rome, Italy (1974)
8.	IIMI:	International Irrigation Management Institute: **Digana** (Sri Lanka).
9.	AVRDC:	Asian Vegetable Research and Development Center, **Taiwan.**
10.	IITA:	International Institute of Tropical Agriculture, **Nigeria, 1968**
11.	CIAT:	International Center for Tropical Agriculture, **Columbia**, 1967
12.	WMO:	World Meteorological Organization, Head Quarters in **Geneva** (Switzerland).
13.	WTO	World Trade Organization, Head Quarters is located in **Geneva** (Switzerland), 1, Jan.,1995

Other Agricultural institutes:

1.	NAARM:	National Academy of Agricultural Research and Management, **Hyderabad** (Andhra Pradesh).
2.	FRI:	Forest Research Institute: **Dehradun** (UK)
3.	NCERT:	National Council of Education Research and Training, **New Delhi.**
4.	NRC-WS:	National Research Center for Weed Science, Jabalpur (MP)
5.	NRC-WT:	National Research Center for Water technology, **Bhubaneswar** (Orissa). Directorate of Water Technology Research, Bhubaneswar (Orissa)
6.	NRC-WAR:	National Research Center for ***Women in Agriculture Research***, **Bhubaneswar** (Orissa). Now Directorate of Woman in Agriculture Research.
7.	PDWMR:	Project Directorate of Water Management Research, **Patna** (Bihar). Now Indian Institute of Water Management
8.	PDRR:	Project Directorate of Rice Research, Hyderabad (Telangana). Now renamed as Indian Institute of Rice Research (IIRR).
9.	PDMR	Project Directorate of Maize Research, Pusa, New Delhi. But it renamed as Indian Institute of Maize Research, shifted at Ludhiana (PB).
10.	PDWR:	Project Directorate of Wheat Research, Karnal (Hariana). Now renamed as Indian Institute of Wheat and Barley Research **(IIWBR).**
11.	NIOF:	National Institute of Organic Farming, **Ghaziabad (UP)**
12.	NCIPM:	National Center for Integrated Pest Management, New Delhi.
13.		
14.	DFR:	Directorate of Floriculture Research, Pune, MH
15.	DIPA:	Directorate of Information and Publication in Agriculture, **New Delhi**
16.	NBB:	National Biodiversity ACT, (2002)
17.	NBA:	National Biodiversity Authority, Chennai (2003)
18.	NASM:	National Agricultural Science Museum, (ICAR),New Delhi, 2004
19.	DRMR:	Directorate of Rape seed & Mustard Research, Sewar, Bharatpur (Raj.), 2009 (Former it was NRC for rape seed & mustard)
20.	APEDA:	Agricultural and Processed Food Product Export Development Authority, New Delhi.
21.	NAFED:	National Agricultural Cooperative Marketing Federation, Oct. 1958
22.	NABARD:	National Bank for Agriculture and Rural Development, 12 July 1982
23.	GATT:	General Agreement on Tariffs and Trade, 1947 (Now it replaced by WTO in Jan, 1995)

24.	**CACP:**	Commission for Agriculture Costs and Prices
25.	**NIAM:**	National Institute of Agricultural Marketing, Jaipur (RAJ.), 1988
26.	**CFQCTI:**	Central Fertilizer Quality Control & Training Institute, Faridabad, HR
27.	**NPPTI:**	National Plant Protection Training Institute, Hyderabad, AP
28.	**DMI:**	Directorate of Marketing and Inspection, Faridabad, HR

ICAR awards:

SN	Award	Field
1	**Hari Om Ashram Trust award**	Published Research in Crop Science
2	**Punjabrao Deshmukh Women**	For woman Agricultural Scientist
3	**Jawaharlal Nehru award**	Best work in Ph. D in the field of Agriculture & Allied Sciences
4	**Jagjivan Ram Kishan award**	Two Innovative Farmers in Agriculture & Allied Sciences
5	**Choudhary Charan Singh award**	For Excellence Journalism in the field of Agriculture Research and development

Important publication related to agriculture:

Journal/magazine	Published by	Place
Kurukestra	Ministry of Rural Development	**New Delhi**
Current science	Current Science Association & Indian Academy of Science	**Bangalore**
Science reporter	**C.S.I.R**	**New Delhi**

Unit-2. Current Status of Indian Agriculture

✓	Geographical area of India is:	**328.74 mha.**
✓	Total reporting area of India is:	**306 mha**
✓	Total degraded land of India is:	**187 mha**
✓	Gross cropped area of India is:	**219.15 mha**
✓	Gross irrigated area of India is:	**112.23 mha**
✓	Net irrigated area of India is:	**75.46 mha**
✓	Net sown area of India is:	**141 mha**
✓	Cropping intensity of India is:	**155.4 %** (2023-24)
✓	Major source of irrigation in India is:	**Tubewell** (48.51%)
✓	Second major source of irrigation is:	**Canals** (22.96%)
✓	Total forest cover in India is:	**80.9mha** (2023)
✓	Percentage forest area of India is:	**21.76 %** (2023-24)
✓	Optimum forest area required:	**33%** of geographical area
✓	State which is having highest geographical area in India	**Rajasthan** (10.41%)
✓	State which is lowest in geographical area:	**Goa** (0.11%)
✓	State which is 2nd in geographical area:	**MP**
✓	Largest forest cover (area) in the state:	**MP**
✓	Largest percent area of forest in the territory:	**Lakshadweep** (91.33%)
✓	Among the states largest percent area of forest in:	**Mezoram** (85.34%)
✓	Lowest forest covering State in India is:	**Haryana** (3.63%)
✓	Agriculture sector in india has a share of GDP in the country at current price:	**18.2%** (2023-24)
✓	State which is 1st in pulses production:	**MP**
✓	State which is 1st in food grain production:	**U. P.**
✓	State which is 1st in total coarse cereals production:	**Rajasthan**
✓	State which is 2nd in food grain production:	**MP**
✓	State which is 1st in wheat production:	**U. P.**
✓	State which is 2nd in wheat production:	**Punjab.**
✓	State which is first in soybean production:	**MP**
✓	State which is first in rapeseed & mustard production:	**Rajasthan**

✓	Water vapour in the atmosphere is constitute:	**0.001 % of the total global water**
✓	Annual World precipitation is:	**1000 mm**
✓	Average annual rainfall of India:	**1194 mm**
✓	Generated rainfall volume:	**400 mha-m**
✓	Maximum area under irrigation is under:	**Ganga basin**
✓	Maximum area under drip irrigation in:	**Maharashtra**
✓	Maximum area under sprinkler irrigation in:	**Rajasthan**
✓	Highest fertilizer consumption rate (Includes States and Union Territories)	**Pundicherry** (1091kg/ha)
✓	Highest fertilizer consumption rate (Among States) :	**Punjab** (254 kg/ha)
✓	Total fertilizer consumption is maximum in :	**U.P.**
✓	Average total fertilizers consumption in India:	**139.8 kg/ha**
✓	Total consumption of fertilizer products:	**60 MMT**
✓	All India current NPK use ratio:	**11.8:4.6:1**
✓	Food grains (cereals + pulses) production in India:	**357.73mt (2024-25)*** ***Final estimate (DA&FW)**
✓	Total cereals production in India:	**332.05mt** (2024-25)*
✓	Total pulses production in India:	**25.68mt** (2024-25)*
✓	Among pulses highest production:	**Gram** (11.11mt)*, followed by moong (4.24mt & tur 3.62mt)
✓	Total oilseeds production in India:	**42.98mt** (2024-25)*
✓	Among the oilseeds which crop is highest in production:	**Soybean** (15.26mt) (2024-25)* followed by Mustard, 12.66mt & ground nut 11.94mt
✓	Rice production in India:	**150.18mt (2024-25)***
✓	**Wheat** production in India:	**117.94mt (2024-25)***
✓	**Maize** production has (3rd position) in cereals:	**43.40mt (2024-25)***
✓	Production of millets (**shree anna**):	**18.59mt (2024-25)***
✓	Production of nutri or coarse cereals:	**63.92mt (2024-25)***
✓	Sugarcane production in India:	**454.61mt (2024-25)***
✓	Cotton production in India:	**29.72million bales***
✓	Total milk production in India:	**247.87 (2024-25)** **(According BAHS, 2025)**
✓	Leading state in milk production (2023-24):	**UP (followed by RJ)**

✓	Milk availability per capita/ day is:	485g (2024-25)
✓	Total fruits production in India:	112.97mt* *Final estimate (2023-24)
✓	Vegetables production in India:	207.20mt (2023-24)*
✓	Potato production in India (2023-24)	57.05mt
✓	Onion & totato production in India (2023-24)	24.26mt & 21.32mt
✓	Leading state in vegetable production	UP
✓	Leading state in potato production	UP
✓	Leading state in onion production	MH
✓	Leading state in tomato production	MP
✓	Total horticulture production in India:	354.74mt (2023-24)* Final estimate (2023-24)
✓	Among fruits highest production from:	Banana (37.61mt)
✓	Total area under horticulture crops in India:	29.08mha
✓	Leading state in fruit production	Andhra pradesh
✓	Leading state in mango fruit production	UP
✓	Leading state in banana fruit production	MH
✓	Leading state in honey production:	UP
✓	Leading state in spices production:	MP
✓	Leading state in cut flower production:	WB
✓	Leading state in loose flower production:	Tamil Nadu
✓	Total wool production in India (2024-25):	34.57million kg
✓	Leading state in wool production:	Rajasthan (47.85%)
✓	Leading state in egg production:	AP (18.37%)
✓	Total egg production in India (2024-25):	149.11billion eggs
✓	Per capita/annum availability of eggs:	106 (2024-25)
✓	Leading state in meat production:	WB (12.46%)
✓	State has second position in meat production:	UP (12.20%)
✓	Leading animal in meat production:	Poultry (51.14%)
✓	Total meat production in India:	10.50mt (2024-25)
✓	Per ha. use of pesticide in India:	0.6kg (2022-23)
✓	Present Union Agriculture Minister is:	Sh. Shivraj Singh Chouhan
✓	Total KVK in India up to Jan., 2026	731
✓	Highest KVK (2026, Jan.) in the State:	UP (89)
✓	Total KVK in Rajasthan State is:	47

Unit-4. Contribution of Agriculture Experts:

1. **Dr. M. S. Swaminathan: (07 August, 1925 to 28 Sept., 2023)**
   - He was eminent wheat breeder and geneticist, worked at CIMMYT, Mexico and IARI, New Delhi.
   - He collaborating with N. Barlaug to bring high yielding dwarf wheat varieties i.e Lerma Rojo and Sonora-64 from Mexico to India which brings green revolution in India (1966).
   - That's why he known as 'Father of **Green Revolution'** in India.
   - He awarded **'World Food Prize** (1987) given by **FAO**
   - He was honored as a **'Millennium prize'** by former Prime Minister A. B. Vajpai in 88th National Science Congress held on 3-7 Jan, 2001 at IARI, New Delhi.
   - He was president of **National Commission on Farmers of India**.
   - The book entitled" **Wheat Revolution**" Written by him.
   - M.S. Swaminathan received 1st World Agriculture Prize at the 11th Global Agriculture Leadership Summit in New Delhi in October 2018.
2. **Dr. N. E. Borlaug: (1914- 2009)**
   - Father of **'Green Revolution'** in World
   - He was American wheat scientist (Plant Pathologist)
   - Norman Borlaug worked at the International Maize and Wheat Improvement Center (CIMMYT) in Mexico, the Rockefeller Foundation and the U.S. Forestry Service.
   - He developed several high-yielding, semi-dwarf wheat varieties in Mexico, including **Sonora 63**, **Sonora 64**, **Mayo 64**, and **Lerma Rojo 64-A** and these varieties, incorporating the dwarfing gene from Japan's Norin 10 wheat, were disease-resistant and produced more grain per head, boosting yields integrating modern farming techniques like fertilizers and irrigation, and introducing these innovations to countries like Mexico, India, and Pakistan, preventing widespread famine and saving millions of live, later sparked the Green Revolution globally.
   - He awarded **Noble Peace Prize** in **1970**.
3. **Verges Kurien: (26 Nov., 1921 to 09 Sept., 2012)**
   - Regarded as 'Father of **White Revolution'** in India
   - He was Ex. Chairman of **NDDB**, Anand (Gujarat)
   - Brain behind the scheme **'Operation Flood'** (1970-1996)
   - He gave idea of **'billion litre'** milk production through cooperative.
   - **Got 'World Food Prize'** in 1989.
   - **National Milk Day in India**, is celebrated on November 26th to honor the birth anniversary of Dr. Verghese Kurien.

- **While** Globally, **World Milk Day** is celebrated on June 1st by the FAO .

4. **S. K. Vashal:**
   - **Maize breeder** working at **CIMMYT**, Mexico.
   - Quality protein maize (QPM) germplasm developed by Dr Vasal is now being used worldwide for developing QPM varieties.
   - Awarded" **World Food Prize**" in the year 2000 for research on **QPM** (maize rich in amino acids, **tryptophan** and **lysine)** along with Dr. E. Villegas.
5. **Dr. K. L. Chadda:**
   - He was 'Pomologist'
   - A renowned scientist and Padma Shri awardee known for his work on mango production and leadership roles in horticulture research and policy.
   - He regarded as 'Father of the **Golden Revolution in Horticulture'** in India.
   - Nirpakh Tutej also recognized as **'Father of the overall Golden Revolution (honey & horticulture)'** in India.
   - Dr. Chadda wrote 'Handbook of Horticulture' book, published by ICAR.
6. **Dr. R. S. Paroda:**
   - He is **forage breeder**
   - Ex. D.G. of ICAR
   - Presently he is 'Chairman of Trust for Advancement of Agricultural Sciences (TAAS)' and Chairman of Haryana Farmers Commission.
   - He got many awards and honours among them top most are- **Padma Bhushan in** 998. The prestigious **Norman Borlaug Award** (2006), **Dr. M.S. Swaminathan Award** (2020): For environment protection and **Dr. APJ Abdul Kalam Lifetime Achievement Award** (2025): Awarded by the National Academy of Biological Sciences (NABS) for his enduring legacy in agricultural sciences.
7. **Sanjaya Rajaram: (1943- 2021)**
   - He was Indian born Mexican **wheat breeder** at CIMMYT.
   - He was awarded **"Chinas Friendship Madel** "in **2001**.
   - Awarded **'World Food Prize'** in 2014 for developing 480 varieties of wheat.
8. **Dr. G. S. Khush:**
   - Rice breeder at IRRI, Manila, Philippines for 35 years.

- He awarded**'World Food Prize'** in 1996 along with Henry Beachell to developing high yielding varieties (e.g. IR-36) of rice often regarded as **father of 'Super Rice'.**
- **Got 'Wolf Prize'** for agriculture, **2000** by Israel for his extra ordinary contribution to **plant breeding** and **genetics** especially in **rice**.

Unit- 5. Achievements of Crop Improvement:

1. Genetically Modified (GM) Crops:

- The crops in which one or many of its **traits** are **altered** or **enhanced** through a process of **genetic engineering** known as **GM crops** or transgenic plants. e.g., ***Bt.* Cotton, GM mustard and golden (GM) rice.**
- In India permission of GM crops for commercially grown are granted by Genetic Engineering Approval Committee **(GEAC)** under **Ministry of Environment and Forestry (MoEF).**

(a) Bt. Cotton-

- ***Bacillus thuringensis*** is a soil bacterium that produces **proteins lethal** to insect **larvae** affecting the **digestive system** of **boll worm**
- The gene responsible for this effect is **Cry-1Ac**.
- **GM cotton** or *Bt* cotton is only crop permitted by Genetic Engineering Approval Committee **(GEAC). In India *Bt* cotton** first permitted on **March, 2002**.
- Punjab Agricultural University, Ludhiana has developed country's first genetically modified (Bt) varieties of cotton- the special advantage over existing MNCs (private sector seed) is the seeds of which could be reused by farmers, resulting in saving the repeated cost they have to bear every season. Because MNCs (private sector seed) are using terminator gene in Bt cotton seeds due to this farmers have to purchase new seed in every year.
- In World first generation of Bt cotton was came in 1996 and known as **Bollgourd-1.**
- **Firstly Mahyco** collaborated with **Monsanto** to developed *Bt.* Variety **Bollgourd-1**, using **Cry-I Ac** gene.
- Recently ***Bt. cotton*** Variety developed as **Bollgourd-II**, having both genes of **Cry I Ac** and **Cry 2 Ab gene**. (From 2006) which is known as 'second generation of Bt cotton'.
- In world first transgenic crop was: ***Bt. Tobacco (1987)***
- Second transgenic crop was: Tomato (Flavr Savr variety) ***(1994)***
- Country having first position for growing of GM crops is: **USA**

- Rank of India for transgenic crops in world (2024) is: **Fourth (USA > Brazil > Argentina > India > Canada)**
- The position of transgenic plants in world (in area): **Soyabean** (48.27%) > **Maize** (32.1%) > **Cotton** (13.5%) > **Canola** (5.3%).
- Mostly GM traits are used as **herbicides resistant** traits followed by **quality improvements** and **insecticides resistant** traits.
- Total area under transgenic in World in 1996 was 1.7 mha and in 2019 it became 270mha.

b. Genetically modified mustard:

- These are **glyphosate** resistance mustard developed by **Pro-Agro Seed Company Ltd.**
- Dhara Mustard Hybrid-11 (**DMH-11**)- it is GM hybrid mustard

c. Golden Rice: (RPSC, AO Exam-2012)

- It is **genetically** modified (GM) rice.
- It having **β-carotene** (precursor of **Vit-A**) rich rice.
- It was engineered from normal rice through incorporating genes from daffodils and a soil bacterium by **Ingo Potrikus** and **Peter Beyer** in 1990s to prevent Vitamin-A deficiency in humans.
- That's why **Ingo Potrikus** is known as **father of golden rice**.

2. Super Rice:

- Developed by **Dr. Gurdev Singh Khush** at **IRRI**, which would bring an improvement of 25% of over present day varieties.

3. Hybrid Rice:

- 1st time in World **China** developed hybrid rice in 1974 by **Prof. Long Ping Yuan, he** known as **'Father of Hybrid Rice'.**
- India is **second** to developed hybrid rice after China, first varieties i.e. **MGR-1** (earlier named as **CoRH-1**) developed first by **TNAU,** Coimbatore and **PRH-1 & PRH-6** developed by **IARI**.

4. Hybrid Mustard:

- Developed by **DRMR** (Directorate of Rapeseed and Mustard Research), Sewar, Bharatpur (Rajasthan)
- Varieties developed by DRMR are: ***NRC-HB-506*** (it is first hybrid mustard in India).
- **Dhara Mustard Hybrid-11** (DMH-11) - it is genetic modified hybrid mustard which developed at Delhi University.

5. Super Wheat:

- The research on super wheat is in progress at **DWR, Karnal**
- It was expected to come by 2005, yield increase **15-20** per cent.

6. Green Revolution:

- The term green revolution was coined by **William S. Gaud** in 1968 of **USAID**; to describe the productivity based improvement in food production particularly in **wheat and rice**.
- Main components in green revolution are **High yielding varieties, use of chemical fertilizers** and **plant protection chemicals**.

7. Ever Green Revolution:

- This term was coined by **Dr. M S. Swaminathan** to denote the green revolution based on sustainable methods of crop **intensification** and **diversification**.

8. Vertical Revolution in Agriculture:

- **Maximizing production per unit land area** per unit of time using **intensive cropping** system cropping system, high production inputs and improved management practices.

9. Other Revolutions in Agriculture:

Revolution:	Related to:
1. White Revolution:	**Milk** Production (operation flood)
2. Blue Revolution:	**Fish** Production
3. Brown Revolution:	**Food** Processing/leather production
4. Grey Revolution	**Fertilizer Production**
5. Yellow revolution:	**Oil** Seeds Production **(mustard)**
6. Red Revolution	**Tomato/Meat** Production
7. Pink Revolution:	**Prawn /Onion** Production
8. Golden Revolution:	**Fruit /honey** production
9. Round Production:	**Potato** Production
10. Silver Revolution:	**Egg/ poultry** production
11. Black revolution:	**Biofuel/Jatropha** production
12. Orange revolution:	**Citrus fruit** production
13. Golden fibre revolution:	**Jute** production
14. Silver fibre revolution:	**Cotton** production
15. Rainbow revolution:	**All sector of agriculture**
16. Almond (*badami*) revol:	**Spices** production
17. Protein revolution:	**Pulses** production
18. Saffron revolution:	**Solar energy**
19. Sweet revolution:	**Bee keeping**
20. Prabhani kranti:	It is a variety of **Okra**

India shares in world:

- Total geographical area: 2.4 %
- Population: 17.5%
- Livestock: 15%
- Forest: 1.5%
- **Water Resources** 4%

India's position in World Agriculture:

1. **First position in:**
   - Mango production
   - Banana production
   - Total irrigated area
   - Milk production
   - Jute & allied fibers
   - Cashew production
   - Coconut production
   - Total pulses production
   - Tea and spices production
   - Livestock production
2. **Second position:**
   - Wheat production
   - Oilseeds production
   - Vegetables production
   - Sugarcane production
   - Rice production
   - Fruits production
   - Silk production
   - Cotton production
3. **Third position:**
   - Tobacco production
4. **Fourth position:**
   - Natural Rubber (share 8.5%)

Position of Agricultural trades in India:

a. **Agriculture export commodities:**
   - **Rice** > Cashew > Wheat > Tea > Tobacco

b. **Agriculture imports commodities:**
   - **Vegetable oil** > Cashew > Pulses > Fruits

UN's International years related to agriculture:

2004: International Year of **Rice** **(OBC, Bank exam-08)**

2006: International Year of **Desert and Desertification**

2007: International Year of **Water**

2008: International Year of **Potato**

2009: International Year of **Fiber**

2010: International Year of **Biodiversity**

2011: International Year of **Forest**

2012: International Year of **Co-operative**

2015: International Year of **Soil**

2016: International Year of **Pulses**.

2021: International Year of **Fruits** and **Vegetables**.

2022: International Year of **Artisanal Fisheries and Aquaculture**
2023: International Year of **Millets**
2024: International Year of **Cammelids.**
2025: International Year of **Glaciers Preservation**.
2026: International Year of **Woman Farmers**
2026: International Year of **Rangelands and Pastoralists.**
2026: International Year of **Volunteers for Sustainable Development**.
2027: International Year of **Sustainable and Resilient Tourism.**
2029: International Year of **Asteroid Awareness** and **Planetary Defence.**

Important days related to agriculture:

SN	Date	Famous day
	24th January	Uttar Pradesh Day (from 1950)
1	15th March	World Consumer Day
2	21St March	World Forestry Day
3	30th March	Rajasthan Day (from 1949)
4	5th June	World Environment Day
5	11th July	World Population Day
6	16th July	ICAR Day
7	16th September	Ozone Day
8	16th October	World Food Day
9	4th December	Women in Agriculture Day
10	5th December	World Soil Day
11	23rd December	Kishan Day, *birth day of Choudhary Charan Singh*

Unit-6.
The major *schemes*/programmes of Agriculture in India:

SN	Scheme	Launch	Objective
1	Kisan Credit Card Scheme **(KCCs)**	**1998**	• Farmers can receive crop loans up to a principal amount of **3 lakh at 7** per cent interest from any nationalized bank of the country. • But Government provided an additional 3 per cent interest subvention to those farmers who timely repaid their short-term crop loan. • Thus the effective rate of interest for such farmers will be 4 per cent per annum. • In the budget speech 2025-26 finance minister announced to increase the loan limit under modified interest subvention scheme from **3.0 lacs to 5.0 lacs.**
2	Kisan Call Centre	**Jan, 2004**	• To provide agricultural information to the farming community through **toll-free telephone** lines • The Kisan Call Centre scheme was launched in 21, **Jan 2004** • A country-wide common 11-digit number—**1800-180-1551**—has been allocated for KCCs. • Replies to the queries of the farming community are being provided in **22 local languages.** • Calls are attended to from **6.00 am to 10.00 pm** on **all seven days** of the week
3	Integrated Scheme of Oilseeds, Pulses, Oil Palm and Maizc **(ISOPOM)**	**2004**	▪ During 2014-15 this scheme is merged in to 'National Mission on Oilseeds and Oilpalm (**NMOOP**). In the financial year of 2018-19 NMOOP has been merged in to **NFSM-Oilseeds**. From October, 2024 new named as 'National Mission on Edible Oils- Oil Seeds' (**NMEO-OS**).
4	Agricultural	**2004**	• To operationalize the extension reforms

	Technology Management Agency **(ATMA)**		with the active participation of farmers/farmer groups, non-government organizations (NGOs), Krishi Vigyan Kendras, Panchayati Raj Institutions and other stakeholders operating at **district level under NATP of ICAR**
5	National Bamboo Mission **(NBM)**	**2004**	• The Mission aims to promote holistic growth of the bamboo sector by adopting an area-based, regionally differentiated strategy and to increase the area under bamboo cultivation and marketing.
6	National Horticulture Mission **(NHM)**	**2005**	• The Ministry of Agriculture has been implementing the NHM for the holistic development of the horticulture sector • NHM upgraded with Mission for Integrated Scheme of Horticulture Development **(MIDH)** from, 1, April, 2014
7	National Food Security Mission **(NFSM)**	Rabi, **2007**	▪ To enhancing the production of rice, wheat, and pulses by 10 mt, 8 mt, and 2 mt, respectively by the end of the Eleventh Plan ▪ It includes rice, wheat, pulses, commercial crops, oilseed crops, nutri-cereals and coarse cereals. ▪ The funding pattern is **60:40** for nation and state (from 2015-16) ▪ From 2024-25 NFSM has been renamed as the National Food Security & Nutrition Mission (**NFSNM**)
8	Rashtriya Krishi Vikas Yojana **(RKVY)**	**2007**	▪ To achieve a **4 per cent growth** rate in agriculture and allied sectors during the Eleventh five year Plan ▪ It is recommendations of the National Development Council of India ▪ The funding pattern is 60:40 for nation and state (from 2015-16)
9	National Mission on Micro Irrigation **(NMMI)**	**2010**	• The Mission is being implemented during the **Eleventh five year Plan** for enhancing water-use efficiency by adopting drip and sprinkler irrigation

			systems in all States and Union Territories for both horticulture and agricultural crops
10	Mission for Integrated Scheme of Horticulture Development (MIDH)	2014	▪ A Centrally Sponsored Scheme of **MIDH** has been launched for the holistic development of horticulture in the country during XII plan. ▪ The scheme, which has taken take off from 2014-15, integrates the ongoing schemes of **National Horticulture Mission, Horticulture Mission for North East & Himalayan States, National Bamboo Mission, National Horticulture Board, Coconut Development Board and Central Institute for Horticulture**, Nagaland
11	National Mission for Sustainable Agriculture (NMSA)	2014-15	▪ In this scheme 4 ongoing national schems have been merged i.e. Rainfed based area development programme, National mission for micro irrigation, National organic farming programme and National soil health & fertility management programme.
12	**Soil Health Card Scheme (SHCs)**	19 Feb. **2015**	▪ This scheme was inuagrated by Prime Minister **Shri Narendra Modi on 19th Feb, 2015 at Suratgarh, Rajasthan** under the National project on management of soil health and fertility ▪ Under the scheme, the government plans to issue soil cards to farmers which will carry crop-wise recommendations of nutrients and fertilizers required for the individual farms to help farmers to improve productivity through judicious use of inputs. ▪ The collected samples are tested & result/suggestion will be displayed in the cards
13	Pradhan Mantri Krishi Sinchai Yojna (**PMKSY**)	2015	▪ The theme is **'Har Khet Ko Pani'** and improving water use efficiency **'More crop per drop'** by MIS.

14	Parampragat Krishi Vikas Yojna (**PKVY**)	17 Jan, 2016	▪ Paramparagat Krishi Vikas Yojana (Traditional Farming Improvement Programme) has been launched by Government of India to support and to **promote organic farming** and thereby improving soil health. This will encourage farmers to adopt eco-friendly concept of cultivation and reduce their dependence on fertilizers and agricultural chemicals to improve yields.
15	Pradhan Mantri Fasal Bima Yojana **(PMFBY)**	18 Feb. **2016**	▪ PMFBY will replace the existing two schemes National Agricultural Insurance Scheme (1999) as well as the Modified NAIS with the objective of. ▪ To provide insurance coverage and financial support to the farmers in the event of failure of any of the notified crop as a result of natural calamities, pests & diseases. ▪ There will be a uniform premium of only 2% to be paid by farmers for all Kharif crops and 1.5% for all Rabi crops. In case of annual commercial and horticultural crops, the premium to be paid by farmers will be only 5%.
16	**Operation Greens Scheme**	2018-19	▪ Ministry of Food Processing Industries (MoFPI) has approved the scheme in bugedt speech of 2018-19 with an outlay of Rs 500 crores to stabilize the supply of Tomato, Onion and Potato **(TOP)** crops and to ensure availability of TOP crops throughout the country round the year without price volatility.

17	Pradhan Mantri Kisan Samman Nidhi (PM-KISAN)	Feb., 2019	▪ To provide an assured income support to the small and marginal farmers, (now all farmers) the governmnet launched the **Pradhan Mantri Kisan Samman Nidhi (PM-KISAN).** ▪ It is 100% central assistance scheme, under the scheme an income support of **Rs. 6000/- per year** is directly provided to the bank accounts of all farmers across the country in three equal installments of **Rs. 2000/- each every four months.**
18	National Beekeeping & Honey Mission (NBHM)	2020	▪ India's key initiative to achieve the **'Sweet Revolution'**, a national push for scientific beekeeping to boost honey production, farmer incomes, and exports under the **Atmanirbhar Bharat Abhiyaan.**
19	PM Programme for Restoration, Awareness, Nourishment and Amelioration of Mother Earth" **(PM-PRANAM)**	2023-24	▪ To incentivize the States and UTS to promote usage of alternative fertilizers and balanced use of chemical fertilizers.
20	**Shree Anna Scheme**	**2023-24**	▪ To promote the cultivation and production of **coarse grains** i.e. Jowar, Bajra, Ragi, Sawa or Sama, Kangni, Kodo, Kutki, Kuttu in the country. ▪ The cultivation of Shree Anna or coarse grains requires less labor and requires less water. ▪ This is such a grain that can be produced **without irrigation and without fertilizers.** ▪ The Directorate of Sorghum Research was upgraded to the status of **Indian Institute of Millets Research (IIMR) during 2014.**

21	Digital Agriculture Mission (**DAM**)	**2024**	▪ This mission is an umbrella program for **digital agriculture initiatives**. ▪ It includes the creation of Digital Public Infrastructure (**DPI**), the Digital General Crop Estimation Survey (**DGCES**), and a nationwide **digital crop survey**.
22	National Mission on Natural Farming (**NMNF**)	**November 25, 2024**	▪ To rejuvenate soil, restore ecosystems, and build climate resilience by encouraging farmers to adopt natural farming practices. It evolved from the Bhartiya Prakritik Krishi Paddhati (**BPKP**). ▪ NMNF focus on farmer training and promoting chemical-free agriculture through practices like using bio-inputs (**Jeevamrut, Beejamrut**) and integrating livestock, aiming to reduce costs, improve soil health, and build a national brand for natural produce. Key actions involve training **'Krishi Sakhis**,' establishing Bio-Input Resource Centers (**BRCs**), developing model farms, offering farmer incentives, and creating simple certification for market access.
23	Pradhan Mantri Dhan-Dhaanya Krishi Yojana (**PMDDKY**)	**Oct. 2025**	▪ The primary goal is enhance overall agricultural productivity and farmer income in targeted districts. ▪ Focusing on the entire agri-value chain, crop diversification, and integrated farm development. ▪ Target area is 100 selected low-productivity districts across the country with outlay of ₹24,000 crore.
24	Dalhan Atmanirbharta Mission (**DANM**)	**Oct. 2025**	▪ Primary goal is make India self-reliant in pulse production to reduce imports by focusing on research, high-yield varieties and increasing the area under pulse cultivation. ▪ Financial outlay of **₹11,440 crore**. ▪ The mission, along with the *PM Dhan-Dhanya Krishi Yojana*, targets 100 agriculturally backward districts across India for intensive development.

UNIT 7.
ORIGIN OF AGRICULTURE AND RELATED TERMS

The literally meanings of agriculture and related terms, which derived from different languages, are follows:-

1. **Agriculture**- Latin word- ***'AGER'*** meaning *'soil'* and ***'CULTURA'*** meaning *'cultivation'*.
2. **Agronomy**- Greek word- ***'AGROS'*** meaning *'field'* and *'NOMOS'* means *'manage'*.
3. **Horticulture**- Latin word- ***HORTUS'*** meaning *'garden'* and *'CULTURA'* meaning *'culture or growing'*
4. **Pomology**- Latin-Greek word- Combination of Latin word ***'POMUM'*** meaning *'fruit'* and Greek word ***'LOGOS'*** means *'discourse*.
5. **Olericulture**- Latin word- ***'OLERIS'*** meaning *'pot herb'* and english word *'CULTURE'* meaning *'cultivation'*
6. **Floriculture**- Latin word- ***'FLORIS'*** meaning flower, concerned with cultivation of flowering & ornamental plants.
7. **Extension**- Latin word- ***'EX'*** meaning 'out' and ***'TENSIO'*** meaning 'stretching'.
8. **Soil**- Latin word- ***'SOLUM'*** meaning *'ground'* or floor of earth crust.
9. **Monsoon**- Arabic word- ***'MAUSIM'*** meaning 'change of season'.
10. **Biodynamics**- Greek word- ***'BIOS'*** meaning *'life'* and ***'DYNAMICS'*** meaning *'energy'*.
11. **Truck gardening**- French word truck- ***'TORQUER'*** meaning *'to barter or exchange of goods'*
12. **Nematode**- Greek word- ***'NEMA'*** meaning *'thread'* and ***'TOID'*** meaning *'form'*
13. **Pedology**- Greek word- ***'PEDON'*** meaning 'soil or earth'.
14. **Edaphology**:-Greek word ***'EDAPHOS'*** means influence soil for plant growth and LOGOS' means Study or *discourse'*
15. **Market**- Latin word- ***'MARCATUS'*** meaning *'place of trade, traffic or merchandise'*.
16. **Evolution**- Latin word- ***'EVOLUTIO'*** meaning 'unrolling or rolling out'.
17. **Enzyme**- Greek word- ***'EN'*** meaning *'in'* and ***'ZYME'*** meaning *'ferment/living'*.
18. **Credit**- Latin word- ***'CREDO'*** meaning *'I trust you'*.
19. **Legume:** Latin word ***'LEGUMEN'*** means 'bean'.
20. **Pulses:** Latin word ***'PULSUS'*** means protein rich edible grains.
21. **Cereals:** Roman word ***'CERES'*** means Goddess of grains or agriculture

22. **Sugarcane:** Sankrit word '***SARKARA*** or ***SAKKARA***' meaning Sugar.
23. **Cotton:** Sanskrit word '***KARPASA***' means cotton tree.
24. **Groundnut:** Greek word '***ARACHI***' means groundnut.
25. **Rabi, Kharif and Zaid:** these are **Arabic words**.
26. **Society**- Latin word- '***SOCIETAS***' which was derived from noun '*SOCIUS*' meaning '*comrade, friendally*'.
27. **Vulgarization**- French word- '***VULGARIZE***' meaning '*popularize*'
28. **Cell** – Latin word- '***CELLA***' meaning 'a small group acting as a unit within a larger organization'.
29. **'In- vivo' and 'In- vitro'** – Latin word- '***IN-VIVO***' meaning '*within the living*' and '***IN-VITRO***' meaning '*within the glass*'
30. **Plastid** – Greek word- '***PLASTIKAS***' meaning '*formed or moulded*'
31. **Ecology**- Greek word- '***OIKOS***' meaning '*house/dwelling place*' and '*LOGOS*' meaning '*study*'
32. **Entomology**- Greek word- '***ENTOMON***' meaning '*insect*' means '*cut into (section)*' and '***LOGOS***' meaning '*discourse*'
33. **Arthropoda**- Greek word- '***ARTHROS***' meaning '*segmented*' and '*PODUS*' meaning '*leg*'.
34. **Phytopathology**- Greek word- '***PHYTON***' meaning '*plant*', '***PATHOS***' meaning '*ailments*' and '*LOGUS*' meaning '*knowledge*'
35. **Fungicide**- Latin word- '***FUNGUS***' meaning 'fungi' and '*CAEDO*' meaning 'to kill'
36. **Meteorology**- Greek word- '***METEOROS***' meaning 'above the earth's surface' (atmosphere) and '*LOGY*' meaning 'indicating science'.
37. **Climate**- Greek word-'***KLIMA***' means weather condition of a region.
38. **Sociology:**-The Latin-Greek words. Latin word ***SOCIOUS*** means ***campanion*** and Greek word *LOGOS* means *speech.*
39. **Economics: -** Greek word '*OIKONOMIA*' with ***OIKOS*** means household and *NOMOS* means management.
40. **Psychology:** Greek words ***PSYCHE*** means soul and ***LOGOS*** means study.
41. **Communication:** - Latin words '***COMMUNIS***' means common.
42. **Entrepreneur:** - French word '***ENTERPREDRE.***
43. **Manure:** French word '***MANOEUVRER***' means work with soil.
44. **Compost:** Latin word '***COMPONERE***' means put together.
45. **Fertilizer:** Latin word '***FERTILIS***' means '**fruitful**' or 'productive'.
46. **Herbicide**: two Latin words: '***HERBA***' (meaning 'plant' or 'grass') and '***CAEDERE***' mean 'to kill'.

Types of Agriculture

The term agriculture is derived from the two **Latin** words:

Ager = Soil
Cultura = Cultivation

- **Cultivation** of soil is called **agriculture**.

Sustainable agriculture/eco-farming:

- Sustainable agriculture is one which makes of use of low cost inputs, lower amounts of chemical fertilizers; maintain soil fertility and ecological harmony **(FAO, 1989).**
- It is also known as **ecological farming/natural farming** or **organic farming/permaculture**.

Shifting cultivation/land rotation or jhuming:

- The **productivity** of soil **is lost** due to cultivation of **same crop** generally rice on the **same forest land** year after year hence, the crop is **shifted** to other **burnt land** which is known as **jhuming**.
- It causes **soil erosion**.
- It practiced in **Jharkhand**, **MP** and **hilly areas** of Northern states of India.

Response farming:

- **Stewart, 1988**, coined this term based on analysis of weather data.
- It refers to the **prediction of monsoon season** (rainfall), based on analysis of weather data of pre-monsoon period and managing crops accordingly.

Precision farming/site specific farming:

- Prof. P.C. Robert is recognized as the father of precision farming.
- It is farm management approach that uses information technology and data analysis to optimize crop and soil condition. Hence the target **specific** use of **inputs** (like seeds, fertilizers, pesticides etc.) for crop production according to crop requirement on localized basis is known as precision farming.

Ideotype (modal plants):

- The concept of Ideotype was given by **Donald** in 1968.
- Ideotype is the model plant which may be defined as "a biological modal which is expected to perform or behave in a predictable manner within a defined environment.

Terra-forming:

- To develop the mars (**planet**) according to the earth's environment is called terra- forming.

- At present **bacteria** and **plants** are being developed by creating the environmental condition of **mars** on the earth.

Hydroponics:

- Growing of plants under **soil less** condition is called **hydroponics.**
- Solution culture is being used for raising **flowers and vegetables** at homes.

Response curve:

- Suggested by **Holliday**, 1960:
- The **relationship** between **plant population** & **yield**

Two types of response curve.

1. **Asymptotic response**:
   - Where entire dry matter is the economic product as in the case of fodder crops or most of dry matter as in tobacco.
   - It means increase yield with increase plant density e.g. **cereal** crops for **fodder**.
2. **Parabolic**: after reaching a level, increase plant density with decrease yield (kept wide space between plants e.g. cereals and pulses for **grain purpose**.

A close **functional relation** between **input** & **output**:

- **Quadratic** response: **y** = $a + bx + cx^2$
- **Linear** response: **y** = $a + bx$
- **Mitscherlish** equation: **y** = $y + a\ (1\text{-}10^{-kx})$.

  Or **y** = $dy/dx = (A\text{-}y)\ ^C$

Zero tillage:

- Father of zero tillage **Glubler B. Triplets** (USA), 1950
- In this primary tillage completely **avoided.**
- Secondary tillage **restricted** to seedbed preparation in the **row zone only**.
- Keeps **50-100% of** residue on soil surface.
- Before sowing, herbicides **Paraquat, Glyphosate** are used for weed control.
- Till planting is adopted.

Till planting: It is the practice in zero tillage including four operations

- ✓ Cleaning the crop row
- ✓ Opening of soil for seed insertion
- ✓ Placing of seed
- ✓ Covering the seed
- These functions are accomplished in **one tillage** operation.

Minimum tillage:

- Can be **defined** as a method aimed at reducing tillage to the minimum necessary for **ensuring a good seedbed**, rapid germination, satisfactory crop stand and favorable growing conditions.
- Started in **USA,** because of **high cost** of tillage due to steep rise in **fuel price in 1974**.
- Tillage operation is done only for **seed bed preparation**.
- **Keep 30-50 %** crop residues on soil surface.
- **It improves soil condition** due to in situ decomposition of plant residue.
- Weed control can be done by herbicide e.g. **Paraquat, Glyphosate**

Global warming:

- Surface of the **earth** is **warmth** from **sun heats.**
- Earth absorbs most of sun energy but **reflects** back some energy in the form of **infra-red radiation.**
- **Major greenhouse** gases are **CO_2** (55%), **Methane** (25%), **CFC** (11%) and **N_2O** (4%).
- These **gases are present** in atmosphere, **transmit the infrared radiation** and **reflect back** to the earth.
- This reflected energy falls on the earth surface and keeps it **warmer**.
- This is called **global warming** or greenhouse effect.

Source of greenhouse gases:

- In **developed** countries: **Emission** from **automobiles** & factories contains **CFCs.**
- In **developing** countries: **Deforestation** cases rise in **CO_2** level, **methane** gas from paddy field & livestock and **nitrous oxide** from **N** base fertilizers.

Effect of global warming on world and agriculture:

- Increase overall **temperature** on earth e.g. earth's surface temperature has increased **1.4° F** in first on **Century** (It has been forecasted that **5 °F** will rise in **next Century**).
- Cause changes in **climate** tremendously.
- Results in **melting of ice** in polar region.
- Increase sea level resulting in **submerging** of coastal areas.
- **Drought** in warmer region.

Important facts:		
✓	Most poisonous pollutant in water is:	**Arsenic.**
✓	Maximum concentration of ozone is found in:	**Lower stratosphere**
✓	Forest conservation Act enacted in the year:	**1980.**
✓	The test has self-purification capacity of water body:	**BOD test.**
✓	Most commonly disinfectant in water purification is:	**Chlorine.**
✓	Gases responsible for acid rain:	**SO_2 and NO_2.**
✓	Pollutant free alternative to petrol for automobiles:	**Propane.**
✓	The gas responsible for Bhopal gas tragedy 1984:	**MIC** (methyl iso cyanate).
✓	MIC was used for manufacturing of insecticide:	**Carbaryl**
✓	Author of famous book 'silent spring' is:	**Rachel Carson (1962).**
✓	Largest ocean:	**Pacific.**
✓	Establishment of NABARD:	**12 July, 1982.**
✓	**R.B.I.** was established in which year:	**1935.**
✓	Proteins are made up from chain of---------- which connected by peptide bond.	**Amino acids.**
✓	Ozone **depletion** is mainly due to:	**CFC** (chloro floro carbon)
✓	One mol. of **CFC** is destroyed about 100,000 molecules of:-	**Ozone.**
✓	Gas filled in fire extinguisher cylinders is:	**CO_2**
✓	Plough which is used to turn over the soil is:	**Mould bould plough**
✓	Carbohydrate which is only present in animal cells:	**Glycogen**
✓	On which crop first all India Co-ordinated Research Project (first **AICRP**) was started	**Maize** (1957)
✓	First National Research Center (**NRC**) for:	**Groundnut** (Junagarh, Gujarat).
✓	The region inhabited by living organism on land, ocean and atmosphere is:	**Biosphere.**

2. AGRONOMY

Important points:

✓	**Mycorrhiza** is a symbiotic association between:	Fungus & roots of higher plants
✓	VAM: *Vascular Arboscular Mycorrhyza* is used as:	P solubilizing fungi in plant roots
✓	Cereals are deficient in which amino acid:	Lysine
✓	Amino acid which is deficient in legumes:	Methionine
✓	Power tiller is most suitable for the cultivation of:	Paddy
✓	Ammonium polyphosphate contains:	15% N & 62 % P_2O_5. (RPSC, AO-09)
✓	tetrazolium (**Tz**) test is done for:	Viability of seeds.
✓	In wet nursery of rice, level of water is maintained:	5 cm
✓	The optimum depth of **puddling** in rice is:	5 cm.
✓	Ultra violet radiations are absorbed by:	Ozone layer.
✓	First Indian director of IARI was:	Dr. B. Vishvanath.
✓	The most critical stage of irrigation in wheat is:	CRI stage (21 days after sowing)
✓	**Jhuming** is a practice of rice cultivation in:	Eastern part of India.
✓	Basic object of sustainable agriculture is:	To maintain ecological balance
✓	The plants growing in salt water are known as:	Halophytes
✓	Crop canopy temperature is measured by:	Infra-red thermometer.
✓	Reclamation disease is due to deficiency of:	Copper
✓	Hormone related to drought tolerance is:	Abscisic acid (ABA)
✓	Contribution of livestock in agriculture is	29.7 %
✓	Hormone used as an herbicide:	2, 4-D, Dicamba
✓	Element available in both anion and cation is:	Nitrogen *(IARI, Ph.D-11)*
✓	Dormancy breaking hormone is:	Gibrellic acid
✓	Ring worm disease is caused by:	Fungus.
✓	Monsoon is responsible to cause major rainfall in Tamil Nadu:	North East monsoon
✓	Stress hardening in plants can be achieved by:	ABA.
✓	Maximum permissible limit of biuret in urea is	1.5%.

✓	Carbon content in organic matter:	**58 %.**
✓	Bio fertilizer used in wheat is:	**Azatobacter.**
✓	Bio fertilizer used in sorghum and grasses is	**Azospirillum**
✓	Vertical mulch is used in which type of soils	**Black cotton soils.**
✓	Criteria of essentiality were given by:	**Arnon & Stout, 1939.**
✓	The optimum spacing for wheat is:	**22.5 cm** (line to line)
✓	The non-traditional area for cultivating wheat is:	**Eastern India.**
✓	Major P fertilizer commonly used in India is:	**DAP.**
✓	State having minimum BPLpopulation is:	**Punjab.**
✓	Maximum number of BPL peoples are in:	**Bihar**
✓	Bio fertilizer more suited for sugarcane is:	**Azatobacter**
✓	Depth of sowing of soybean seed:	**3 cm.**
✓	First variety of rice introduced in India is:	**TN-1 (1964-65)**
✓	Indian mustard is known as:	***Brassica juncea.***
✓	Explosive fertilizer is:	**Ammonium nitrate.**
✓	Most prominent soil group of India:	**Alluvial soil.**
✓	In **SSP** sulphur contains:	**12 %.**
✓	Which micro nutrient deficient in Indian soils:	**Zn.**
✓	Which macro nutrient mostly deficient in Indian soils	**Nitrogen**
✓	**Spike tooth** harrow is a type of implement:	**Secondary tillage**
✓	Most widely used material for correcting soil acidity is:	**Lime.**
✓	**Diara** cultivation method is followed in:	**Cucurbits.**
✓	**C: N** ratio of organic matter is:	**10:1**
✓	National Seed Act was passed in:	**1966.**
✓	National Seed Corporation (NSC) in:	**1963.**
✓	Root promoting hormones:	**IBA**
✓	Flower and fruit setting hormone:	**NAA**
✓	Bench terracing is done when slope is more than:	**15 % (16-33%)**
✓	Mass per unit volume is called:	**Bulk density.**
✓	Disk plough is used when the soil is:	**Tough.**
✓	First man made cereal is:	**Triticale.**
✓	Cheapest N contains fertilizer is:	**Urea** (Rs. 6/kg & 46% N)
✓	First product of urea hydrolysis:	**Ammonium**

		carbamate.
✓	**NABARD** was set up on the recommendation of:	Sivaramman committee
✓	Free living nitrogen fixing organism is:	Azatobacter
✓	The fruit of mustard is known as:	Siliqua.
✓	Supplemental irrigation is known as:	Life saving irrigation
✓	Major agricultural importing commodity in India is:	Edible oil (65%)
✓	Lunishree is a variety of:	Super rice.
✓	Plants with separate male and female flower on same plant are known as:	Monoecious
✓	National Biodiversity Board is situated at:	New Delhi.
✓	***Beaufort scale*** is used to measure:	Wind force/speed
✓	One standard atmosphere is equal to	1013.25mb
✓	Die back of shoots occur due to deficiency of:	Copper
✓	Occurrence of grey spots on leaves is due to deficiency of:	Manganese (Mn).
✓	Most destructive disease of sugarcane is:	Red rot of sugarcane.
✓	Potatoes are borne on:	Stolon
✓	Potato tubers are a modified form of:	Stem
✓	Sowing of sugarcane in trench method is helpful to prevent:	Lodging
✓	For seed purpose carrot is grown as:	Biennial
✓	Boron is harmful to plants when concentration is:	More than 3 ppm
✓	Which crop has maximum productivity among the cereals:	Maize
✓	**Kaolin** is a type of anti-Transpirants:	Reflecting type
✓	**Cycocel** (CCC) is a:	Growth retardants
✓	Family of sesame:	Pedaliaceae
✓	Family of jute:	Tilliaceae
✓	Exhaustive family of crops is:	Graminae (poaceae)
✓	Origin of maize:	Mexico
✓	Origin of soybean:	China
✓	Origin of tobacco:	Mexico
✓	State having highest area under summer maize is:	Bihar
✓	Indian Journal of Agriculture Sciences is	ICAR

	published by:	
✓	Yellowing of tea leaves may be due to the deficiency of:	**Sulphur**
✓	The greenhouse gas that released from paddy field is:	**Methane**
✓	The edible part of sweet potato (***Ipomoea batata***) is:	**Adventitious root**
✓	Carbon dating technique is used to determine age of:	**Fossils**
✓	The irrigation method which is suitable for saline soils is:	**Flooding method**
✓	Element contributing to the disease & drought resistance is:	**Potassium**
✓	Element involved in energy transfer and storage in plants is:	**Phosphorus**
✓	At present how many nutrients are termed as essential elements for plants:	**17 nutrients**
✓	According to Arnon & Stout how many nutrients are termed as essential elements for plants:	**16 nutrients**
✓	Apical bud dominance is caused by due to:	**Auxin hormone**
✓	Highest consumption of k fertilizer in:	**Maharashtra**
✓	Highest production of cereal crop in world is	**Wheat**
✓	Seed rate of sorghum is:	**9-12 kg/ha**
✓	De-suckering is a important practice of :	**Tobacco.**
✓	Name the weed, which was first biological controlled:	***Lantana camera***
✓	State, where seed law are adopted:	**Karnataka**
✓	Soil moisture tension directly measured by:	**Tensiometer**
✓	Evapo-transpiration (ET) measured by:	**Lysimeter**
✓	Parshall flame is used for measurement of:	**Water flow**
✓	Flood irrigation method is suitable in the crop:	**Rice & jute**
✓	Fertilizer application through irrigation is known as:	**Fertigation**
✓	One of the most important cultural practice in rice field is:	**Puddling**
✓	Post-harvest losses in cereals accounts for:	**10%**
✓	Saltation is a process of:	**Wind erosion**

✓	Irrigation efficiency of loamy soils:	**70% (highest)**
✓	Sowing pattern used in dry land is:	**Broadcasting**
✓	Most efficient method of irrigation is:	**Drip irrigation**
✓	Extensively grown pulse crop in India is:	**Chick pea**
✓	Relative humidity measured by:	**Psychrometer**
✓	Crops growing for conservation of soil moisture is known as:	**Mulch crops**
✓	Example of erosion permitting crop is:	**Maize**
✓	Example of erosion resistant crops are:	**Cow pea, Groundnut**
✓	The herbicide used as pre-plant incorporation (PPI) is:	**Fluchloralin**
✓	Example of C_3 weed is:	**Bathua** (lamsquarter)
✓	The botanical name of **two rowed** barley is:	***Hordeum distichon***
✓	The name of **six rowed barley** is:	***Hordeum vulgare***
✓	In which crop first **AICRP** was started	**Maize (1957)**
✓	What is the optimum plant population for **Bt**-cotton:	**10,000 plants/ha**
✓	**Dropsy** disease in human being is caused by the weed:	***Argemone mexicana*** (Satyanasi)
✓	Crop is suitable for soil having P^H 4-6:	**Rice, tea and potato**
✓	Red hemoglobin in root nodules is responsible for:	**Transport of O_2**
✓	Annual rainfall in Rajasthan is about:	**500-575 mm**
✓	Aminophilic plants:	**Sugarcane and rice**
✓	Country having 100% of the cropped area under irrigation:	**Egypt**
✓	Country having highest irrigation efficiency:	**Israel**
✓	Country having highest irrigated area:	**India**
✓	Optical wave lengths in **remote sensing ranges:**	**0.3** to **15 um**
✓	India is the **first** country to develop and use of hybrids in bajra, and the first variety was:	**HB-1(1965)**
✓	In India which crop is having maximum area under irrigation:	**Wheat**
✓	Most extensively grown fruit crop in the world is:	**Mango**
✓	**Pusa giant** is the variety of:	**Berseem, Napier grass**

		& Papaya
✓	Type of soil water which is most useful for plants is:	Capillary water.
✓	At **equator** day and nights are always:	Equal.
✓	Atmospheric pressure always **decreases** with:	Altitude.
✓	**Horizontal** heat transfers from warmer area to cooler area, the effect is known as:	Cloth lineeffect
✓	In which effect **vertical** heat transfers from warmer to cooler crop area:	Oasiseffect
✓	The enzyme of **nitrogen reduction** is:	Nitrogenase
✓	The element of **nitrate reduction** is the:	Molybdenum.
✓	Most of field crops require the pH range for optimum growth is:	6.5-7.0
✓	Hence; cultivated mostly crops are:	Acidic in nature (IFFCO Exam- 09)
✓	Depth of water required by a crop is known as:	Delta
✓	Area irrigated by one cusec discharge of water is:	Duty
✓	Period in days for which irrigation is supplied to a crop is:	Base period
✓	Rancidity in sunflower oil is due to:	Oxidation.
✓	Varieties of black gram are:	T-9, Pant U-19
✓	Famous variety of moth bean is:	RMO-40, RMO-257
✓	Elite varieties of moong bean are:	Pusa baisakhi, SML-668
✓	**0.1%** solution in **500 liters** of water is:	500 ml (RPSC, AO-09/SRF-09)
✓	**Bioclimatic** law was proposed by:	Hopkins
✓	Chemical safeners are also known as **herbicide antidote** e.g.	NA (it used first in maize)
✓	**Crop** which has **highest** K^+ up take:	Potato
✓	The low pressure area near the equator is called:	Doldrums/ITCZ
✓	**Triacontanol** is a growth stimulant obtains from:	Leaves of Lucerne

Important points-

- ✓ **Tillage increases** the **water holding capacity** and **infiltration** of soil but it **reduces** bulk density.
- ✓ **Wet land plough** is **smallest** among the wooden ploughs it is used to **puddle soil** & efficiency of **0.1 ha** for 8 hours.
- ✓ Blade harrow is known as ***Bakhar*** used for weed control before sowing.
- ✓ The **sprinkler system** of irrigation provides **protection from cold** by the **release of latent heat of fusion.**
- ✓ **Splitting** of a **pod** or **fruit** in characteristic manner at **maturity** is known as ***Dehiscent*** e.g. pods of pulse.
- ✓ Continuous consumption of Lathyrus causes a disease known as ***lathyrism*** as its grains contains **BOAA** (Benzo-oxalin acitic acid), aneurotoxin.
- ✓ Fertilizer which is preferred to grow succeeding crop in green manured soil: ***Phosphatic***
- ✓ The cattle consuming grasses of **low** magnesium content suffer from **hypomagnesaemia**, commonly called as ***'Grass Tatany'***.*(AAO Exam-2012)*
- ✓ The practice of grazing of animals on public lands is called as: ***Ranching***
- ✓ **Weather** is a state or condition of atmosphere at a given place at a given instant of time.
- ✓ Recently scientists separated germs in maize corn for manufacturing: ***Contraceptives***
- ✓ India's rank **first** in production and consumption of **sugar** in the world.
- ✓ The tobacco contributes the maximum **excise revenue** among agricultural commodities in India.
- ✓ India is the only country to have developed and grown **hybrid cotton** commercially.
- ✓ **Karnataka** is the largest producer of **coffee** in the country.
- ✓ **China** is the largest producing and consuming country of **tobacco** in the worldwhereas, India has third position in production of tobacco.
- ✓ Phenyl mercuric acetate (**PMA**) is a chemical used in agriculture crops in order to: ***Reduce transpiration***
- ✓ Jowar is known as: ***Camel crop/Kings of coarse grains***
- ✓ India is the **second country** after china to have commercialized the **hybrid rice** technology.
- ✓ **HD 2329** variety of wheat contributed maximum production during green revolution in India.
- ✓ Plant rectangularity is: **Plant to plant distance/row to row distance.**
- ✓ The direction from which the winds are coming called as: **Wind ward side.**

- ✓ Agronomic measures are used to reduce erosion where slope is: ***Less than 2%.***
- ✓ **Bench terracing** is usually a practiced on slopes ranging from: ***16-33 %***
- ✓ The highest award presented to an agricultural scientist in the country: ***Rafi ahmad kidwai award.***
- ✓ Name the plants which produce food themselves through photosynthesis: ***Phototrophs/autotrophs or producers.***
- ✓ **Sui generis**" used in **GATT** proposals is a Latin word means ***'a system of its own'.***
- ✓ In northern India, sometimes **rain takes** place in **winter** which is due to: ***Western depression*** **(JRF-97/RPSC, AO-09).**
- ✓ In total agricultural credit the contribution of **commercial banks is highest (74%)**, followed by co-operative banks (17%) and rural banks (9%) in 2009. **(Pre-PG-2013)**
- ✓ **Agriculture** came into existence in: **7500 B.C.** when **cultivation of wheat** and **barley** firstly **started.**
- ✓ Minimum support prices (MSP) of various crops are recommended by *Commission on Agricultural Cost & Price* ***(CACP),*** at present 22 mandated crops (14 kharif and 6 Rabi crops and two commercial crops i.e. jute & copra) are covered under MSP while sugarcane is covered under fair and remunerative price (FRP).
- ✓ Accumulation of nutrients in a lake or pound due to human intervention or natural process is called as: ***Eutrophication***
- ✓ India is meeting **85 per cent** its urea requirement through **indigenous** productionbut depends heavily on **imports** for its **phosphatic and potash fertilizer** requirements. **(RAS-2012)**
- ✓ Nitrogen fixation in rice field is carried out by blue green algae with: water fern ***Azolla***.
- ✓ Cultivations of such crops which have different natural habit and zero competition is known as: ***Parallel cropping***
- ✓ The variety of wheat which is resistant to all the three rusts is: ***Choti lerma***
- ✓ How many times a **ratoon** crop of sugarcane is advisable: **One time**
- ✓ Cultivation of two or more than two crops of **different heights** simultaneously on a certain piece of land in a certain period is called: ***Multistoried cropping***
- ✓ Nursery area required for seedling of one hectare rice field is: ***0.10** ha*
- ✓ **Red** and **purple** color of maize leaves is due to the deficiency of: ***Phosphorus.***

Center of orign:

- Vavilon (1924) identified **8 centers of origin** and **3 sub centers** of crops.

Major crops and their origin place:

Crop	Origin	Crop	Origin
Rice	S. E. Asia (Burma)	Tea	China
Sugarcane	New Guinea	Soybean	China
Wheat	S. W. Asia (Turkey)	Mustard	China
Gram	S.W. Asia	Cow pea	Africa
Maize	C. America (Mexico)	Cotton	India
Tobacco	C. America	Brinjal	India
Groundnut	Brazil	Arhar	Africa
Moong	India	Jowar	Africa
Potato	S. America (Peru)	Bajra	Africa

Terms and related agronomists:

	Terms/field	Agronomist
✓	Irrigation Agronomy	Dr. N.G. Dastane
✓	Cropping System	Dr. S.S. Bains
✓	Irrigation Engineering	A.M. Michele
✓	**SRNF**(slow release nitrogen fertilizer)	Dr. Rajendra Prasad
✓	CRI stage of irrigation in wheat:	B. L. Bhardwaj
✓	Functional nutrient	Nicholus, 1963
✓	Essentiality of nutrient (criteria)	Arnon & Stout 1939
✓	Agrobiology	Wilcox
✓	IW/CPE Ratio	Parihar *et al* 1974
✓	Transplanting in pearl millet	Dr. Gautum
✓	Coefficient of effectiveness	Nichiporovic (1950)
✓	H-4 hybrid of cotton	Dr. C.T. Patel (GUJ.) 1970
✓	Jaya (first Indian variety of rice)	Dr. Shastry
✓	Law of minimum	J.V. Liebig in 1840
✓	Law of limiting factor	Black man (1905)
✓	Increase yield N low	Wilcox (1929)
✓	Leaf area index	Watson (1947)
✓	Nutrient mobility concept	Brays R.H
✓	A" value concept	Dean & Friend (for P)
✓	Harvesting index	Donald

✓	Photoperiodism concept	Gardener & Allard 1920
✓	Thermoperiodism	F. Went
✓	Vernalization	Lysenko
✓	Antibiotics (Penicillin)	Alexandrian Fleming 1929
✓	***Bacillus radicicola*** (*now**Rhizobium***)	Beijerinck 1890
✓	Rothamsted experimental station	J.B. Lawes & J.H. Gilbert
	Founders/father of different fields:	
✓	Father of field plot experiment:	J. B. Boussingault
✓	Father of pedology:	V.V. Dokuchaiv
✓	Father of golden revolution:	Dr. K.L. Chadda
✓	Father of tillage/weed science:	Jethro Tull
✓	Father of soil testing technique:	M. L. Troug
✓	Father of soil microbiology:	S. N. Winogradsky
✓	Father of agriculture chemistry:	J. V. Liebig

Unit-1
AGRONOMY

The term derived fromof two **Greek** words:

- ***Agros*** = **Field**
- *Nomos* = To manage

Definition:

- Can be defined as a branch of agricultural sciences that deal with methods which provide favorable environment to the crop for ***higher productivity.***
- Indian agriculture is predominantly of the **subsistence type**
- **Peter Decresenzi** (1230-1307 A.D.) collected many literatures related to agronomy in his book "***Opus Ruralium Kamo Daram***" for the first time. He is known as **'Father of agronomy'**.
- Van Helmont (1577-1644 A.D.) concluded that the main principle of vegetation is water.
- **Jethrotull** (1674-1741 A.D.) published a book ***'Horse Hoeing Husbandry'*** and the words ***'Weeds'*** and ***'Zero tillage'*** were firstly given by him. He is known as **'Father of weed science'**.
- **Aurthor Young** (1741-1820 A.D.) Conducted pot culture experiments to increase the yield of crops by applying several materials like poultry dung, litter, gunpowder etc. published his work in 46 volumes as' ***Annals of Agriculture'***.

- A truly **scientific approach** to farming was started in 1840 by ***Justus von Liebig***.

Important concepts:

1. Leaf area index:

- Conceptwasdeveloped by **Watson, 1948**
- **LAI** = $\frac{\text{Total leaf area (m}^2\text{)}}{\text{Ground area (m}^2\text{)}}$

2. Land equivalent ratio (LER):

- It means the relative land area occupied by a sole crop that is required to produce the yield achieved in intercropping.
  - **LER** = (Y12/Y11) + (Y21/Y22)

Here,

- Y12 = yield of crop 1 when intercropped with crop 2
- Y11 = yield of crop 1 when grown as a sole crop
- Y21 = yield of crop 2 when intercropped with crop 1
- Y22 = yield of crop 2 when grown as a sole crop

When **LER > 1**: Intercropping is **beneficial.**

- E.g. - **LER: 1.50** means mixture gives 50% more yield.

3. Cropping intensity:

- C.I. = $\frac{\text{Total cropping area}}{\text{Net cultivated area}} \times 100$
- Cropping intensity of India is nearly: **150-%.**

4. Crop rotation:

- In crop rotation **land is fixed** but **crops are rotated** year after year on **same land**. i.e. after cultivation of **cereals** should be taken **pulse** crop in next year.
- It **maintains** and **improves** soil **fertility**.
- It checks the soil **erosion** and **conserves** moisture.
- The **incidence** of harmful pathogens, insect-pests, nematodes and weeds which are associated with particular crop can be **managed**.

5. Crop rotation intensity:

- C.R.I. = $\frac{\text{No. of crops in crop rotation}}{\text{No. of years in crop rotation}} \times 100$
- Crop rotation for 1 year: Rice - Pea (200%).

- Crop rotation for **2 years**: Groundnut - Wheat - Sugarcane (150%).
- Crop rotation for **3 years**: Paddy - Potato - Sugarcane - Arhar - Wheat (166%).

6. **Cropping system:**

- It refers to the cropping patterns used on a farm and their interaction with farm resources, associated farm enterprises and the available component technology.

7. **Harvesting index**:

- The efficient utilization or assimilation of CO_2 in the form of photosynthesis is expressed in terms of harvest index. **H.I.** is also called **coefficient of effectiveness.**
- $$\text{H.I.} = \frac{\textbf{Economic yield (grain)}}{\textbf{Biological yield}\text{ (grain + straw)}} \times \mathbf{100}$$

- H.I. is similar to ***partitioning coefficient***.

H.I. of major filed crops-	
Potato = 0.82 (82%)	Wheat= 0.4
Rice = 0.5	Maize= 0.5
Pulses = 0.3	Arhar = 0.19
Oilseeds = 0.3	Cotton= 0.3
Sugarcane = 0.27	

8. **Coefficient of effectiveness:**

- **C. E.** = **Sink/Source** x 100.
- *Source*: Parts of the plant where **photosynthesis** occurs, generally **leaves.**
- *Sink*: Where photosynthesis **accumulates**, generally **grain is the sink** but may be **exceptions** *viz.* **stem inpotato, roots** in **sugar beet** and **leaves in Palak** is the *sink*.

9. **Sorghum sickness**:

- This is harmful effect on succeeding crop due to unavailability of Nitrogen.
- The **stubbles of sorghum** takes a long time to decompose due to **wide C: N** ratio hence, during the process of decomposition soil nitrogen gets blocked or temporally **immobilized,** affecting the growth of succeeding crop by showing early nitrogen deficiency symptoms due to wide C/N ratio of sorghum residue.

Unit-2

Dry land agriculture:

- Means cultivation of crops entirely under **Rainfed** condition.

a. **Dry farming**:

- Cultivation of crops in areas where annual rainfall is **less than750** mm.
- **Crop failure** due to prolonged dry spells during crop period is most common.
- Dry farming is practiced in **arid regions** with the help of moisture **conservation** practices.
- **Alternate land** use system is suggested in this region.

b. **Dry land farming**:

- Cultivation crops in areas where annual rainfall is **more than 750** mm but **less than 1150** mm is called dry land farming.
- The soil moisture conservation measure is the **key for dry land farming** practice in **semi-arid** regions.
- **Drainage** facility may be required especially in **black soils**.

c. **Rainfed farming**:

- Means cultivation of crops in regions where annual rainfall is **more than 1150 mm** is called as **rain fed farming.**
- This farming is practiced in **humid regions**.

SN	Constituent	Dryland farming	Rainfed farming
1	Rainfall (mm)	< 800 mm	>800
2	Moisture availability to crop	Shortage, PET > P	Enough, PET< P
3	Growing season	< 200 days	>200 days
4	Growing region	Arid, sub arid & sub humid	Humid & sub humid
5	Cropping system	Single or mixed cropping	Double or Intercropping
6	Constraints	Wind & Water erosion	Water erosion

Unit -3

1. **Agro-climatic regional planning**:

- It was initiated in **1988** by the **Planning Commission** under the **Seventh Five Year Plan**.
- Regional planning was come into force on the basis of agro-climatic factors for balanced regional growth.
- Planning commission (now NITI Aayog) is headed by **Prime Minister** of India.

- **NITI** Aayog- National Institution for Transforming India (2015).

2. **Agro-climatic zones:**
   - Agro-climatic zones are classified on the basis of criteria of homogeneity in agro-characteristics such as rainfall, temperature, soil, topography, water resources and cropping system.
   - According to planning commission India has been divided into **15 agro-climatic zones**.
   - NARP/ICAR has divided into **131 agro climatic zones**.

The trans-gangetic plain zone:
   - It consists of Punjab, Haryana, Union Territories of Delhi and Chandigarh and Ganganagar district of Rajasthan.
   - It has highest irrigated area, highest net sown area, high cropping intensity and high ground water utilization.

3. **Agro-ecological region**:
   - Agro-ecological region is the land unit cut out of agro climatic region when superimposed on land form and the kind of soils and soil conditions that act as modifiers to climate and length of growing period.
   - According to **NBSS&LUP** India has been divided into 21 agro ecological zones (**at present only 20**).
   - According to **ICAR** agro-ecological zones are divided in to **8 zones**.

4. **Drought**:
   - As a situation occurring in any area in a year when annual rainfall is **less than 75** % of the normal is known as drought.
   - When deficiency of rainfall is **above 50%** of the normal, called **severe drought**.

5. **Moisture Availability Index** (MAI):
   - It is the ratio between rainfalls (weekly/monthly) at 50 % probability level to potential evapotranspiration of the corresponding period.
   - MAI = Rainfall at 50 % probability/PET.

6. **Evapo-transpiration (ET):**
   - Soil moisture is the most limiting factor in dry land agriculture.
   - About **60-70%** of the rainfall is lost through evaporation.
   - The evapo-transpiration is the evaporation from the soil surface and transpiration from the plant surface.
   - ET losses can be reduced by use of **mulches, anti-transpirants** and **wind breaks.**

a. Anti-transpirant:

- Materials are applied to plant surface **to reduce** transpiration from the plant.
- **Anti-transpirants reduce** photosynthesis.
- Hence, their use should be limited to save the crop from death under severe moisture stress.
- They have some practical use in **nurseries** and **horticultural** crops.
- Anti-transpirants aremainly four types:

(i) **Stomata closing type**:
- They reduce water loss through **closing stomata.**
- E.g. Phenyl mercuric acetate (**PMA @ 10^{-4}**)
- **Atrazine** and **CO_2** are also acts as effective anti-transpirants.

(ii) **Film forming**:
- Checks transpiration loss of water due to **formation of thin film** which acts as physical barrier.
- E.g. **Mobileaf, hexadecanol, silicone, oils, waxes.**

(iii) **Reflecting type**:
- These chemicals **reflect the radiation** and reduce leaf temperature.
- Application of kaolin @ 5% reduces transpiration losses.

(iv) **Growth retardants**:
- These chemicals **reduce shoot** growth and increase root growth and thus enable the plants to resist drought.
- They also induce **stomata closing e.g. Cycocel** (CCC).

b. Wind breaks and shelterbelts:

- **Wind breaks** are such structures which break the wind flow and **reduce wind speed**.
- While **shelterbelts** are rows of trees or shrubs planted across the wind direction for protection of crop against wind.
- Wind break & shelterbelt provides the protective shelter against desiccating winds to the extent of **5-10 times** the height of tall tree on **windward** side and up to **30 times** on **leeward** side.

7. Agro-forestry:

- Agro-forestry is a **sustainable land management** system which increases the use of land **combines** the production of crops, fruits, forest plants and or animals.
- It is also known as tree-crop planting system.

Components of agro forestry:

a. **Agri-silviculture**: The growing of agricultural crops along with the forest trees is known as Agri-silviculture.
   - Agriculture crops + Forest crops (silviculture).

b. **Sylvo-pastoral system**: Forests are managed for the production of wood as well as for rearing of domesticated animals.
   - Silviculture + Pasture management.

c. **Agro-sylvo-pastoral system**:
   - Agriculture crops + Forest crops + Pasture management.

d. **Agri. Horti- silvicultural system**:
   - Agriculture crops + Horticulture + Silviculture.

e. **Agri-Silvi-Aquaculture System:**
   - Agriculture crops + Forest crop + Fish Production.

Unit-4
Soil conservation measures:

1. **Agronomic measures**:
   - Adopted where slope is **less than 2%** and erosion problems are not severe.
   - It includes contour cultivation, tillage, mulching, and strip cropping.

a. ***Contour cultivation***:
   - Cultivation of crops by sowingacross the slope.
   - It conserves **soil, soil fertility** and water.

b. ***Mulching***:
   - Mulch is any material applied on the soil surface to **check evaporation** and improve soil water.
   - E.g. C**rop residues**, leaves, manures, **polythene films** etc.
   - It is most important tactics to **conservation of soil moisture** and reduces **erosion** for agriculture in dry regions.

c. ***Strip cropping***:
   - System of cropping in which long and narrow strips of erosion resisting crops (close growing crops) are alternated with strips of erosion permitting crops.

2. **Mechanical measures**:
   - Adopted to supplement the agronomical practices and when land slope is **more than 2 %.**
   - It includes following practices.

a. *Contour bunding*:

- Contour bunding is adopted up to **6% slope** of land and used in **arid and semi-arid** areas with high infiltration and permeability.

b. *Graded bunding*:

- Recommended in areas receiving rainfall of **more than 800 mm per year**.
- In general **bunding** is suitablefor lands having slopes from **2-10 %.**
- **Contour** and **graded** bunding are mostly used in India.

c. *Sub-soiling*:

- Breaking of hardpan of soil by the sub soiler at 30-60 cm depth and 90-180 cm interval.

d. *Broad Bed and furrow* **(BBF) system**:

- Suitable for managing rainwater in **deep black soils** where surface drainage during the monsoon period is a problem.

e. *Bench terracing:*

- Adopted in lands where slope ranging from 16-33% on steep slopping and undulating land.

f. *Zing terracing*:

- Adopted in lands with **3 -10** % slopes.
- Zing terraces are constructed in medium to deep soils in moderate to high rainfall areas.

g. *Vegetative barriers*:

- These are closely spaced plantations, usually a few rows of grasses or shrubs grown along contours for erosion control.
- **Khus** (*vettiveria zelanica*) is the most suitable plant for this purpose.

h. *Vertical mulching*:

- Digging narrow trenches across the slope at intervals and placing the straw or crop residues in these trenches.
- It is mostly practiced in **coffee gardens**.

Unit -5
CLIMATE

Weather: a state or condition of atmosphere at a given place and at a given time.

Atmosphere:

✓ It is a colorless, tasteless and odorless mixture of gases that surround the earth.

✓ It extends up to a height of about ***1600 km.***

✓ However 99 per cent of total mass of the atmosphere is within a height of 40 km from the earth surface.

Structure of atmosphere:

- Based on ***vertical temperature differences*** atmosphere is divided in to four layers.

1. **Troposphere**:
   - It is the lower most layer of atmosphere near to earth surface
   - Extending up to a height of 8 to 18 km from the earth surface. **(RPSC, AAO-O9).** (8km from polar and 18km from equator)
   - **Temperature decreases** with altitude at the rate of **6 °C /km.**
   - It influences **earth climate**.
   - It is dense part of atmosphere, contains **85 per cent** of the atmosphere's mass.
   - All weather phenomena like clouds, rain, mist, fog, dew etc. occur in troposphere. **(IARI, Ph. D -09)**
2. **Stratosphere**:
   - It is a second major atmosphere layer which lies beyond the height of 8 to 18 km and extending up to 50km.
   - **It is warmest layer, increase** temperature with altitude.
   - It is the actual seat of most of the photochemical reactions in air.
   - Due to presence of ***Ozone layer***, UV rays are absorbed, hence it is also called ***ozonosphere***
   - Most of concentration of ozone is present about a height of 25 km from the earth surface.
3. **Mesosphere**:
   - It lies above the stratosphere; a height of **50 km** extending up to **80 km** from the earth surface.
   - It is coldest layer of atmosphere av. temperature is being -90^{0}C.
   - Strong decrease in temperature with increase in altitude.

4. **Thermosphere:**

- In this layer atmospheric densities are extremely low
- From earth surface it is extending a height of 80 km up to 1600 km.
- The lower layer of thermosphere (80-300km) is called ***ionosphere***.
- Long distance radio communication is made possible through this ionized layer.
- It is outermost layer is called ***exosphere.***
- Exosphere is the layer above ionosphere.
- In this layer temperature **increase** with altitude.

1. **Composition of atmospheric air** or principle gases of atmosphere:

SN	Gases	Volume %
1	Nitrogen (N)	78.08
2	Oxygen (O_2)	20.95
3	**Argon (Ar)**	**0.93 (RPSC, AAO-09)**
4	Carbon dioxide (CO_2)	0.03 (300-360ppm)
5	Neon (Ne)	0.0018%

2. **Composition of soil air**:

- **Nitrogen** percentage is more or less **same** as in atmosphere.
- **Oxygen** levels slightly **decrease** as compared to atmosphere.
- **CO_2 levels** will be nearly **8-10 times more** (0.25-3.0%) as compare to atmosphere.

Solar constant:

- Defined as the energy falling in one minute on a surface area of one square centimeter at the outer boundary of the atmosphere, held normal to the sunlight, at the mean distance of the earth from the sun.
- Solar constant value is: **1.94 Cal/cm^2/min** or **SI** unit is: **1353 Watts/m^2**
- The solar radiation is received in the form of electromagnetic waves/particles are called ***quanta*** or ***photons.***

PAR (photosynthetic active radiation):

- Photosynthesis in green leaves use solar energy in wavelengths from **0.4 to 0.7 um or** the radiation **400-700 nm** is most active in photosynthesis is known as Photosynthetic Active Radiation.
- PAR or Photons are measured by **Quantum sensor**.
- Unit is ***Einstein*** or Watt/m^2
- ***Lux:*** it is the oldest unit to measures intensity (brightness) of solar radiation.

- *Spectrometer-* measures solar radiation in narrow wave bands.

Lapse rate:

- The **vertical temperature decrease** or gradient is expressed as ***lapse rate***.
- Normally lapse rate is **6.5 °C/km**.

Adiabatic lapse rate:

- The rate at which the temperature changes as **air rises** or **falls** is called ***adiabatic lapse*** rate:
- **Dry adiabatic** lapse rate: **10 °C/km**
- The condition in which there is abrupt rise instead of fall in temperature is called ***inversion.***

Albedo:

- It's the ratio between **incoming** and **outgoing** solar radiations on the earth surface.
- It is the percentage of the incident solar radiation reflected by a surface is known as ***Albedo.***
- It is expressed in percentage (%).
- The albedo of crop surface is ranging between 23-30 %.
- The albedo for fresh snow is ranging from 75-95%.
- The albedo for sea surface or water is ranging from 6 -10%
- The albedo for land area of earth is ranging from 10-40% (av. 30%)

Cyclone:

- A **cyclone** is a roughly circular low pressure area whose diameter may be from hundred to a thousand miles and wind movement towards the center. Atmospheric pressure is **always lowest** in the center of this region and increases rapidly outward e.g. *Tornado, Typhons, Herrikens* are names of cyclone.

Climate:

- Climate is a generalized weather or **summation of weather condition** over a given region during a comparatively longer period.

Classification of climate:

a. **De Candolle** (1900 AD):

- The first person to attempt the classification of climate **based on vegetation.**

b. **Koppen** (1936):

- Divided the world climate into **5 groups**, based on **weather elements**.

c. **Troll** (1965):

- This classification has been found satisfactory to explain vegetation zones of tropical **Africa & S. America**.
- This classification is suitable for agricultural purpose
- The **ICRISAT** (Hyderabad) adopted this method for classifying semi-arid tropics in **India**.

d. **Thornthwaite:**

- Based on PE index, TE & Season distribution of rainfall
- PE index = **P/E x 100**

Where,

- **PE** = Precipitation of effectiveness
- **P** = Annual precipitation.
- E = Annual evaporation.

e. **Moisture Deficit Index (MDI):**

- AICRP on dry land (ICAR) adopted classification based on MDI.
- MDI = **P-PET/PET ×100**

Types of climate	MDI values
Sub humid	**0 to 33.3**
Semi-arid	-33.3 to -66.6
Arid	**> - 66.6**

Measures of humidity:

- Humidity means water vapor content of the atmosphere.

a. **Relative humidity:**

- $$\text{RH} = \frac{\text{Water vapor present in the air}}{\text{Water vapor required for saturation}} \times 100$$

- Relative humidity is expressed in: **Percentage** (%)
- It is simplest in measurement, mostly useful for agriculture.

b. **Absolute humidity:**

- Unit of absolute humidity is: ***g/m^3*.**

c. **Specific humidity:**

- Unit of specific humidity is: ***g/kg*.**

Forms of precipitation:

- **Rain**: Drops of > 0.5 to 6 mm
- **Drizzle**: Drops of < 0.5 mm
- **Snow**: Ice crystal resulting from sublimation
- **Dew**: Moisture condensed in small drops upon cool surface

- **Frost**: A feathery deposit of ice
- **Fog:** A thin cloud of varying size, droplet of 1-50μ
- **Mist**: A very thin fog, droplet size is **50-150 (μ) microns**.
- **Vapor:** Water vapor in the atmosphere. It constitutes only about **0.001 per cent** of the total global water.

Temperature:-

- It is the most important variable of weather.

Measurement of temperature:

- The S.I. unit for temperature is degree **Celsius** (°C)

a. **Celsius:** in Celsius scale, the melting and boiling point of water are **0** and **100 ° C** respectively.

  - **C = (F-32) 5/9**

b. **Fahrenheit: I**n Fahrenheit scale,the melting and boiling point of water are **32 ° F** and **212° F.**

  - **F = C (9/5) + 32**

c. **Kelvin:** ***absolute temperature*** is measured in Kelvin units.

  - When there is no activity of molecules of substances, then it is said to be at **0 °K.** The melting point of water in Kelvin scale is **273 °K** and steam point is **373 °K**
  - **°K = °C + 273.**

Instruments:

a. **Maximum thermometer**: It is used to measure the highest temperature attained during the day. It consists of hollow glass tube filled with ***mercury.***

b. **Minimum thermometer:** Itis used to measure lowest temperature. It consists of a glass tube with ***alcohol*** in bulb.

c. **Soil thermometer:** It is a mercury thermometer with a ***bend*** at the bulb end.

d. **Thermo-couples:** these are used for measuring soil temperature **and** they are two types, viz., ***thermo-electric*** and ***resistance.***

Phenomenons:

- **Contour**: Lines joining equal ***elevations.***
- **Isopleths**: Line joining uniform values of *any* ***phenomena.***
- **Isobar**: Imaginary lines connecting points of equal ***atmospheric pressures.***
- **Isohyets**: Lines of connecting points of equal ***amounts of rainfall.***
- **Isobaths**: Imaginary lines connecting points of equal ***water table elevations.***

- **Isohels**: Lines connecting points of equal ***sunshine hours.***
- **Isophene**: Lines connecting points of similar ***seasonal phenomenon***.
- **Isotherm**: Lines connecting points of equal ***temperature.***
- **Isotope**: Elements having equal ***number of neutrons***.
- **Isopluvial**: Lines joining equal ***depth of rainfall.***
- **Isotech**: Lines connecting points of equal ***wind velocities.***
- **Isostere**: Lines connecting points of constant ***moisture contents.***
- **Isodemic**: Lines which join the places of similar ***disease severity.***

Instruments and their use:

Instrument	To measue or Uses
Altimeter	Height
Aneroid barometer	Atmospheric pressure
Anemometer	Wind speed
Beaufort scale	Wind force in Sea
Crescograph	Growth of plant
Dynamometer	Convert mechanical energy to electrical
Cambel stokes	Duration of sun shine
ECG	Movement of heart
Wind wane	Wind direction
Tensiometers	Soil moisture tension
Galvanometer	Electric currents
Porometer	Stomal behavior
Hydrometer	Measures relative density of liquids
Hygrometer/Psychrometer	Relative humidity
Lactometer	Density of milk, milk purity
Lysimeter	Evapo-transpiration/leaching loss of water
Manometer	Root pressure
Planimeter	Area of an irregular figure
Pyrheliometer	Measures amount of direct solar radiation
Pyranometer	Measure total incoming (global) solar radiation
Pycnometer	Specific gravity of soil
Peizometer	Depth of water table
Potometer	Transpiration

Classification of irrigation projects:

Project	CCA (ha)	Cost (Rs).
Major	>10,000	> 50 Crores
Medium	2000-10,000	5 Crores to 25 Lacs
Minor	< 2000	< 25 Lacs

Unit-6
Monsoon:

- The word monsoon has originated from the **Arabic word "*mausim*"** which means **"Season"**.
- Monsoon is often referred to seasonal reversal of winds over the Indian Ocean especially in Arabian Sea.
- India is situated in the north-east trade wind zone and these trade winds continue throughout the year
- Total amount of precipitation in country is: **1194 mm**.
- Or total rainfall generated volume in India is: **400 mha-m**
- Average rainy days in country: **130 days/year**.
- **Rainy day**: > 2.5 mm rains during 24 hours called rainy day.
- **Crop rainy day**: >5.0mm rains during 24 hours called crop rainy day.

1. The South –West monsoon:

- S. W. monsoon reach south India (at coast of Kerala) around **1st June** of every year and in the middle of July it covers whole India under the grip of rainfall.
- It withdrawal start from western Rajasthan in **1st September** to complete withdrawal from whole country by **15 October**.
- It moves ataverage speed **30 km/hour**
- It's called as **'Grand period'** of rainfall in India.
- It contributes around **> 75 %** of total rainfall in India.
- Tamil Nadu get lowest rainfall 20-40% during south-west monsoon since most of the rainfall is received during North-East Monsoon.
- In Jammu & Kashmir the concentration of rainfall during south-west monsoon is about 35-50% and remaining portion is fulfills from western disturbances during November to April.

South-West monsoon has two branches:

a. **Arabian Sea branch:**
   - Contributes **770** mm rainfall (80 % of SW monsoon).
   - This moves northward (in **Delhi)**.

b. **Bay of Bengal:**

- Contributes 340 mm rainfall (about 20 % of SW monsoon).
- This move up to **Assam**.

2. The North-east monsoon:

- This monsoon is onset nearly 15 October and it is limited mostly to southern states (**TN, AP & some part of KA**) and withdrawal almost by mid-December.
- Tamil Nadu gets highest annual rainfall (600-800mm) during north-east monsoon followed by Andhra Pradesh & Karnataka.
- This monsoon is also known as **retreating monsoon**.
- It contributes about **15 %** of total rainfall in India

Western Disturbances:

- It moves from west to east direction
- It defined as a low pressure or cyclonic circulation in the wind above the surface which moves from western countries to Himalayas Mountains (J&K).
- December to March is the main period when western disturbances carry rainfall in north and central India which are very useful for Rabi crops and in Rajasthan locally known as '***Mavath***'.
- **EL Nino** is a phenomenon that leads to rise in the ocean temperatures during mid December (start from cost of Peru) and negatively affect the Indian monsoon, consequent **dry weather and drought in many States** of **India.**
- **LA Nino** this phenomena is opposite of EL Nino that lead to cold sea surface temperature in the central and eastern pacific Ocean which is positively affect the Indian monsoon, lead to above-normal rainfall in india during the south west monsoon season.
- River which utilize maximum flow: **Ganga**

Artificial rain making or Cloud seeding:

- Application of foreign material to clouds to induce precipitation is called cloud seeding.

a. ***Cold seeding***: Silver iodide (AgI) or dry ice is applied at one nuclei for a liter of cloud air. In dynamic cloud seeding, massive quantities of 100-1000 nuclei for a liter of cloud air are applied.

b. ***Warm clouds:*** Generally is used Sodium chloride (NaCl).

Weather forecasting:

- Short range: **< 3 days** or up to 72 hrs.
- Medium range: 3 -10 days
- Long range: **> 10 days**

- The short range forecasting is issued twice a day and especially valid for 36 hrs.
- The long range forecasts are issued thrice in a year and are useful for choosing cropping patterns.
- The prediction of monsoon is based on statistical observations of Long Range Forecasting.
- While Synoptic method of forecasting is used for short range forecasts.

✓ **NC-MRWF:** National Center for Medium Range Weather Forecasting, **New Delhi.**

Properties of water:

✓ Latent heat of vaporization: **580 Calories/g at 0 ºC**
✓ Heat of sublimation is: **680 Cal./g at 0 ºC**
✓ Heat of fusion is: **80 Cal/g at 0 ºC**
✓ Highest density of water at: **4 ºC**
✓ Water expands on freezing: **9% by volume.**
✓ Dielectric constant of water is: **- 80**
✓ Water has highest **specific heat1 Cal/g** of any known substances **except liquid ammonia** i.e. 1.03 Cal/g.

Water logging:

- A situation where all the pores both macro and micro are filled with water temporarily or permanently.

Demerit of water logging:

✓ Reduces **ODR** (oxygen diffusion rate).
✓ Reduce Eh (redox potential): Eh of **well aerated** soil is: **400-700mv**
✓ **Eh of waterlogged soil**: **-250** to **-300 mv**
✓ **Denitrification** loss more in waterlogged soils.
✓ In submerged rice SO_4 are reduced to sulphide that is further reduced to $\mathbf{H_2S}$ that causes '***Akiochi***' in rice due to toxicity of H_2S.

Approaches of irrigation scheduling:

a. **Transpiration ratio** approach:

- Amount of water transpired by a crop to produce unit amount of dry matter.
- This is also known as **absolute approach**.

b. **Soilmoisture deficit approach:**

- **For maize & wheat 25%** depletion of available soil moisture.
- **For drought resistant** crops **50%** depletion of available soil moisture.

c. **Climatologically** approach or **IW/CPE** approach:

- Known as soil **moisture index approach.**

- The ratio IW/CPE serve as soil moisture stress index.
- Lower the ratio will be more stress.
- It is scientific approach but not practical utility.

d. **Critical stage approach**:

- The state of growth when plants are most sensitive to shortage of water and most responsive to correction of deficiency.

Crop	Critical stage	IW/CPE ratio
Rice	Booting, Panicle initiation, flowering	**1.2** (highest)
Wheat	**CRI**, tillering, jointing, flowering, milking, daugh	0.9
Maize	**Tasseling**, silking	0.9
Cotton	Flowering, boll formation	0.7
Gram	**Pre-flowering**, pod development	0.6
Groundnut	Flowering, **pegging**, pod filling	0.6
Pigeon pea	Flower initiation, **pod formation**	0.6
Safflower	Flower initiation	**0.4** (lowest)

Percentage irrigated and water requirement of major field crops:

Crop	**% irrigated area**	**Water requirement** (mm)
Sugarcane	**93%**	1500-2500
Wheat	**85%**	650-800
Rice	50%	1500
Oil seed	25%	500-700
Pulses	10%.	450-700

Measurement of irrigation water:

- 1 Cusec: **28.3 litres**
- 1 Cumec: 1000 litres or 10^3
- 1 ha mm: 10, 000 litres or $\mathbf{10^4}$
- 1 ha cm: 100, 000 lit. or $\mathbf{10^5}$
- 1 ha m: 100, 000, 00 lit. or **10^7 or $1000m^3$**

Measures of water height:

- **1 Atmosphere** = 1036 cm of water
- **1 Bar** = 1023 cm of water

Conversion units of:

- One acre = 0.405 **ha,**
- One ha = 2.47 **acre.**

Ploughing is classified (according to **CRIDA**, Hyderabad 1985) as:

Type	Depth (in cm)
Shallow	5- 6
Medium	15-20
Deep	25-30

Watershed

- It may be a **natural unit of land** whose runoff collects and flows out of the area through a single common outlet in to a river or other water body.

Classification of watershed:-

1. Macro- watershed : > 50,000 ha
2. Sub-macro watershed : 10,000-50,000 ha
3. Milli watershed : 1000-10,000 ha
4. Micro watershed : 100-1000 ha.
5. Mini watershed : 10-100 ha.

Unit-7. Irrigation

- Irrigation is the artificial application of water to partially meet the crop evopo-transpiration requirement.

Irrigation methods:

1. Surface irrigation:

- **Most common** method of irrigation in India.
- It includes following methods.

a. **Flood irrigation**:

- Water is applied to the crop by flooding it on the soil surface.
- This method is practiced where irrigation water is abundant.
- Mostly adopted in wetland rice.
- Useful in uniform surface soils with good water holding capacity.

b. **Check-basin**:

- Most common among surface methods, suitable for **close growing** crops e.g. wheat, groundnut, pearl millet etc.
- It consists in running water in to relatively level basins surrounded by small ridges.
- A limitation of this method is that it has too many ridges which occupy large area of land and require too much labour to formed beds.

c. **Ring basin**:

- Suitable for **fruit** crops or orchards.

d. **Border strip method:**

- In this method the field is divided into number of strips, which are separated by ridges.
- Soils with a **slope** of **0.5-1%** are suitable for this method.
- Suitable for ***all close growing crops*** on medium to heavy texture soils but not suitable for sandy soil.

e. **Furrow** method:

- This method is generally used to irrigate row crops and vegetables.
- It is ideal for slopes varying from 0.2-0.5 %
- The length of furrows may vary from 30m for sandy soils at lesser slope to 300m for heavy soils at greater slope.
- Furrow irrigation is suitable to most type of soils except sands that have a very high infiltration rate.
- Furrows are mostly suitable for root and tuber crops

f. **Corrugation**:

- Small and shallow furrow are known as corrugation, suitable for close growing crops like wheat, ground nut etc.

g. **Surge irrigation**:

- The term 'Surge irrigation' is the intermittent application of water to the field surface under gravity flow which results in a series of ON & OFF modes of constant or variable time span.
- This process of ON- OFF water supply and cut off results in minimized deep percolation and runoff losses (hardly exceeding 20%). Not popular in India

h. **Cab legation**:

- It's an **automatic method** of surface irrigation.

2. Sub-surface or sub soil irrigation system:

- Water is applied into a series of field ditches through underground pipe lines, generally followed in **Western Countries**.
- This method is practiced in **Kerala** for coconut garden and in **Gujarat** and **Kashmir** on sandy loam soils for vegetables.
- Drip system can be used in sub surface method.

3. **Sprinklers**:

- Water is applied with pressure to the surface of any crop or soil in the form of a thin spray, somewhat resembling rainfall.
- This system consists of sprinkler nozzles, which are mounted on risers in lateral lines taken from main line, which is further connected to a pumping unit.

- The rate of spray of water can be regulated and natural rainfall can be simulated.
- Sprinkler irrigation can be used for all most all crops, (except rice and jute) and on most soils (except heavy clay soils).
- It is especially suited for fields with steep slopes or undulating topography.

Advantages of sprinkler irrigation:

- Suitable for **undulating topography** and **sandy soils.**
- Saving of water from **25-50%** for different crops
- Discharge rate is more than **1000 lit/hr**.
- Sprinkler pressure **2.5-4.5 kg/cm^2**.
- Water use efficiency can be as high as 60 % much higher to surface method of irrigation.
- Increase 40 % in irrigated area with same amount of water as compared with surface method of irrigation.
- About 40-60 % saving in labor compared with surface.
- It can be used to protect crops against frost and high temperatures.

4. **Drip irrigation**:

- Introduced from **Israel.**
- Drip or trickle irrigation is one of the latest methods of irrigation which is becoming increasingly popular in areas with **water scarcity** and **salt problems**.
- This irrigation is defined as the **precise** but slow application of water as discrete drops or continuous drops through mechanical devices, called **emitters** located at selected points along water delivering lines.
- This system involves the slow application of water, drop by drop to the root zone of a crop.
- In this method water is used very economically, since losses due to deep percolation and surface evaporation are reducing to the minimum.
- Drip irrigation is best suited in water scarcity area where water quality is marginal, topography is undulating or steep, soil depth is restricted, labour is expensive and crop value is high.

Advantages of drip irrigation:

- Suitable for water scarcity area water saving **50-70%** as compare to surface.
- Irrigation efficiency about 90 pecent campare to surface method

- Discharge rate of water per dripper is generally **1- 8lit/hr.** at 1.5-2.5**kg/cm^2** pressure.
- Fertilizers (Fertigation) and herbicides (Herbigation) also possible to apply with drip (through venture)
- This method of irrigation is mostly suitable for **widely spaced crops**, **orchard** trees and in **green houses** (protected cultivation of vegetables & flowers).

5. **Typhoon system** of drip irrigation:
    - In **sugarcane** modified drip, in this method a particular water depth maintained at a pa rticular growth stage.

Drainage:

- Agricultural drainage is the removal of excess water known as free water or gravitational water from the surface or below the surface of the farm land.
- So as to create favorable soil conditions for plant growth.

Drainage methods: 1. Surface drainage 2. Sub surface drainage.

1. Surface drainage:

- It is **simplest and cheapest** method of drainage
- Most common in India by digging open drains at suitable intervals and depth

a. **Drainage of flat areas:**
- Where slope **less than 2%**

b. **Broad bed and furrow system (BBF):**
- Field is laid out into **120-150 cm** width beds and **45 cm** wide furrows across the slope.
- About **0.5** % slope is provided for free drainage.
- Crops are sowed on beds having two or more than two rows each. The beds are **15 cm** raised.
- BBF methods are widely practiced in **groundnut (clay** soil).

2. Sub-surface drainage:

- Purpose is to lower down the ground water level **below root zone.**

Sub-surface drainage is includes:

a. **Tile drains**: Including **perforated pipes**.
b. **Mole** drainage: Suitable for **clay** soil.
c. **Vertical** drainage: Drainage **by digging wells** is called vertical drainage. This practiced in coffee gardens.

Methods of water measurements:

1. **Orifices**:
   - To measure water in comparatively small streams like flow into border stripes furrows/ check basin.
2. **Weirs**:
   - Used to measure the flow in an irrigation channel or the discharge of a **channel** outlet at the source.
   - Basic equation for discharge through weir is: $Q = \mathbf{CLH^m}$
   - Discharge relationship for suppressed weirs is = **0.0184** $LH^{3/2}$

90° -V" notch weir: is an excellent device for measuring small flows. The discharge may be computed using the formula.

- $\mathbf{Q = 0.0138\ H^{25}}$

3. **Parshall flume** (venture flume):
   - Combination of weir and the submerged orifice
   - Can be used even in relatively shallow channels like irrigation furrows with flat grades.

Unit-8
Soil Moisture Constants

Metric potential:

- The total water potential that is attributed to the solid colloidal matrix of the soil system

Capillary potential:

- The energy with which water is held by soil is defined in terms of capillary potential.

Seepage:

- Horizontal flow of water channel is called seepage.
- Water loss from the irrigation channel or canal is mainly due to seepage.

Percolation:

- Down ward movement of water through saturated or nearly **saturated soil** in response to the gravity.

Infiltration:

- Entry of water from the upper layer of soil is called infiltration.
- It occurs in **unsaturated soil**.

Leaching:

- Downward movement of nutrients and salts from the root zone with the water is called leaching.

Saturation capacity:

- When all the soil pores are filled with water at a tension of zero "o" bars is termed as saturation capacity.
- It is called as maximum water holding capacity.

Field Capacity (FC):

- The soil moisture content after 2-3 days of irrigation and after drainage of gravitational water has become very slow and soil moisture content had become relatively stable.
- At the field capacity, the large pores filled with air and the micro pores are filled with water.
- It is considered as the **upper limit** of water availability to plants.
- The tension at field capacity ranges from **-0.1 to –0.33** (1/3) Atm.
- It is determined with the help of **pressure plate apparatus**.

Permanent wilting point (PWP):

- The concept permanent wilting point (PWP) proposed by **Briggs and Shantz** in 1912.
- They utilized dwarf **sunflower** as an indicator plant.
- Soil moisture content at which plants can be no longer obtains enough moisture to meet their requirement and remained wilted unless water is added to the soil.
- It is the **lower limit** of available water to plant.
- Although plants not dead, but they are now in a permanently wilted condition and die if water is not added.
- Soil moisture tension at PWP is about **-15 bars** or 15 atmosphere

Wilting coefficient:

- The percentage of moisture in root zone at the **permanent wilting** of plants is called the wilting coefficient or critical moisture point.

Available water:

- Concept of available water given by **Veihmayer and Hendrickson** 1981.
- It is the soil moisture available for maximum plant use.
- It is the portion of capillary water held between **Field Capacity** to **PWP** (i.e. –0.3 to -15 bars).

Ultimate wilting point:

- The moisture content at which **wilting is complete** and the **plant die** is called **UWP**.
- At **UWP** the soil moisture tension is as high as **-60 bars**.

pH: the term pH given by **SPL Sorenson** (1909):

- Defined as the negative logarithm of hydrogen ion activity.
- Log 1/ (H^+)

- Range: 0-14
  - Acidic: < 7
  - Neutral: 7
  - Basic: > 7

P^F Value: it is defined by **Schoefild** (1935).

- The logarithm of centimeter height of a water column to give the necessary suction.
- Expressed in **centimeter** of water (based on the height of a water column above free water level in cm).

Soil condition	P^F values	Pressure (atm/bars)
Saturated soil:	0	0.001
Field capacity:	**2.53**	**1/3**
PWP:	**4.18**	**15**
Hygroscopic point:	4.50	31
Oven dry soil:	7.0	10000

P^E given by **Silen:**

- It is negative logarithm of electron activity.
- P^E = **Eh**/0.0591
- It measure intensity of **oxidation-reduction.**

Measurement of soil moisture:

- **Indirect method** (on the basis of tension):-
- **These are followings:-**

a. **Tensiometer/irrometer**:

- This method is based on principle that the capillary tension increases with decrease in soil moisture contents and vice versa.
- **Tensiometers** are sensitive up to **0.85 bars of soil moisture.**
- Suitable for **sandy** soil at field capacity or below it in which most of the available water **below 1 bar** tension.
- **Not suitable for clay** soil.
- Tensiometers are very useful in scheduling irrigation to **frequently** irrigated crop.

b. **Electrical resistance/bouyoueous** moisture meter/**gypsum block**:

- Gypsum blocks are containers made up of gypsum, fiber glass etc.
- It based on the principle that electrical resistance increases with decrease in moisture content.
- Based on conductivity of electricity low resistance at field capacity and high at **PWP**
- Gypsum block perform satisfactory measurement in ***dry soil***

- **Not suitable for saline** soil as salt in soil increase conductivity.
- It is used for measuring soil moisture tension up to **0-15 bar** and for scheduling irrigation of crops of delayed intervals.

c. **Neutron moisture meter**:

- Rapid method used for **Field** or **in situ** measurement of soil moisture from large volume of soil based on the measurement of number of H^+ nuclei present in a unit volume of soil.
- This is **best method** at present available for quantities estimation of changes in soil moisture. ***(ARS-05)***

d. **Pressure member and pressure plate apparatus**:

- It is used for laboratory measurement of soil moisture potential.
- Generally used for estimating FC, PWP and moisture content at different pressures.
- It is suitable for measuring soil moisture tension: 0-15 bar.

Soil moisture characteristic curve:

- **A curve** showing relationship between the energy status of water i.e. tension and amount of moisture in soil.

Soil moisture characteristic curve:

- For sandy soils= **L** shaped
- For clay soils= **I shaped** (almost a straight line).

ET (evapo-transpiration):

- Total amount of water lost due to transpiration by crop and evaporation from the soil surface during a specified time from a particular area.
- Evapo-transpiration constitutes nearly **99 %** of the total water uptake.

PET (potential evapo-transpiration):

- Concept of **PET** was given by **Thornthwaite** (1948)
- The ET occurs when the ground is fully or completely covered by actively growing vegetation and no limitation of soil moisture.
- PET is the **upper limit** of ET for a crop in a given climate.

Measurement of ET:

a. **Lysimeter**:

- It consists growing of crops in large containers installed in a crop field measuring their water loss and gains.
- ET and Leaching losses can also be measured by lysimeters.

b. **Soil moisture depletion studies:**

- Depletion (D) = **FC-PWP x D. Db.**

Where,

FC = Field capacity, PWP = Permanent wilting point

D = Depth, Db = Bulk density.

c. **ET can be measured by evaporation data:**

(i) **USWB class 'A' open pan evaporimeter**:

- It isstandard evaporimeter most widely used by agro-meteorological observatories.
- It having diameter of **120** cm with a depth of **25** cm and it filled with water up to **20** cm
- It is painted **white**. Evaporation is expressed as mm/day

(ii) **Sunken screen** evaporimeter:

- Developed by **Sharma** and **Dastane**, 1968
- It is used to measure evaporation (ET) under field conditions.
- It is determination of **PET/ET** ratio which according to them varies from **0.95** to **1.05**.

(iii) **Portable evaporimeter:**

- It was developed in Israel in 1984 to measure evaporation for ***very short periods.***

Consumptive use (CU) of water:

- It is used to designate the losses due to evapo-transpiration and the water used by the plant for its metabolic activities.
- **CU = ET + Water used in metabolic activities**.
- About **99** percent of water loss in plant due to evapo-transpiration, only 1 % of water is used in actual metabolic process.

Estimation of PET/ET:

a. **Penman method**:

- Developed by penman, 1950 for estimating evaporation from free water surface,
- It is most accurate estimation of ET_0.
- As it utilize all meteorological parameter responsible for ET.

b. **Water use efficiency (WUE)**:

- It is the yield of a marketable crop produced per unit of water used in evapo-transpiration. **OR**
- **WUE** = the dry matter produced per unit of water used and it expressed as kg/ha-mm.
- Water use efficiency are **two types:**

(i) **Crop water use efficiency**:

- It is the ratio of crop yield (y) to the amount of water depleted by the crop in the process of evapo-transpiration (ET).
- WUE (Crop) = **Y/ET**

(ii) **Field water use efficiency:**

- It is the ratio of crop yield (y) to the total amount of water used in the field (WR).
- WUE (Field) = **Y/WR**

Water use efficiency of major field crops:

S.N	Crop	WUE (kg/ha mm)
1	Finger millet	**13.4**
2	Wheat	12.6
3	Groundnut	9.2
4	Sorghum	9.0
5	Pear millet, maize	8.0
6	Rice	**3.7** (lowest)

Unit-8. Important Terms

Terms	Crop
Curing:	Tobacco, tea, onion
Stripping:	Jute
Nipping:	Gram, cotton
Wrapping:	Sugarcane
Propping:	Banana
Trashing:	Sugarcane

Oil percentage in field crops:

Crop	Oil (%)
Coconut:	**60**
Sesame:	**46-52**
Ground nut:	**45**
Castor:	**35-58**
Safflower:	**24-36**
Rape seed and mustard:	**33-39**
Soybean:	**20**

Protein content in cereals:

- Rice: **6-7%** Wheat: **11-12%** (**Gluten**)
- Maize: **10%** Sorghum: **10-12%**
- Bajra: **11-12%** Barley: **11.5%** (**Albuminoids**)

Protein content in pulses:

- Gram: **21.1%** Lentil: **25%**

- Pea: 22.5% Arhar: 21-25%
- Moong: 25% Urd: 24%
- Cow pea: 23.4% (Known as **vegetable meat**).

Protein content in oilseeds:

- Soybean: 42%
- Groundnut: 26%
- Sesame: 18-20%
- Linseed: 36%
- Safflower (Cake): 40-45%

Classification of crop based on families:

1. Graminae (new name is **Poaceae**): it is largest cultivated crops family known as grass family including **Cereals**, **Napier grass.**
2. Leguminaceae (=Papilionaceae): e.g. Pulses, groundnut, **sunhemp**, berseem, alfalfa, dhaicha, methi etc.
3. Solanaceae: Brinjal, tomato, potato, chilies, **tobacco.**
4. Cruciferae (Brassicaceae): Mustard, **radish**, cabbage, **cauliflower**.
5. Cucurbitaceae: **All gourds**, cucumber, **pumpkin**.
6. Malvaceae: Cotton, **bhindi**.
7. Asteraceae (Compositae): Sunflower, safflower, niger.
8. Chenopodiaceae: **Spinach**, beet, sugar beet.
9. Euphorbiaceae: Castor, topica.
10. Convolvulaceae: Sweet potato.
11. Umbelliferae: Coriander, cumin, carrot.
12. Aliaceae: Onion, garlic
13. Zingiberaceae: Ginger, turmeric.

Unit-10
Crop Production

1. Paddy (*Oryza sativa*): 2n=24

- ✓ Rice is the stable food of **more than 60%** world population and most of the people of S.E. Asia, about **90%** of rice grown in the world is produced and consumed in **Asia**.
- ✓ It is a crop of wet tropical climate and also grown in humid regions of subtropics.
- ✓ In India, rice occupies the first position among the cereals in respect of both area and production.
- ✓ Rice contains **6-7 %** protein (lowest among cereals) & contains 2-2.5 % fat which is lost during milling.
- ✓ **Hulling** percent in rice is **66%** or **2/3** of paddy.
- ✓ In general cereal crops required about **400-500 liters** of water for the production of one kg of plant dry matter.
- ✓ But rice consumes ten times more water **i.e. 5000** liters of water to produced 1 kg rice grain. **(OBC, Bank-09).**
- ✓ In rice field to prevent **de-nitrification** and **leaching** losses N- fertilizer should be incorporated in **reduced zone** so that the process of nitrification is delayed.
- ✓ Rice field submergence in 5cm deep water during the reproductive and grain formation stages is beneficial. **(NET-2001)**
- ✓ The main protein in rice is: ***Oryzenin***

Area & Distribution:

- Area: **India** > China > Indonesia
- Production: **China** > India > Indonesia
- Productivity: **USA** > Japan > China

In India:

- Area: **West Bengal** > UP > Bihar
- Production: **West Bengal** > UP > Andhra Pradesh
- Productivity: **Punjab** (34 q/ha).

Classification:

Genus ***Oryza*** have 24 species out of which two are cultivated.

1. *Oryza sativa*, Diploid, **2n=24**, it has three main geographic races.
    a. **Indica: -** Grown throughout tropics and sub-tropics in India and characterized by **tall** stature, weak stem, elongate grains, droopy leaves, low productivity and **irresponsive** to high input conditions.

b. **Japonica**: - Grown in temperate and sub-tropical regions of **Japan** and characterized by **short** stature, narrow erect & dark green leaves and **responsive** to intensive input condition having highest productivity, grain shape is oval to round in shape.

c. **Javanica/Balu rice**: - Grown in **Indonesia**, few tillers, **wild form** of rice, has many awns and long duration crop period.

2. *Oryza glaberrima*: it is grown in **West Africa** only.

Origin: Indo-Burma region (S.E. Asia).

- ✓ Rice is **self-pollinated** crop, fruit type is ***caryopsis.***
- ✓ The grain or caryopsis is tightly enclosed by the **lema** and **palea** collectivelycalled the ***hull.*** **(IARI, Ph.D-1997)**
- ✓ Test weight (wt. of 1**000 grains**) is = **25g**
- ✓ Rice stem is commonly called the ***haulm*** or ***culm*** which made up of a series of nodes & internodes.
- ✓ Inflorescence of rice is ***panicle*** which is a group of spikelet.
- ✓ **Climate: C_3, short day plant** and required hot & humid climate.
- ✓ **Soil: clay** and **clay loam** are best suited for rice cultivation.
- ✓ Acidic soils are good for rice cultivation, most suitable pH 4- 6.
- ✓ Average temperature for plant growth is 21-37°C
- ✓ In India mostly rice cultivated on low land condition.
- ✓ Rice is semi aquatic plant grown best under submerged condition, under these conditions the atmosphere oxygen is transported by **aerenchymatous tissues** of leaves.

Seasons:

- **Aus**: Sowing time **April-May.**
- **Aman** (Kharif): 84% of total area, sowing time **June-July.**
- **Boro** (spring): Sowing time **November - December.**
- In North-West part of country only Aman (**kharif**) crop is taken due to falling of temperature in winter.

Varieties:

- First dwarf variety of rice developed in world is **Taichung Native-1 (TN-1)** in **Taiwan in 1964-65.**
- First high yielding dwarf variety developed in world is **IR-8**, by cross between ***Peta×Dee-gee-woo-gen.*** It was the first known photo insensitive variety to be released by International Rice Research Institute in **1966.**
- Literally meaning of Dee-gee-woo-gen is: ***Brown tipped short legged.***
- **First introduced** variety of rice in India was TN-1 in 1964-65.

- **IR-8** was the second introduced variety of rice in India from the IRRI, Manila (Philippines) **in 1966**.
- First Indian high yielding earliest semi-dwarf variety ***Jaya*** released in **1968**, evolved by **Dr. Shastry**, and it's a cross between ***TN-1 × Type 141***. It gives better yield as compare to TN-1 & IR-8 both; hence called '***Miracle rice in India***'.
- Cuttack (odissa) evolved two varieties: ***Malinga & Mahsuri***
- Mutant dwarf varieties of rice: ***Jagannath*** and ***Sattari***.
- Varieties suitable for direct seeding: ***Bala, Cauvery, Nagina-22, Pusa 2-21***
- BLB (bacterial leaf blight) tolerant: ***TKM-6***
- Salinity tolerant: ***IR-8, Lunishree***
- Suitable for waterlogged area: ***Jalmagan, Jalmani***
- First hybrid of basmati in world: ***PRH-10.***
- First hybrid of rice in India is: ***CORH-1 (1994)***
- Variety released from **IARI**: ***Norin-8, 18, pusa 2-21*** *and basmati-370, Pankaj, IR-20.*
- **Pusa Basmati-1**: Its world's first high yielding dwarf variety under **quality** rice has been developed by **IARI** through convergent breeding.
- **Two genome edited** varieties of rice by ICAR: DRR Dhan 100 (kamala), DST Rice-1
- **Other varieties**: Padma (for upland), Pankanj (waterlogged), Ratna, Chambal, Sita, Sabarmati etc.

Seed rate:

- Broadcasting: 100 kg/ha
- Drilling: 60 kg/ha
- Hybrid rice: 15 kg/ha
- **Dapog method**: 1.5-3kg/m^2
- **RxP distance**: 20 x10cm.

Fertilizer management:

- NPK ratio **100:60:60** and 25 kg zinc sulphate (at last puddling) is generally recommended in rice for high yielding varieties.
- The water logged puddle soil develops two zones; the upper zone (**1 to 10mm**), a thin surface layer which contains oxygen is called ***"oxidized zone"*** and immediately below the oxidized layer lies a reduced layer which contains no oxygen is called ***"reduced zone"***. **(RPSC, AAO-2012).**

- Use of nitrate fertilizer in paddy field should be **avoided** due to more susceptibility to loss of nitrogen through **leaching** and **denitrification**.
- Hence, in low land the deep placement of the ammonical form of nitrogenous fertilizers in the reduced zone has been proved to be efficient method of nitrogen application than other methods.
- If we have to use nitrate fertilizer in low land then we should be incorporated in oxidized zone only.
- Nitrogen use efficiency in rice: **30-40%**. **(IARI, Ph.D-1997)**

Biological Nitrogen Fixation:

- It has been estimated that the total nitrogen requirement of rice could be brought down by about 25-30 per cent through bio fertilizer (algae).
- In addition to **blue green algae** (BGA), the association of the water fern, **Azolla** with the blue green algae, ***Annabaena azollae*** also helps in biological nitrogen fixation.
- **Azatobacter**, a non-symbiotic nitrogen fixing bacteria also fixes atmospheric nitrogen biologically in upland rice field.

Khaira disease:

- **Zn** deficiency causes ***Khaira*** disease in rice usually at **nursery** stage.
- The tip of the zinc deficient leaves first become yellow and then dry and plant remained stunted.
- Zinc deficiency is marked by the appearance of **rust-coloured** or brownish red coloration on the surface of the outer leaves.
- First reported by **Y.L.Nene** in **1966** at **Pantnagar**, Uttrakhand.
- It is managed by spray of @ **5 kg zinc sulphate at** nursery stage.
- or foliar spray of 0.5 per cent zinc sulphate in early stage of crop.

Browning of rice:

- Iron toxicity in rice is known as ***browning of rice***
- **Iron toxicity** is the major problem to rice production in water logged acid-sulphate soils of ultisols & oxisols soil orders.
- The beginning of iron toxicity is characterized by the appearance of purple **coloration** at the **leaf tips** and finally has a **scorched** appearance.
- **Phalguna** variety is tolerance to **Fe-toxicity**.

Silicon in rice:

- Silicon is regarded as essential nutrient in rice production.
- Silicon deficient plants are soft and droopy.

- Silicon supply mostly required during the panicle initiation stage because during this stage root activity is relatively reduced and the rate of transpiration is high.
- It has been noticed that silicon provides a **thick cuticle-silica** layer which functions as a barrier against attack by insects-pests.
- Silicon deficiency can be corrected by application of **basic slag** @ 1.5 2.0 t/ha. Because basic slag is the most common source of calcium silicate.

Weed management:

- ***Echinochola crusgalli*** (barn yard grass) and ***E. colonum*** (wild rice) are major weeds in rice.
- The weed problem is **less in low land transplanted** rice compared to upland direct seeded.

Herbicides used in rice crop:

1. Benthiocarb (**Saturn**) @ **2** kg a.i. /ha is applied as pre-emergence of weeds.
2. Propanil (**Stam F-34**) @ 3 kg a.i. /ha is applied at 1 week after transplanting or post emergence of weeds.
3. Butachlor (**Machete**) @ 2-3.0 kg/ha is used as pre-emergence i.e. weeds have not emerged in field. It is effective to control both annual grasses and broad leaved as pre emergence of weeds because emerged weeds are not killed.
4. Anilophos: - it is a pre-emergence and early post-emergence selective herbicide used for control of annual grasses, sedges and some broad-leaved weeds in transplanted and direct seeded rice.
5. **Pyrazosulfuron** (Saathi):- it is new selective and systemic pre and post emergence herbicide of rice can be used before and early post emergence of weeds.

Beushening:

- It involves **cross ploughing** the young crop, 4-6 weeks after sowing with a light country plough in 5-10 cm standing water, ones or twice.
- Purpose behind this to **control weeds**, **optimize** crop stand and provide soil **aeration**.

Puddling:

- It is a process of destruction of soil aggregates by mechanical force in soils with excessive moisture content.
- Puddling destroys the soil structure which **hampers** root penetration and the development of the succeeding upland crops grown in rotation with rice.

- Puddling **increases the bulk density** of soil about from 1.4 to 1.7 g/cc.
- It is done with aequipment, known as **puddler**.

Objective of puddling:

- To **reduce percolation** losses of water. (main object)
- To reduce **weeds infestation.**
- To increase **nutrients availability**.

Critical stages for water.

- **Tellering stage**: it comes 2-20 days after transplanting.
- **Booting** or late panicle initiation stage: Swelling of the flag leaf sheath is called **booting.** It is most critical stage for water in rice crop.

Dapog method of rice cultivation:

- This method was introduced from **Philippines**.
- In this method, seedling becomes ready for transplanting on **12th** day after seeding.
- It comprises rising of a dense stand of seedlings without contact of soil.
- Rate for nursery sprouted seeds @ **3kg/m²** on wooden planks, trays, concrete floor, and seedbed covered with polythene sheet.
- Hence, 2**5-30 m²** nursery area is sufficient for one hactare.

Hybrid rice:

- In the last decade of 1970, hybrid rice was firstly developed by **China** by using **cytoplasm male sterility-fertility restore gene system**.
- After China, India is the **second** country to commercialized hybrid rice of **APRH-1, APRH-2** (AP), **KRH** (Karnataka), and **MGR-1** (TN).
- **PHB-71** is the only hybrid rice which is released by private organization.

Golden Rice-

- Term coined by **Ingo potrykus.**
- **beta carotene** rich gene inserted in rice through genetic engineering
- That fortified the vitamin A content in rice
- It is transgenic or GM rice.

SRI (System of rice intensification):

- This methodology was developed by Henri de Laulanie in Medagaskar in 1980s.
- It is an agro-ecological approach for increasing the productivity of irrigated rice by changing the management of plants, soil, water and nutrients and cultivating of rice with as much organic manure as possible.

- Basic principles in SRI are significantly reducing irrigation quantities, plant population & improving soil conditions by using less or nil herbicides or chemicals.
- Benefits include 20-100% or more increased yields, up to 90 % reduction in required seeds and up to 50 % water saving.
- With SRI, average yields of rice are **6-7 tons** per hectare, compared to conventional method (2-3 tons/ha).
- SRI mainly adopted for irrigated upland or rainfed rice as well as other crops such as wheat, sugarcane, millets, mustard etc.
- SRI is more feasible for the farmers who have small landholdings and need to get the highest yields possible from their available lands.
- In SRI, wide spacing planting (**25x25cm**) are adopted and transplanting of young seedlingsat two leaf stage (8-12 days old), nursery area is **100 m^2 enough** with seed rate of **2kg/acre** or **5 kg/ha.**
- **Demerit of SRI is higher labour costs** in the initial years (for individual seedling transplanting and weeding).

Processing of rice:

1. **Milling:** The aleurone layer and embryo are removed during the polishing of rice grain. There is a loss of **vitamin B$_1$**, particularly thymine, the deficiency of which causes the disease called ***'beri-beri'*** in those persons who continuously eat polished rice. **(UPPCS Exam-2010)**
2. **Polishing of rice**: The removal of very fine bran clinging to milled rice and smoothening of the surface is called ***polishing.***
3. **Parboiling:** It is a hydrothermal proccss in which the crystalline form of starch is changed in to an amorphous form on account of irreversible swelling and fusion of starch.

Important questions:

✓	Sowing time of Aus/autumn rice:	**April-May**
✓	Sowing time of Boro rice:	**November-December**
✓	Rice inflorescence is known as:	**Panicle.**
✓	Rice is a type of plant:	***Short day plant***
✓	Cardinal temperature of rice is:	**30-32 degree**
✓	Hulling % of rice is:	**65-66 %.**
✓	Fruit of rice is known as:	**Caryopsis.**
✓	Widely used nitrogenous fertilizer in rice:	**Ammonium sulphate**
✓	Optimum pH of soil required for growing rice is:	**4-6 pH.**
✓	**Mat type** nursery is related to crop:	**Paddy.**

✓	Test weight of basmati rice:	**21 g.**
✓	Most critical stage of water for rice:	**Booting stage.**
✓	Biofertilizer for rice is:	**Azolla.**
✓	Seedling ready for transplanting in Dapog method after sowing…..days:	**11-14**
✓	Most dominant weed species in rice:	***Echinochola sp.***
✓	**White eye** of rice due to:	**Fe deficiency.**
✓	The gas emits from rice field is:	**Methane.**
✓	Akiochi disease in rice due to:	**H_2S toxicity.**
✓	Highest nitrogen losses in the rice crop by:	**Denitrification.**
✓	Paira & Utera cropping system is closely related to which crop:	**Rice**
✓	Parboiling of rice conserve the vitamin:	**Vitamin B_{12}**
✓	Seed rate in Dapog method for rice:	**3-4 kg/m^2**
✓	Area required for seedling preparation in Dapog method:	**25-30 m^2**
✓	Miracle rice of India is:	**Jaya**
✓	Gene responsible for dwarfness in rice:	**Dee-gee-woo-gene**
✓	Spacing for rice sowing of seedling:	**20 × 10 cm**
✓	Khaira disease of rice is due to:	**Zinc deficiency**
✓	Reclamation disease of cereals due to:	**Copper deficiency**
✓	A rice grain is said to be extra-long if it is:	**7 mm or more**

2. Wheat (*Triticum aestivum*): 2n = 42

- ✓ Wheat **ranks first in the world** among the cereals both in respect of area and production so wheat is called as ***'king of cereals'***.
- ✓ In India wheat is the **second** most crop in area and production after rice
- ✓ The word "***Cereals***" derived from a **roman word *'Ceres'*** which means the ***Goddess of grain.***
- ✓ Wheat contains spongy **protein** i.e. ***Gluten*** (this is colloidal complex of **gliaidine** and **glutenin** in equal amounts) which is essential for ***baking** (bakery products).*
- ✓ Gluten is somewhat harmful to children's health.
- ✓ Wheat protein is rich in niacin & thiamine content.
- ✓ **Perling index** in wheat: to determine the **hardiness** of grain, a **higher value** of P.I. means **lower kernel hardness**.

- ✓ In wheat the **permanent adventitious roots** develops below the soil surface after the appearance of the first leaves (at CRI stage) is called ***clonal roots***. **(NET-2001)**
- ✓ Wheat **inflorescence** known as ear or **spike** and seed is **caryopsis**.
- ✓ Mustard is very common crop usually intercropped with wheat.
- ✓ **Test weight** of wheat grain is about **40g** while test weight of ***phalaris minor*** is only **2g**.
- ✓ Gene responsible for dwarfness in wheat is **Norin-10** which is a Japanese variety isolated by **Dr. Borlaug** in 1960 at **Mexico**.
- ✓ In India introduced dwarf, photo insensitive high yielding Mexican varieties in 1963 like **Lerma Rojo, Sonaro 63, Sonaro-64** and **Mayo 64** were imported by then Minister of Agriculture, Shri C. Subramaniam and released for commercially cultivation in **India (1965)** these varieties were responsible for **green revolution** in India.

Area & Production (India):

- Area: U P > M P > Punjab
- Production: U P > Punjab > Haryana
- Productivity: Punjab (45q. /ha) > Haryana
- The **highest productivity** of wheat in **Punjab** due to **nearly 100 % area** of wheat is under **irrigated** conditions and **higher doses** of fertilizer consumption.

Classification:

1. Common bread wheat or bread wheat (***T. aestivum***):
   - Its **hexaploids, 2n =42** (imported species but mostly grown in **India).**
   - It is occupying about **87%** of wheat area. It is good for chapati making & bakery products.
   - Introduced in India by N.E. **Borlaug** from **Mexico**, hence called **Mexican** dwarf wheat and responsible for green revoluation.
2. Durum/Macaroni wheat (***T. durum***):
   - Tetraploids, 2n=28.
   - Used for ***Suji, Semya*** preparation.
   - This occupies about **12%** of wheat area, particularly in **Central** and **Southern** India in rainfed condition.
   - Durum varieties:- *Jairaj, Malavika*
3. **Emmer wheat (*T. diccocum*)**:
   - Tetraploids, 2n=28
   - Occupies about **1%** wheat area confined to GJ, **MH, AP, KN** and **TN,** Suitable for preparation of south Indian dish"***Uppumav***"

4. **Diploid species**:-
   (i) ***Triticum monococcum***:-2n = 14, diploid wheat
   (ii) ***Triticum aegiloploid*, 2n = *14*,** diploid wheat.

Origin of place: **S. W. Asia** (Turkey).

Irrigation stages of wheat (given by **R.B.L. Bhardwaj**)-

1. **CRI** (crown root initiation) stage: it comes **21st days** after sowing
   - It is most critical stage of irrigation.
   - Transition zone between root and shoot is known as **crown root initiation**.
2. **Tillering** stage: Up to 40-45 Day after sowing.
3. **Jointing** stage: Up to 60-65 Day after sowing.
4. **Flowering** (Booting) stage: Up to 80-85 Day after sowing.
5. **Milking** Stage: Up to 100-105 Day after sowing.
6. **Dough** stage: Up to 115-120 Day after sowing.

✓ Dough stage is refers to when **milky fluid** in wheat **grains turns harden**.

✓ If water is available to provide only one irrigation then it should be applied at **CRI** stage.

✓ If water is available to provide two irrigations, then it is preferred at: **CRI** and at flowering.

Climate: C_3, Long day plant. The most ideal conditions for desirable wheat growth are:-

1. Cool and moist weather during the vegetative growth period.
2. Warm and dry weather during grain formation. **(ICS/RAS)**

Booting stage:

- It is reproductive stage in wheat characterized by the swelling of the upper most internode.
- The spike which is inside is held by the last leaf called ***flag leaf***. This is the stage of the maximum leaf areas & responsible for contributing 50% of total yield.

Varieties:

- Norin 10 cultivar has created a new chapter in the history of wheat breeding.
- The dwarfness of norin was the result of two particularly recessive genes, which marked with the symbols of ***Sd 1*** and ***Sd 2*.** **(ICS Exam-2002).**
- Maxican varieties were red coloured grains while Indian & its cross are amber coloured.

1. **Single gene dwarf:**
    - **Lerma Rojo 64-A**
    - **Sujata**
    - **Girija:** suitable for hills of north India
    - **Sonalika:** developedfrom selection of ***HD 1553*** and ***RR 21***
2. **Double gene dwarf:**
    - **Kalyan Sona:** developed from ***Kalyan 227*** and ***Sona 227***
    - **Sonaro-64**, **Chhoti Lerma**
    - HD-2009 (**Arjun**), HD-2329, **Janak.**
3. **Triple gene dwarf**: (developed in 1970)
    - Heera, Moti, Lal Badshah.
4. **Mutant varities**: released from IARI, New Delhi
    - **Sharbati sonora** developed from: **Sonoro 64**
    - **Pusa lerma** from: Lerma Rojo 64-A

Other popular varieties:

- **Lok-1**: Suitable for late sowing.
- **C-306** (Farmy wheat): tall variety, suitable for rainfed, excellent in chapatti making quality.
- **K-65**: Tall variety, suitable for late sowing and saline soils.
- **Shailaja**: Suitable for Himachal Pradesh.
- **Malavia**-12: protein content is about 15%
- WH-147: Suitable for making good quality of '***Chapatti'***
- **PBW-502** (2004):Good yielder under irrigated, cool climate
- **RAJ-3765**: Late sowing variety, resistant to lodging
- **RAJ-4037** (2004): Suitable for warm climate
- **RAJ-3077**: Suitable for saline and late sowing.
- RAJ-4120 (2009): Resistant to ***UG-99 race*** of rust, yield 48-58 q/ha.

New varieties developed at IIWBR, Karnal-

- Karan Vandana (DBW-187)- biofortified, iron rich variety.
- Karan Vaidehi (DBW-370)
- Karan Vrinda (DBW-371)- heat tolerant variety (2023)
- Karan Varuna (DBW-372)
- DBW-327- yield 80qt/ha.

New varieties released by IARI-New Delhi

- Pusa Gautami (HD-3086)-
- Pusa Amulya (HD-3090)-
- Pusa Yashasvi (HD-3226)

Seed rate:

- For normal sowing: **100kg** seeds/ha.
- For late or broadcasting: 125kg seeds/ha.
- For dibbling required: 25-30 kg seeds/ha.

Spacing:

- R x P: - **22.5** x 10cm for irrigated.

Depth of sowing-it is depends on **length of coleoptiles**.

- Single gene dwarf varieties: 5-6cm depth.
- Double gene dwarf varieties: 4-5cm depth
- Triple dwarf varieties are sown at: 3-4cm depth.
- An average 4-**5cm depth** is ideal for wheat sowing.

Sowing date: 1st fortnight of November is the best time of wheat sowing, while rainfed wheat is sown in second fortnight of October

Sowing methods:

1. **Drilling:** Seed-cum -ferti drills is best and most common method.
2. **Broadcasting:** Required more seeds
3. **Dibbling:** sowing with dibbler, required less seeds than other methods
4. **Behind the plough method**: these are two types
   (a) **Kera method**: When seeds are dropped by hand in the furrows made by country plough is known as ***Kera method.***
   (b) **Pora method:** When seeds are dropped in furrow by ***funnel or nai***attached to theplough is known as ***Pora method.***

5. **FIRB Method**: New method of sowing is furrow irrigated raised bed system: it gives average to higher yield than other methods.

Fertilizer and Manures:

- Nitrogen: 120 kg/ha **(IARI, Ph.D 1998)**
- Phosphorus: 60 kg/ha
- Potash: 40 kg/ha

Dose:

- **Half dose** of nitrogen and **full dose** of P & K applied as basal and remaining **half dose** of nitrogen applied as **top dressing** at **1st**irrigations.
- Under rainfed condition full dose of nitrogen, P and K should be placed in to soil before sowing.
- Under rainfed condition spray of thiourea @ 0.5% in standing crops gives 10-15% higher yield.

Weed management:

a. **Monocot weeds:**

- ***Phalaris minor*** (for the control of *Phalaris minor,* applied Isoproturon @ 0.75kg a.i. /ha at 30-35 days after sowing.
- Phalaris minor has showed resistance against isoproturon reported by **Dr. R.K. Malik** at **Hissar**, hence **Sulfosulfuron** @ 20g a.i./ha or **Metasulphuron** @ 4g a..i./ha can be used instead isoproturon against ***Phalaris minor***.
- ***Avena fatua*** (wild oat): to control ***Phalaris minor*** and ***Avena fatua*** a **post emergence** application of Clodinafop-propargyl 60g a.i/ha.

b. **Dicot weeds:**

- ***E.g. Chenopodium album***, ***Anagallis arvensis***, ***Melilotus spp.*** etc. Applied 2, 4-D Eater Salt as post-emergence i.e. **30-35** DAS @ **500ml**/ha to control **dicot** (broad leaved weeds).
- In dwarf varieties application of 2, 4-D should be sprayed at 30-35days after sowing because one week earlier spray may cause ear malformation in wheat.
- (**2, 4-D**: 2, 4-dichlorophenoxy acetic acid).

Important questions:

- Botanical name of (common bread) wheat is: ***Triticum aestivum.***
- **Origin** of wheat is: ***South West Asia*** *(turkey)*
- Flowering portion of wheat is called: ***Spike or ear***
- Macaroni wheat is known as: ***Triticum durum***
- Emmer wheat is known as: ***T. diccocum***
- Suitable late sown variety of wheat: ***Lok-1, Raj-3765, Sonalika.***
- Spacing of row to row in wheat: ***22.5cm***
- Most critical stage for irrigation: ***CRI (21 DAS)***
- Gene responsible for dwarfness in wheat is: ***Norin-10***
- Important mimicry weed of wheat is: ***Phalaris minor***
- Test weight of wheat seed: ***40g***
- Test weight of Phalaris minor is: 2g **(IARI, Ph.D -1998)**
- Drift hazard in to adjoining pulse crop is due to spray of: **2,4-D**
- **Triple** gene dwarf varieties of wheat: ***Heera, Moti, Arjun***
- Central **zig zag axis** of wheat spikelet is called: ***Rachis***
- Fruit type of wheat is: ***Caryopsis***
- First **man made** cereals: ***Triticale***
- **Triticale** is a cross of: ***Wheat x Rye***
- **DT-46** is a variety of: ***Triticale***

- Temperature for germination of wheat seed: ***20-25 °C.***
- **Protein** content in wheat: ***8-11%***
- World staple food grain crop is: ***Wheat***
- Most of the present days Indians wheat varieties contain the gene: ***Rht 1 & Rht 2***
- The wheat protein which is essential for good bread quality and chapatti making is: ***Gluten***
- For **safe storage** moisture content of grain should be ranges from: ***10-12%***
- ***Triticum aestivum*** was first introduced in the India by: ***NE Borlaug of Mexico***
- Moisture content at harvesting stage in wheat is: ***25-30%***
- Triple gene dwarf varieties were released in the year: ***1970***
- The state which has **highest production** and cultivation **area** of wheat: ***UP***
- Export quality of wheat is influenced by: ***Karnal bunt***
- Dwarfing agent for wheat crop is: ***Cycocel***
- Objectionable weed of wheat is: ***Convolves arvensis***
- Variety of wheat , susceptible to 2, 4-D where spike deformity occur: ***HD-2009 (Arjun)***
- Primary roots in wheat that support and nourish the plants at initial stage is called: ***Seminal roots***
- In wheat hybridization method most common use: ***Goojo method***
- Absolute weed of wheat is: ***Chenopodium album***
- Recommended ratio for wheat + mustard intercropping is: ***9:1***
- Biofertilizer for wheat is**: *Azatobacter***
- Multilines in wheat are produced by: back cross breeding.

3. Barley: *Hordeum vulgare*, Graminae, 2n = 14

- The earliest type of barley is two-rowed but now days most barley growing groups are six-rowed barley.
- Barley is comparatively more tolerant to saline and alkali soils than other cereals.
- The barley flour does not contain gluten.
- The leaves of barley are usually broader, light green and roughly surface than wheat.

- The barley grains are mostly used in the production of **malt** (carbohydrates) which is used in breweries to make beer, alcohol, malt, vinegar and yeast.
- Protein content (11.5%) in barley in the form of **albuminoids.**

Classification of varietal groups of barley:-

1. ***H. Vulgare:***
   - It is **six-rowed** barley & all the three spikelets are fertile.
   - Spikelets in six distinct rows are arranged at a uniform distance around the rachis
   - Lemma of all the florets are awned or hooded
   - This group is most commonly cultivated in India. **(ICS/NET-2002)**
2. ***H. Irregulare:***
   - In this group ear may be **six-rowed** and **two-rowed.**
   - Lemma does not bear awns or hoods
3. ***H. Distichon:***
   - It consist of only **two-rowed** barley
   - Only central spikelet is fertile

Varieties:

- Azad, Amber, Kailash **(JRF-1997)**, **Dolma** (Huskless)
- Clipper, Ranjit, Ratna, Jyoti, RS-6, RD-2552, Karan maltsona
- Molya nematode resistant: RD-2052, RD-2035, Rajkiran (RD-387)
- Salt tolerant: RD 2794, BL-2, NDB-1173
- Pusa sheetal (NHS-400) - released by ICAR for northern hills zone.

4. Maize: *Zea mays. 2n=20*

- Origin: Central America
- Maize is a **non-tiller** plant.
- Maize is usually a cross-pollinated crop.
- In maize yellow color due to presence of: ***Cryptoxanthin***
- Maize possesses the **highest yield potential** among the cereals that is why known as **queen of cereals**.
- Maize grain contains 10% protein, 4% oil and 70% carbohydrates.
- Maize protein is called ***'zein'*** and is **deficient** in ***tryptophan and lysine***.
- Maize is particularly sensitive to **water logging.**
- **Seminal roots** present in maize to nourished seedling plants.
- Maize is ***monoecious*** plant having two different types of inflorescences.

- Maize is a warm weather plant and requires warm night temperature for higher yield.
- Water requirement for maize is about: 500-800mm
- If maize is sowing in east-west direction it helps to augment the grain productivity due to greater availability of light. **(NET-2001)**
- In the persons who wholly depended on maize lead to a disease called '***pellagra***' due to low content of vitamin B complex, niacin and tryptophan in maize.
- The terminal male flower clusters are called ***tassel,*** it is a branched panicle bearing of pollen or staminate inflorescence.
- The **style** is a very long silky filament, bears the female hairy cluster of which is known as ***"silk"***.
- The female pistillate inflorescence also called the ear or cob comes out from a node generally half way on the stem.
- Idea of **hybrid** maize was first conceived by **E.M. East** and **G.H. Shull** in 1910 by **single cross technique** among the inbred lines.
- An inbred line is a **'pure line'** developed by self-pollination and selection until apparently homozygous plants are obtained. **(ICS Exam - 1993)**
- **D.F. Jones** (1920) proposed the use of **Double cross technique** for hybrid seed production.
- At present double cross techniques are **mostly** used in India.

Classification of maize on the basis of endosperm characters: Shurtevant (1899) classified into **seven groups:-**

1	**Dent corn**	*Zea mays indentata*	The kernel of the dent corn appears like a tooth so named as dent corn, **extensively grown in the USA.**
2	**Flint corn**	*Zea mays indurata*	Its kernel is hard and smooth, **commonly cultivated in India.**
3	**Sweet corn**	*Zea mays saccharata*	Known as sweet corn, the sweet corn kernels present a **translucent and horny appearance** before maturity. This corn is chiefly cultivated to produce kernels for the **canning industry**.
4	**Flour corn**	*Zea mays amylacea*	Kernels almost entirely consist of soft starch so given name ***soft corn.***
5	**Pop corn**	*Zea mays everta*	When popcorn grains are heated, the endosperm cracks open due to the vapor pressure consequently releasing the loose floury mass. It is chiefly used to prepare the food products like ***corn flakes, corn chips*** and

			popcorn etc.
6	**Waxy corn**	*Zea mays* ***ceratina***	The waxy corn caryopsis has two-layered endosperm, and it is used to **produce starch**.
7	**Pod corn**	*Zea mays* ***zunicata***	**Husked, primitive** type of corn, no significantly importance.

Varieties:

- First time in India (1961), four double cross hybrids **Ganga-1**, **Ganga-101**, **Deccan**, **and Ranjeet** were released under all India co-ordinated research project on maize.
- **Mahi Kanchan** and **Mahi Dhavan** are varieties of maize originating from Rajasthan.
- **Top cross** varieties: **Ganga-2** and **Hi-Starch**.
- A top cross hybrid is obtained by combing an inbred line and an open-pollinated variety.
- In 1967, six composites were released: **Vikram, Vijay**, **Amber, Kisan, Jawahar** and **Sona.**
- Composites for early duration: **Pratap, Agety-76, Kanchan, Laxmi, Tarun etc.**
- In 1971, **Lysine** rich Opaque-2 composites **Protina** (**4%** lysine), **Shakti** and **Rattan** were released (these are very rich in lysine and tryptophan amino acids).
- **Synthetic variety**: Means their advanced generations of a multiple hybrid increased by open-pollination e.g. **Amber**
- **New varieties- Pusa Vivek Hybrid 27 Improved (2020), DMRH 1301, 1308, DHM-121,Pusa HM-9, Pratap Hybrid Maize-3**
- New varieties of quality protein maize (**QPM**) have released by using ***Opaque-2 gene*** e.g. **Shaktiman-1**, **Shaktiman-2** and **HQPM-1, 4, Pusa Vivek-QPM-9**
- Optimum plant population in Kharif is **65,000-70,000** plants/ha.

Seed rate:

- For composites: 15-20kg seeds/ha (in **kharif** crop).
- For hybrids: 25 kg/ha.
- For fodder purpose: 40-50 kg/ha
- RxP: 60 x 20cm.

Critical stages for water:

- For moisture and nutrients point of view **tasseling** and **silking stages** are critical. **(IARI, Ph.D-1998)**.

Weed management:

- **Atrazine** is a selective pre emergence herbicide can be used @ 0.75kg/ha.
- **Metolachlor 50EC @ 1kg a.i. /ha.-** it can be used for pre emergent control of grasses and broad leaved weeds of maize, sorghum and soybean.
- **Mesotrione @ 75-150g a.i. /ha** - it is new selective herbicide of maize can be used as **post emergence** control of broad leaved and grass weeds in maize.
- **Halosulfuron methyl (Sempra) 75%WG @ 67.5 g a.i./ha** is systemic & selective herbicide can be used as post emergence (15-20DAS) to control of broad leaved and grass weeds in maize.

Important question:

- Suitable temperature for growth of *Zea mays is:* ***32 ^{0}C.***
- Commonly herbicide used in maize is: ***Simazine*** *(0.5-1.0 kg)*
- Hybrid varieties of maize: ***Sangum*** *and* ***Ganga safed.***
- Double cross techniques of production of maize introduced by: ***D.F Jones (1920).***
- On maize plant, first appearance is: ***Tassels*** *(male flower).*
- Water requirement for the maize crop is: ***500-800mm.***
- Protein content of maize: ***10%***
- **Rabi** maize is extensively grown in**:** ***Bihar***
- ***White bud*****:** In **maize** due to deficiency of zinc the apical leaves become white.
- **Phosphorus deficiency** in maize is characterized by stunted growth, leaf tips & margins are purplish and abnormally dark green leaf colour.

5. Bajra: *Pennisetum americanum. 2n= 14*

- Bajra is known by different names like pearl millet, cattail millet, candle millet, dark millet, spiked millet etc.
- Highest quantity of minerals (2.7%) is found in bajra among the cereals.
- Rajasthan having **first position** both in area and production
- In pearl millet the roots coming out from the second and third nodes provide support to the plant and are called ***prop*** or ***brace roots***.
- **Highest drought** tolerant crop among cereals & millets.
- Among the cereals, the water requirement for pearl millet is lowest (250-500mm) but sensitive to water logging.

- In pearl millet the inflorescence is predominantly **protogynous** and therefore stigmas are seen a few days before the appearance of anthers which greatly facilitate natural cross-pollination.

Seed rate: -for direct sowing: **4-5kg/ha**, for transplanting: 2 kg/ha.

Hybrid varieties:

- Tift 23-A was the first male sterile line of pearl millet.
- First hybrid: **HB-1** released in 1965 from PAU, Ludhiana. Its cross of Tift 23-A (male sterile) x BIL-3B.
- HHB-67, Pusa moti, BJ-104 (from IARI).
- MPMH-17, 21, MH-179, RHB-233, 234,
- **Dhanshakti**- it is first **iron** & zinc rich, **bio-fortified** variety of pearl millet.
- PHB-10:- Hybrid released from PAU, Ludhiana in 1975
- **Napier grass**: For development of forage variety **Pusa napier-1;** the first inter-specific hybridization was carried out in 1947 between *Pennisetum purpureum* x *Pennisetum glaucum.*

6. Jowar: *Sorghum bicolor*. 2n= 20

- First hybrid of sorghum: ***CSH-1*** (released in 1964)
- Jowar is native of: ***Africa***
- Witch weed ***(Striga spp.)*** a partial root parasite weed, which is a major problem in sorghum and Pearl Millet production.
- Seed rate of Jowar: ***9-12 kg/ha***
- Seed rate for fodder purpose: ***40-50 kg/ha.***
- The best high yield variety of Rabi Jowar is: ***M 35-1***
- **SSG-58-8** is a multicut fodder variety of sorghum.
- Drought and salinity tolerant variety of sorghum is: **CSH-9**
- Sorghum is considered as camel crop because of: ***Resistance to drought.***
- Basic races of Sorghum***: bicolor, guinea, cardatum, kafir and durra.***
- Most important male sterile variety: ***Combine kafir-60.***
- Sorghum crop for fodder purpose should be used after 50% flowering because sorghum leaves at knee stage possess a high amount of hydrocyanic acid ***(HCN)*** or ***dhurin*** or ***prussic acid*** which is poisonous to the animals. **(ICS Exam -1993)**
- Under drought condition if animal is grassed of sorghum leaves with 0.5g of HCN may be dying.
- Sorghum is called the ***camel crop*** because it well thrive in arid and drought-prone environment.

- The inflorescence of sorghum is a ***panicle*** known as ***heads***
- The sorghum grain poor in Lysine content but rich in Leucine.

7. Sugarcane (*Saccharum officinarum*)

- ✓ Sugarcane - *Saccharum officinarum* is known as **Noble cane.**
- ✓ In India great research on hybridization of cane is done by **T.S. Venkataraman** hence he known as **sugarcane wizard.**
- ✓ The word "***Saccharam***" derived from **Sanskrit** word **'sarkara'** which means **sugar**.
- ✓ Sugar industry is the **largest agro** based processing industry after **textiles** in India.
- ✓ In India (2006) there are about **582 sugar factories** maximum in **Maharashtra (134).**
- ✓ The maximum sugar factories (317) are in cooperative sector followed by private sector (203 factories).

Area and distribution:

- Area: **India** >Brazil > Cuba
- Production: **Brazil** >India> China
- Productivity: **Peru** (115t/ha) > Australia

Area & Production (India):

- Area: **UP** > Maharashtra > Tamil Nadu
- Production: **UP** > Maharashtra > Karnataka
- Productivity: **Tamil Nadu (99.3 t/**ha.)

Botany:

- Family: **Graminae** (poaceae).
- 2n=80, **perennial** plant.
- **Inflorescence** of sugarcane known as ***Arrow*** which is an open branched panicle.
- **C_4**, ***intermediate short day*** plant (requires 12-14 hrs for flowering).
- Sugarcane is tropical plant and gives more yield in southern part of country so South India is ideal for sugarcane cultivation.

Origin: **New Guinea**

Classification: there are **three cultivated** and **two** wild species.

1. ***S. officinarum*: Noble** cane, this is tropical cane, thick and **juicy**, low fiber content: 2n = **80.**
2. ***S. Barbari***: origin is India known as **Indian** cane, low sucrose and high fiber content,with short and thin stalks, 2n = **82-124**.
3. ***S. sinense***: **Chinese** cane, medium sucrose, high fiber content, long and thin stalks, 2n = **118.**

Wild species:

1. *S. spontaneum*: Very **low sugar** and very **high fiber** content.
2. *S. robustum*: Very low sugar

Seeds and sowing:

- Material used for sowing is known as seed pieces or **sets**.
- One year old top **1/3** to half **portion** of cane being comparatively **immature** having **high glucose** content good for germination.
- **Three budded**: **35,000-40,000 setts/ha** (obtained from **75-80 qt.** of cane).
- Setts should be treated with **0.5%** solution of **Agallol** or Aretan.

Time of planting:

1. **Autumn planting:** planting in October in N. India.
2. **Adsali planting:** planting in June-July and takes 15-18 months for harvesting. It is commonly grown in Maharashtra (South India).
3. **Eksali or spring planting:** Planted in Jan.-Feb. and takes about one year time for harvesting. It is commonly growing in North India.

Planting methods:

1. **Flat planting**:
   - Planted **shallow** 8-10 cm deep furrows at **75-90 cm** distance
   - It is followed in **N. India.**
2. **Furrow method**:
   - Followed particularly in **heavy soils.**
   - Adopted **10-15** cm deep furrow in N. India
   - About **20 cm** deep furrow in S. India.
3. **Trench method**: **(IARI, Ph.D-1997)**
   - Followed in **tall growing** cane areas such as **coastal** areas to **protect** from **lodging.**

Manures and fertilizer:

- **Heavy feeder** crop but higher doses of **nitrogen** during ripening stage **decrease sucrose** content in **maturing** canes.
- **Blanked** application of nitrogen (120-150 kg): phosphorus (80 kg): potash (60 kg).
- The best source of nitrogen in normal soil is **Ammonium sulphate**, in saline soil is urea and in acidic soil is used CAN.

Irrigation stages:

	Phase	**Duration**
1.	Germination	0-60 DAP
2.	**Formative**	60-130 DAP (max. requirement of water in this stage)

3. Grand growth 130-250 DAP
4. Maturity 250-365 (**DAP** = Days after planting)

- **Furrow method** of irrigation is **most common** method of irrigation in sugar cane.
- Under high temperature, sucrose gets converted into glucose and quality of the produce becomes poor.
- **Ratooning** of sugarcane: it is economical to take **only one ratoon** crop, in ratoon crop **20% more nitrogen** required as compare to planted crop.
- Artificial ripening of sugarcane by use of: **Polaris, Glyphosine @ 5 kg** a.i. /ha at maturity.

Blind hoeing:

- Animportant practice in sugarcane, hoeing after planting and before crop emergences for the purpose of weed control.

Harvesting: cane maturity can be measured by followings:

a. **Brix ratio:**

- Sugarcane considers optimum matured if **brix** value or **refractometer** reading is about **18-20.**
- For reading juice should be taken from **middle** portion of stalk

b. **Fehling test**:

- For maturation of cane **Fehling test solution** reading should be less **than 0.5% glucose.**

Crop logging:

- **H.F. Clements** in Hawaii developed the concept of crop logging.
- It is the **foliar diagnosis** comparing the **nutrient status** of comparable leaves of high and low yield crop plants generally used in **sugarcane** in **Hawaii.**

Crop lodging:

- The **falling down** of crop due to **strong wind**,
- It prevented by trench planting, earthling up, wrapping and propping of plants.

Water lodging:

- The falling down of crop due to **flooding.**
- That occurs when saturation of both macro and micro pores.

By-products of sugarcane:

1. **Bagasses**: Used for **fuel** and **paper** making. **(RAS-2013)**
2. **Molasses**: For alcohol production.
3. **Press mud**: (Contain lime), **reclamation** of **acidic** soils.

- In **Brazil**, sugarcane is also used as a source of energy.

- **Gasohol** is prepared from **80% petrol** + **20%** ethanol or **alcohol** (from **sugarcane**) which are used in **automobiles**.

Diseases of sugarcane:

Diseases	Causal organisms	Remarks
Red rot	*Colletotrichum falcatum*	Reddning of the stem pith
Ratooning stunting disease (RSD)	*Clavibactor xyli*	
Grassy shoot/ Albino	*Phytoplasma*	
Gumosis	*Xanthomonas vascularam*	

Insect-pests of suger cane:

Insects	Scientific name	Remarks
Shoot borer	*Chilo infuscatellus*	Forms **dead heart** on growing shoot, it can be managed by foliar application of cartap hydrochloride (Padan 4G) @ 25 kg/ha & use of *Trichgramma chilonis* @ 1.5 lacs/ha (7 trichocards)
Pyrilla	*Pyrilla perpusilla*	Nymphs& adults suck juice from underside of the leaves, for biological control its impotant parasitoid is a lepidopetra insect '*Epiricania melanoleuca*'
Top borer	*Tryporyza noviella*	Damage at later stage cause **bunchy top** of sugar cane. Apply carbofuron 3G @ 20 kg/ha
White fly	*Aleurobolus borodensis*	
Mealy bug	*Saccharicoccus sacchari*	
Root borer	*Emmalocera depressella*	
Internode borer	*Chilo sacchariphagus indicus*	
Gurdaspur borer	*Acigona steiniella*	
Termite	*Odontotermus obesus*	

Important questions:

- Inflorescence of sugarcane is called as: ***Arrow***
- Higher dose of nitrogen decrease the: ***Sucrose content***
- Most critical stage for irrigation is: ***Formative stage***
- Most popular planting method in the north India: ***Flat bed planting***

- Brix reading should be for the proper maturity: ***18-20%*** **TSS** (total soluble solids)
- Soils which are unsuitable for sugar cane are: ***Saline soils***
- Scientific name of noble cane is: ***Saccharum officinarum.***
- Adsali sugarcane planted in: ***June-July*** (18 month crop).
- State, have highest productivity of sugarcane: ***TN***
- **Earthing** in sugar cane should be done in: ***June-July.***
- **Earthing** in sugar cane is important practice in: ***Maharashtra.***
- **Tying** should be done in the month of: ***August***
- Spacing row to row: ***90 cm***
- Name of wild cane: ***Saccharum spontaneous***
- Burning of canes is done for: ***Improve sucrose & juice quality.***
- Indian cane indigenous to north and eastern India: ***Saccharum barberi.***

8. Chick-pea/Bengal gram/gram: *Cicer arietinum*

- ✓ Gram is rich source of **calcium**, **iron** and **niacin.**
- ✓ Used as **blood purifier** and germinated seeds are recommended to cure **scurvy** disease.
- ✓ India is the **largest** producer of gram in the world sharing **65%** area & **70%** of total global **production.**
- ✓ Major pulse crop in India and share largest acreage and production among the pulse crop in the country, followed by pigeon pea.
- ✓ Most of legumes fulfill their 75% N- requirement through root nodules (symbiosis).

Origin of place: South West Asia

Area and distribution:

- Area: **India**> Pakistan > Turkey
- Production: **India** > Turkey > Pakistan

In India:

- Area: **MP** > Raj > UP
- Production: **MP** > Raj > UP

Classification:

1. **Desi or brown gram** (***C. arietinum***), 2n = **14 or 16**
   - ✓ It is most **widely cultivated** in India.
   - ✓ It has good branching habit.
   - ✓ Yellow to dark brown seed
   - ✓ Seeds are small sized
   - ✓ Cosmopolitan in nature.
   - ✓ **Test weight**: **140-260**g

2. **Kabuli/white** gram (***C. kabulium***), 2n = **16**
   - ✓ **It has bold** attractive **white** seeds
   - ✓ Plants are taller
   - ✓ **Test weight: 270-375** g
   - ✓ Its **poor yielder** as compare to desi gram.

Botany:

- Family: **Fabaceae** (Leguminaceae), sub family: **Papilionaceae**
- The **sour taste** of **leaves** and **pods** is due to the presence of **Mallic** acid **90-96%** and **Oxalic** acid **4-10%**, about 4-10 kg of these acids can be obtained from one ha. crop.
- **Tap root** system, **cloddy** and **rough seed bed** is required to provide adequate aeration for proper germination of seeds because gram is highly sensitive for soil aeration.
- **C_3, long day** plant.
- Gram is **highly susceptible** to **frost** at **flowering** stage.

Varieties of desi gram:	
Variety	**Features**
Avrodhi:	**Wilt resistant**
Pusa-256/BG 256:	Most common and **best for Rainfed** condition.
C-235:	Widely adopted for **dry land**, tolerant to **Ascochyta blight**.
Gaurav:	Resistant to **Ascochyta blight**.
RS-10:	Drought resistant, suitable for Ajmer & Kota districts of Rajasthan
RS-11:	Mutant variety, drought resistant, suitable for heavy soils of Rajasthan
ICCC-2:	Short duration variety
Aparna (2006):	**Grains** remain in **green colour**
Type-2	Suitable for Rajasthan **(RAS Exam-1998)**
Pusa-209	Seeds are medium, test wt. 135g/1000 seeds
Other varieties:	GNG-2171, GNG-1551, 1958, Chaffa, RSG-2, GNG-16, Radhey. (IARI, Exam-1998)

Kabuli:

- **C-104, K-4, L-144**

Preparation of Seed bed:

- **Rough** seed bed is required for gram to provide good **aeration**.
- **Sowing**: **2nd** fortnight of **October** is optimum time.
- **Seed rate for desi**: **60-80** kg seeds/ha.

- **Spacing**: Row to Row **30 ×** 10cm
- **Seed rate for kabuli- 80-100 kg /ha.**
- **Delayed** planting **increases** the incidence of (pod borer), ***Helicoverpa armigera.***
- **Early** and **shallow** sown crop is more liable to be damaged by **wilt**.
- Hence, Seed should be sown at least 8cm deep to prevent wilt disease.

Irrigation management:

- **Pre-flowering** and **pod development** are the mostcritical stages.
- In gram no irrigation should be provided at the time of flowering.

Practices in gram:

Nipping and **topping**:

- Pruning of **top** branches of the plants to encourage reproductive growth, this practice is done **30-40** days after sowing.
- It can be done through rapidly passing of a **flock of sheep** in field.
- Chemical for **nipping**: **TIBA @ 75 PPM** i.e. Tri-iodo-benzoic Acid.
- **Yield** of gram **20-25q**/ha in irrigated field.

Important points:

- Root system of gram is: ***Tap root system***
- Sour test of gram leaf is due to: ***Maleic & oxalic acid***
- Best variety for dry land is*: **C-235***
- Sowing time of gram is: ***15 October- 20 Oct.***
- Gram fruit is known as: ***Pod***
- Protein content of gram: ***21 %***
- Deep sowing of gram is protect from: ***Wilt disease***
- Critical stage for irrigation: ***Pre flowering, pod development stage.***
- Herbicide application in gram is: ***Basalin @ 1kg before sowing.***

9. Pigeon pea: *Cajanus cajan*

- Known **as red** gram/arhar/tur
- **Most droughts tolerant** among pulses.
- In India it is second important pulse crop after gram.

Origin: Africa

Classification:

1. *C. cajan* var. **bicolor**:
   - ✓ **Perennial**, tall, much branched
   - ✓ **Late maturity**, 4-5 grains/pod
   - ✓ Commonly grown in **North India.**
2. *C. cajan* var. **flavus**:
   - ✓ **Annual**, Short plant

- ✓ **Early maturity,** 2-3 grains/pod.
- ✓ Mainly grown in **South India** as a *seed crop*.

Area and distribution:

- Area: **India**> Malari > Uganda
- Production: Same trend in area and production

In India:

- Area: **Maharashtra** > UP > Karnataka
- Production: **UP**> MH > Gujarat
- Productivity: **Bihar**> UP
- In India **90% area** and **85% production** on **Global** basis.

Botany: Features

- ✓ **Deep** rooted, **self**-pollinated.
- ✓ **Hypogeal** germination.
- ✓ C_3, **short day** plant.
- ✓ **Hardy** crop, **most drought tolerant** crop among major pulses.

Varieties:

a. Short duration:

- **UPAS-120**: It takes about 120-125 days for maturity.
- **Prabhat**: Extra-early maturity variety, takes 110-120 days
- **ICPH-8**: 1st hybrid of arhar in world developed at ICRISAT, Hyderabad, 1991.

b. Medium duration: 150-180 days

- Mukta (wilt resistant).
- Sharda, Basant, Paras

c. Long duration: 180-300 days.

- PUSA-9.

Seeds and sowing:

- Seed rate: **12-15 kg/ha.**
- Spacing: **60 × 15 cm.**
- Plant population for kharif is: 55,000 plants/ha
- Plant population for winter crop: 3.33 lakh plants/ha
- Sowing time: **1st fortnight of June**
- Harvesting Index is: **19% lowest** in among **pulses.**

Important points:

- Botanical name of Arhar: ***Cajanus cajan***
- Origin of arhar: ***South Africa.***
- Sowing of arhar: ***15 June***
- Highest production of arhar in: ***UP***
- Protein content of arhar is: ***25%.***

- **Highest** productivity of arhar: ***Bihar (1115 kg/ha).***
- Most important disease of arhar is**:** ***Steirlity mosaic (Viral)***
- The vector for sterility mosaic is**:** ***Eriophyid mite***
- Specific pest of arhar is***: Tur pod fly***

10. **Cotton, (*Gossypium sp.*)**

- ✓ It known as **'White gold'**
- ✓ **King** of **appraisal fiber.**
- ✓ ***Bt*** cotton is effective against **boll worm** pest complex of cotton.
- ✓ In India ***Bt.*** **Cotton** is grown about 11.27**Mha** (94.68% of total cotton area) in 2023.

Origin: **India**

Area and distribution:

- Area: **India**> USA > China
- Production: **China** > USA > India

In India:

- Area: **MH** > A P > Gujarat
- Production: **MH** > A P > Haryana

Botany:

- ✓ Family: **Malvaceae.**
- ✓ Cultivated cotton is **annual.**
- ✓ **Tap root** system.

Two type of branches:

1. **Monopodial**: these are auxiliary buds known as **vegetative branches**
2. **Sympodial**: these are **reproductive branches**

Classification:

Out of **20** species of cotton only **4** are cultivated.

1. *G. arborium*: 2n = **26**, diploid
2. ***G. herbaceum***: 2n = **26**, diploid
   - These above two are known as **desi cotton/old world/Asiatic** cotton.
3. ***G. hirsutum***: 2n= 52, tetraploids. Generally called American cotton.
4. ***G. barbadense***: 2n= 52, tetraploids. Highest fiber length 3.6-5.0 cm.
   - *G. barbadense* is known as **Egyptian/**sea Island cotton.

India: is the only country in the world where **all four** species of cotton are grown on commercial scale besides hybrid, present status follows as-

- ***G. hirsutum***: it representing **90 % of hybrid cotton.**
- ***G. arborium***: nearly 2.5% of total area
- ***G. herbaceum***-: 2.5% area
- ***G. bardadense***: 5% area.

Varieties:

- *G. arborium*: **Lohit**, Virnar
- *G. herbaceum*: **Digivijaya**, Sanjay
- *G. hirsutum*: **Ganganagar agety**, **Bikaneri nerma**, **Mahalaxmi**
- *G. barbadense*: **Suvin** and **Sujata**
- **Sujata**: it is the first spinning variety of **Egyptian** cotton released in India).
- **MCU-5:** It is extra-long staple variety of hirsutum cotton (SRF-09).

Hybrid varieties:

a. **H-4**:

- It was **first** commercially **hybrid** variety of cotton.
- Developed in the world by crossing **G-67** x **American Nectorless** by **Dr. C.T. Patel** of **GAU, Surat** in **1970**.
- It's a cross of **intra hirsutum**.

b. **Varalaxmi**:

- **It is 1st interspecific hybrid** between ***G. hirsutum*** and ***G. barbadence.***
- Released from **UAS, Dharwad (Karnataka)**. **(RPSC-AO-2012)**

c. **Maru Vikas**/Raj. HH-16:

- It's **1st hybrid** of **hirsutum** in Rajasthan **(JRF-09)**.

d. **Savita, Surya:**

- **These are intra specific** hybrid of hirsutum.

e. **Dhanlaxmi:**

Delinting/deffusing:

- Done with concentrate **sulphuric acid** in a ratio of acid to cotton seed about 1:10.
- This helps in grading the seed
- Kill **hibernating** insect stages
- **Destroy** the diseased **pathogens**
- It makes seeds **easy** to sowing & **germinates** rapidly.

Seed rate:

- **Desi cotton**: 10-18 kg/ha. at 67.5 x 30 cm spacing
- **American** cotton:15-25 kg/ha at 67.5 x 30cm spacing
- **Hybrid** cotton: **2-3kg/ha.** at **100 x 60** cm spacing.
- For *Bt.* cotton: **1-1.5 kg/ha/ha.** (av. 10,000 plant population)

Sowing time:

- **North India**: **1st** fortnight of **May.**
- **Central India**: Last week of **June** to first week of July.
- In **Tamil Nadu:** September - October

Ultra Narrow Row:

- The concept of **ultra-narrow-row** (UNR) was developed in **USA** in which the Row to Row and Plant to Plant spacing kept **19 x 19 cm^2**.
- Keeping a plant population of **2, 77,000 plants/ha**.

Square in cotton:

- The appearance of **flower bud** in cotton is known as square and fruit is boll or capsule.

Topping:

- Removal of **terminal growing** point once from each plant at a height of 1-1.2 m (80-90 DAS) to protect further terminal growth and to encourage **sympodial** branching and good **boll** development by **diverting** the energy flow.

Ginning %:

- Separation of fiber from the seed cotton is known as **ginning**,
- In general its ranges **30-35%**.

Weeds management:

- Important weed flora-***Trianthima***, ***Echinochola, Diseria***etc.
- Pre-emergence/post emergence spray of **diuron** @ 0.5 kg a.i. /ha.
- Cotton is **highly sensitive** to **2, 4-D** & **glyphosate**, injured by a **spray drift.**
- 2, 4-D **resistant** cotton- **gene-tfd-A.**

Diseases-

1. **Bacterial blight**- caused by *Xanthomonas malvecerum* and most serious disease. And this disease has four stages which are angular leaf spot (leaf infection)/black arm (stem infection) / boll rot (boll infection) seedling infection (seedling infection).

Physiological disorders:

1. **Tirak**/bad ball opening/**pre-mature** defective opening of balls due to **early sowing**, **water stress** on light sandy or in **alkaline** soils.
2. **Little leaf** in cotton: due to **Zn deficiency.**
3. **Crinkle leaf** of cotton: due to **Mn toxicity.**

Insects-pests of cotton:

1. **Sucking pests**: Jassids, white fly, aphids, thrips (these are now regarded as major pests of *Bt.* cotton).
2. **Bollworm complex** (still lepidopterous larvae are most destructive insects of non Bt. cotton.
   (i) Spotted bollworms (*Earias insulana*) attack on terminal shoots during pre-flowering.

(ii) Pink boll worms (*Pectinophora gossypiella*) attack on mature cotton bolls.

(iii) American bollworms (*Helicoverpa (Heliothis) armigera*) damages to squares flowers and boll.

Bt. **Cotton:**

- At present (2024-25) *Bt* cotton grown in India about 94.68% (11.29mha) of total cotton areas.
- Developed by biotechnology by inserting lethal genes from soil bacterium (*Bt.*) in cotton, which is resistant against cotton boll worms.
- *Bacillus thuringensis (Bt.)* is a soil bacterium that produces proteins lethal to insect larvae affecting the digestive system of boll worms.
- The first generation of Bt. cotton was approved in 1995 and planted for commercial production in 1996
- The best example of transgenic crop protection in developing countries is Bt. cotton containing Cry-1Ac gene from Bacillus thuriengensis, which is grown commercially in India and china
- GM cotton is only crop permitted by Genetic Engineering Approval Committee (GEAC) under Ministry of Environment and Forestry (MoEF) first permitted in India on March, 2002 and commercial cultivation in 2002-2003.
- Mahyco collaborated with Monsanto to developed first Bt. Variety Bollgourd-1, using Cry-I Ac gene.
- Recently (2006) *Bt.* varieties developed as Bollgourd-II, having both genes of Cry I Ac and Cry 2 Ab gene. Which known as second generation of Bt. cotton.
- There are reports of more than 50% reduction in insecticides application on Bt. cotton

Refugia strategy:

✓ The 20% area is kept for non Bt cotton around the field of main *Bt* cotton.

✓ Purpose behind this is to prevent resistance to boll worms against *Bt* cotton.

Important points:

- **Fiber length** and fiber fineness are: ***Genetic traits.***
- Origin of cotton**:** ***India***
- Lint per cent in cotton is**:** ***33 %***
- **Fibermaturity** measured by: ***Arealometer***
- Fiber **thickness** by: ***Nepiness***
- **Spinning** performance measured by: ***Number of counts***

- **One bale** of cotton: ***170 kg.***
- **Oil content** in cotton seed varies from: ***14.5 to 25.6%***
- **Toxic** pigment/phenol compound present in cotton seed is: ***Gossypol.***
- **Cotton fiber** is an elongation/**outgrowth** of an **epidermal cell** of **seed coat**.

11. Jute (*Corchorus spp.*): F. Tilliaceae

- ✓ Also known as **golden fibre** because of its colour.
- ✓ Most important **fiber** crop in India.
- ✓ India's **first rank** in both **area** and **production** of jute in world.
- ✓ In India **West Bengal** has **first position** both in area and production.
- ✓ ***Corchorus olitorius*** known as meth-pat/**tossa**. Grown on **well drained** high lands only.
- ✓ ***C. capsularis*** known **white** jute/**tita,** due to presence of glycoside **'Corcharin'**. Covers **70% area** of total jute cultivation.
- ✓ Weight of one bale of jute is: ***180 kg***
- ✓ Seed rate of jute: ***6-8 kg/ha***

Harvesting:

- Optimum time of harvesting is about **90 DAS**, but this stage the grain yield low.
- So to obtain **quality fiber** and **good yield** jute is harvest at **small pod stage** i.e. **135-140 DAS.**

Ribboning:

- This practice is very common in **China** and Taiwan, but **not popular** in India.
- It consists of **peeled out** of raw bark from the green plant, immediately after harvest and bundles of the ribbons thus obtained are retted.

Retting:

- It is a **microbial process** in which **aerobic** and **anaerobic** bacteria and fungi **loosen** the **fiber** by decomposing & dissolving the pectin, hemicelluloses and other cementing agents.
- It completes within **8-30 days**.
- Optimum temperature required for retting is about **34°c.**

12. Potato (*Solanum tuberosum*)

- ✓ **Known as King** of vegetable.
- ✓ **Cash crop** (high marketable value per unit area).
- ✓ Rich source of starch i.e. **farina** and vitamin especially **B_1**

Origin: Peru (**South America**)

True potato seed (TPS):

- Developed by **Ramanujan**
- **100-150 g seeds**/ha (for **nursery**) is sufficient.

Area and distribution:

- Area: **Uttar Pradesh** > West Bengal > Bihar
- Production: **Same trend** as area
- Potato occupies **largest** area under any **single** vegetable in the World.

Botany:

✓ **Perennial** plant but as a **crop treated** as an **annual** plant.

✓ Propagation by **tubers** i.e. an enlarged underground stem produced on the end of a **stolon** and not on the root properly.

✓ C_3, short day plant from **tuberization** point of view while **flowering** point of view long day plant.

✓ Fruit type is **berry** and potato requires temperature for satisfied tuber growth **17-19** °C.

✓ P^H requirement about **5.0-6.5**, as **acidic** condition **reduces** scab disease incidence.

Varieties:

a. **Early maturity**: Takes only **75-90 days**

- Kufri chandramukhi
- K. ashoka (**extra early**)
- Kufri jyoti
- K. alankar, K. lavkar and K. navtal

b. **Medium duration (90-110 days)**

- K. sheetman (Frost resistant variety)
- K. chamatkar
- K. badshah

c. **Late maturity:**

- K. deva
- K. sinduri

d. **Suitable for processing**:

- K. chipsona-1, K. chipsona-2, 3

Seed sowing:

- **25 g** cut (**piece**) of tuber with **3-5 buds/eyes** @ 15-20 q/ha
- As getting high yield @ 20-25 q/ha for 15g/seed
- Spacing: 45-60 x 15-20cm

Seed plot technique (SPT):

- Developed by Pushkarnath *et al.* in 1967

- It is a technique of multiplication of disease free (virus free) seed in Northern plains of Country by growing of healthy seeds in October and remove upper haulms (tops) between mid-January and leave tubers underground for maturity.

True potato seeds (TPS)-

- Sexual reproduction in potato and given by Ramanujam in 1957
- TPS is sown in nursery for raising seedling which are later transplanted in field.
- Disease free potato seed can be produced and prevention of disease to new areas can be checked.
- Cost of cultivation can be reduced as compared to traditional method of planting by tubers
- 100-150 g seeds/ha in nursery (75 m^2) is sufficient.

Disorders:-

a. **Black heart:** Due to **lack of O_2** in stored potato the internal tissues Break down and become black. Avoid storage temperature above 35 °C and create good ventilation during storage.

b. **Hollo heart:** due to excessive use of N-fertilizer, an irregular cavity in the center of tubers develops.

c. **Chilling injury:** Due to prolonged storage of tubers at temperature of about 0 °C.

Diseases:

- ✓ **Late blight of potato**: caused by *Phytophthora infestance,* fungus, more destructive & responsible for Ireland (Iris) famine in 1845
- ✓ **Early blight of potato** is caused by Fungus, *Altenaria solani* and more common than late blight.
- ✓ **Black scurf** caused by Fungus, *Rhizoctonia solani*
- ✓ **Scab** is caused by Fungus, *Streptomyces scabis*
- ✓ **Wart disease** of potato is caused by Fungus, *Synchitrium endobioticum*
- ✓ **Charcol rot** is most serious storage disease of potato.
- ✓ To control black scrub and common scab disease tubers should be treated 3% borric acid.

Important points:

- **Dormancy breaker**: May be used **1%Thiourea** with I ppm of GA_3 for 10 minutes.
- Fertilizer **KCL not used** in **solanaceae** crops because **chloride** has **adverse** effect on dry matter and **starch** content of potato.

- The **stolen formation** (20-50 DAP) and **tuber elongation** (50-80 DAP) are **most critical** stages of irrigation.
- Insect of potato which infect the tubers in field as well as storage is : Potato tuber moth (*Gnorimoschema operculella*)
- Potato leaf roll viruses are transmitted by: Aphids (*Myzus persicae*).
- Wart disease of potato is due to: Fungus (***Synchitrium endobioticum***)
- ***Domestic quarantine*** is adapted:- transport of planting materials i.e. tubers from **Darjeeling area** (WB) to other parts of the country is restricted to prevent the spread of **wart disease** and transport of planting materials from **Nilgiri hills** (TN) to other parts of country is restricted to protect from **golden nematode** (*Globodera rostochinensis*).

13. Groundnut: *Arachis hypogea*

- Known as Peanut/earthnut/monkey nut
- Contains 45% oil and 26% protein in kernels
- Oilcake of ground nut contains about 7.3% Nitrogen (highest in edible cakes)

Origin: **Brazil**

- Area: India > China > Nigeria
- Production: China > India > Nigeria

In India:

- Area: **AP** > Gujarat > TN.
- Production: **Gujarat** > AP > TN
- Productivity: **Tamil Nadu**

Classification:

- Bunch type/erect type: ***A. hypogea*** var. ***fastigiata***
- Spreading/trailing type: ***A. hypogeal*** var. ***procumbens*.**

Botany:

- Family: **Leguminaceae**
- **Self**-pollinated
- C_3, short day plants
- Fruit known as **pods.**
- Pegging: after pollination ovary becomes stalk like structure known as **gynophore** which enter in to soil as ***peg*** to form pods.

Dormancy:

- Bunch type **non dormant** while spreading type is **dormant** which require a **resting period** of 2-2.5 months for germination,

- It can be broken by **high temperature, storage** (for 2-3months) and use of ethylene chlorohydrin @ **0.7%** and seed soaking in **5-10 ppm** solution of **NAA, IAA.**

Important variety:

- Bunch type: **Jyoti, TMV-2**
- Spreading type: **Chandra, TMV-1, RS-1**
- Semi-spreading type**: AK-10, RG-425, RG-578**

Seed rate:-

- For bunch type: 80-100kg/ha at 30x10cm
- For spreading type: 60-80 kg/ha at 45x10cm spacing
- Depth of sowing: 5cm

Soils:

- Light texture, sandy loam soil is excellent.
- **Ammonium sulphate** and **SSP** are excellent source of **N & P**, respectively.
- **Popping** in Groundnut: **pods without kernels** or unfilled grains due to **Calcium** deficiency.
- Application of 250 kg **gypsum**/ha at **pre-flowering** stage (30 DAS), which is essential for **pod formation** and **development**.
- Generally sowing time is **last week of May** to **1st fortnight** of **June**.
- **Aflatoxin** contamination in Groundnut (**fungal infection** of Groundnut kernel) caused by ***Aspergillus flavus*** due to high moisture content in stored kernels. **(RPSC-2009)**
- **Management** of Aflatoxin is **storage** of kernels at **8 %** moisture level.

Important points:

- Botanical name of ground nut: ***Arachis hypogea***
- Origin of ground nut: ***Brazil***
- State which is largest producer: ***Gujarat***
- Shelling percentage: ***70 %***
- Oil percentage: ***45%***
- Nitrogen % in groundnut cake:*7.3 %*
- Best soil for ground nut production: ***Sandy***
- Spacing of ground nut: ***30 x 10 cm***
- High yield type of ground nut: ***Spreading type***
- Pegging stage in spreading varieties comes after: ***55 DAS***
- Chemical used for floral initiation: *NAA* ***@ 40 PPM.***
- Fruit of ground nut is known as: Pod or lamentum.
- Herbicide used for groundnut is: Pendamethalin (PE/PPI)
- Early post emergence herbicide used in ground nut is: ***Imezathapyr***

- Bitterness of kernel due to: ***Aflatoxin (fungal toxin)***
- Protein content in ground nut: ***26 %***
- Gynophores of ground nut are known as: ***Peg.***
- Yellowing of groundnut leaves due to: ***Iron deficiency***
- Rosette disease is due to: ***Virus (vector aphids)***
- Major pest of ground nut: ***White grub.***
- Bud necrosis in ground nut due to: ***Thrips.***

14. Soybean: *Glysine max*

- ✓ Soybean is basically pulse crop but in India mostly grown as oilseed crop.
- ✓ known as **wonder crop** or **yellow jewel**
- ✓ World first rank crop as a source of vegetable oil.
- ✓ It provides approximately **60%** of **vegetable protein** & **30%** of **oil** in the World.
- ✓ Contains **42** per cent **protein (rich** in **lysine)** and 20 per cent oil.

Origin: **China**

Area and distribution:

▪ Area:	**MP**> MH < Rajasthan
▪ Production:	**MP**> MH < Rajasthan
▪ Productivity:	Andhra Pradesh (> 40 q/ha)

Botany:

✓ Family:	**Leguminaceae.**
✓ Inflorescence:	**Raceme**
✓ Fruit:	**Pod**
✓ Soybean is:	**C_3, short day plant.**

Varieties:

a. **Introduced from USA are:**

▪ Bragg	Lee
▪ Kent	Clark-63
▪ Blackheart	

b. **Indian varieties:**

- Shilajeet, Alankar, Ankur, Punjab-1, Monato
- New varieties: JS-95-60, JS 20-94, JS 20-116, MACS 1407, Swarna Vasundhara.

Seeds and sowing:

✓ Sowing time is:	**3rd week** of **June** to **1st fortnight** of **July.**
✓ Spacing:	**45-60** x **15-20 cm.**
✓ **Depth** of sowing: **3** cm.	

- ✓ Seed rate: 60-**70kg**/ha
- ✓ Plant population: **300,000-400,000plants**/ha.

Weed Management:

- **PPI** application (before sowing)-Trifuralin or **Fluchlorin** @ 1kg/ha.
- **Pre-emergence** -pendamethalin or **Alachlor** @ 1-1.5 kg/ha.
- Early post emergence - application of Quizalofop-ethyl or **Imazethapyr** @ 750g/ha 15-20 days after sowing is effective.

15. Mustard/rape seed: *Brassica spp.*

- ✓ **Pungency** in mustard is due to **isothiocynate**, enzyme hydrolysis product of glucosinolates (in all crucifers).
- ✓ Due to presence of toxic **glucosinolates** in its cake, it is **unsuitable** source of **protein** for both **human & cattle**. It causes **goiter** and effect growth and development. The **upper limit** is fixed at **0.5 %** by the govt.

Varieties:

a. **Rapeseed:**

- **Brown sarson** (*Brassica com.* var. **sarson**): ***Pusa kalyani***
- **Yellow sarson** (*Brassica com.* var. **sarson**): ***Benoy***

b. **Mustard/rai/Indian mustard (*B. juncea*):**

- Kranti — Varuna (T-59)
- Rohini — Pusa bold, RH-30

New released varieties of mustard:

- Laxmi — Vasundara
- Ashirwad — Basanti
- Pusa jai Kisan — NRC-HB-506
- **Pusa jai Kisan (Bio-902):** Its first variety ***Brassica juncea*** developed in the world by **Prof. V.L. Chopra** *et. Al.* with the help of biotechnology through ***somatic hybridization*** in India.
- **NRCHB-506: ** It is **the first hybrid of mustard** in India or World developed from Directorate of Rape seed and Mustard Research, Bharatpur, Rajasthan in Feb., 2009.
- **Pusa Double Zero Mustard 33** **(2021)**: contains < 3% erusic acid and < 30 u mol/g glucosinolates.

Genetically Modified (GM) Mustard:

- **Dhara Mustard Hybrid-11** (DMH-11) is genetically modified hybrid variety of mustard developed by Centre for Genetic Manipulation of Crop Plants at Delhi University.

- But researchers at Delhi University have created hybridized mustard DMH-11 using **"barnase / barstar" technology** for genetic modification.
- It is an herbicide tolerant (HT) crop. In February 2016; the Genetic Engineering Appraisal Committee **(GEAC)** had allowed the commercial productionof another GM crop viz. **Mustard DMH-11**, it is yields about 30% more than a traditional reference mustard variety.

Seed rate: 5 kg/ha

Diseases of Mustard:

SN	Disease	Causal organism
1	Alternaria blight /black spot	*Alternaria brassicicola*
2	Clubroot	*Plasmodiophora brassicae*
3	Bacterial blight/ black rot	*Xanthomonas campestris*
4	Damping-off	*Fusarium* spp., *Rhizoctonia solani*
5	Downy mildew	*Peronospora parasitica*
6	White rust	*Albugo candida* (the most visible symptom is the ' stag head' formation)

Insect-pests:

- **Mustard saw fly** (*Athalia lugens proxima*) important pest at seedling stage. In severe case at the seedling stage, the crops have to be resown.
- **Mustard aphid** (*Lipaphis erysami*): very serious pest and major limiting factor in mustard production
- **Painted bug** (*Bagrada cruciferanum*),
- Leaf minor and diamond black moth are other insects in mustard crop.

16 Til: *Sesamum indicum*

- **Family**: Pedaliaceae
- Popularly know as queen of oilseed due to its high degree of resistance to oxidation and rancidity.
- The anti-oxidative agents (sesamin, sesamolin, sesamol), their glucosylated forms sesaminol glucosides and tocopherol make the oil very stable and therefore it has a long shelf life.
- Sesame is one of the oldest oilseed crops.
- Sesame has one of the highest oil contents (50-60%) of any oil seed crop and contains high amount of PUFA, mainly linoleic acid (37-47%).

- Traditional uses: - Sesame seeds are used in Hindu culture as a **'symbol of immortality'** and used as substitute of butter (Ghee) hence it is known as **'Butter of the Middle East'**.
- Sesame oil can trigger allergic reactions in some peoples.
- Sesame crop is drought-tolerant but most commercial cultivars of sesame are intolerant of water-logging
- Sesame seeds are protected by a capsule which only bursts when the seeds are completely ripe. This is called **dehiscence**
- The sesamum seeds need to be harvested as dry as possible and stored at 6% moisture or less.

Area and production:

- India stands second in production after Myanmar (2016)
- In India Gujarat first in both area and production, followed by WB.
- Inflorescence is known as: raceme
- Fruit is known as: Capsule

Seed Rate: **3-4 kg/ha** for row sowing, 4-5 kg/ha for broadcasting and 1 kg/ha in mixed cropping or inter cropping

- Varieties: RT-46, Pratap, TC-25 for PB, WB, HR, Raj, UP and for Gujarat-Purva and Gujarat til-1
- Gauri, Madhavi and Chandana popular varieities in Andhra Pradesh

Major diseases:

- **Phyllody**: Most destructive disease caused by **Phytoplsma** and transmitted by leaf hopper, the disease apparent at the flowering stage, All floral parts are transformed into green leafy structures followed by abundant vein clearing in different flower parts, Finally, plants look like witches broom. It can be managed by vector control through spray of diamethoate (Rogor) or Malathion @ 1.0ml/litre of water.

Insect-pests of sesame:

- Leaf hopper (*Orosius albicinctus*): it transmits phyllody disease
- Til hawk moth/sphinx moth (*Acherontia styx*): it is major pest
- Gall fly

17. Linseed: *Linum usitatissimum*

- **Family:** Linaceae
- It is dual purpose oil seed crop and when it is grown as fiber purpose called as **Flax** (in abroad) but in India it is grown as oil purpose, seed contains 33 to 47% of oil.
- Linseed oil is chief source of **Omega-3-fatty acid** and this is good for heart patient.

- The fibre received from linseed is called ***Linen*** fibre
- **Seed rate**: 30-35 kg/ha
- It is Rabi season crop
- Oil contains high **Linolenic acid (>66%),** it is a perfect drying oil, so oil can't be stored for longer time and used in manufacturing of paints, oil cloths, printed ink and varnish.
- In India major growing state is MP, followed by UP
- **Paira** (in Bihar & WB) and **Utera** (in MP): sowing of linseed crop (seed rate 10 kg/ha) in rice field before harvesting of rice crop are known as paira & utera.
- **Important varieties** are: Jawahar-1, 17, Neelum, Sheetal, Sheela, Mukta, Garima etc.

18. Castor: *Ricinus communis*

- **Family**: Euphorbiaceae
- Oil is non edible
- Plants are monocious in nature.
- **Recinine** is an alkaloid found in castor
- **Area and production**: Gujarat stand first in both area and production
- It is used as a lubricant in high-speed engines and aeroplanes, in the manufacture of soaps, transparent paper, printing-inks, varnishes, linoleum and plasticizers. it is also used for medicinal and lighting purposes.
- Oil also used as purgative for cattle and Dying industry
- **Seed rate**: 8-10kg/ha for dibbling, 12 kg/ha for drill, hybrid 5 kg/ha
- **Spacing:** 90-120 x 45-60 for normal and 120 x 90cm for hybrid

Varieites:

- GCH-1, 4, GAUCH-1, TMV-5, RHC-1
- Aruna (NPH-1) is a first hybrid of castor
- Oil contains in **hybrids 51.7 %** and attains maturity at 160-170 DAS.

Pests and diseases of castor crop:

- The castor semi-looper and the capsule-borer are the two most serious pests of castor.
- **Management:** Dusting quinolphos 1.5 % @ 20 kg/ha in early stages or spraying @ 0.1% monocrotophos on the standing crop provides an effective control of these pests.
- Seedling blight and alternaria blight are serious diseases of castor.

19. Sunflower: *Helianthus annus*

- **PUFA** (linoleic acid) is 68% in sunflower oil so suitable for heart patients
- **Origin**: Mexico
- **Family**: Compositae/Asteraceae
- **Inflorescence**: Capitulam (Disk like flower head)
- **Fruit**: Achene
- **Modern** is dwarf and short duration variety
- Day neutral plant and can grow in any season (i.e. Kharif, Rabi, Zaid)
- Seed rate 8-10 kg/ha at spacing of RxP= 60x20 cm^2
- Pollination mediated by honey bees.

20. Safflower (Kusum): *Carthamus tinctorius*

- Highest PUFA (linoleic acid) 78.50% among oil seeds and this is best suited oil for heart patients (reduce cholesterol level in blood)
- **Family**: Compositae
- **Inflorescence:** Capitulam (Disk like)
- **Fruit:** Achene
- **Seed rate**: 15-20 kg/ha
- Most drought tolerant crop among field crops
- Traditionally, used for coloring and flavoring foods, in medicines and making red (**carthamin**) and yellow dyes, especially before cheaper anilinedyes became available.
- Safflower flowers are occasionally used in cooking as a cheaper substitute for saffron, sometimes referred to as **'bastard saffron'**.
- But for the last fifty years or so, the plant has been cultivated mainly for the vegetable oil extracted from its seeds.

21. Cumin: *Cuminum cyminum*

- Also known as Zeera or Jeera, Family: **Apiaceae**
- Its dried seeds are used in the cuisines of many different cultures, in both whole and ground form.
- Cumin seeds contain **2.5 to 4.0% volatile oil** and aldehyde or cuminol which attributes for the aroma and special medicinal properties (Aiyer & Narayan 1950).
- Cumin seed is common ingredient of flavouring of zal-zira, a common drink during summer.

- The main producer and consumer of cumin is India. It produces 70% of the world supply and consumes 90% of that (which means that India consumes 63% of the world's cumin)
- In India, Gujarat (first) and Rajasthan (second) are chief producer of cumin
- Soaking the seeds for 8 hours before sowing enhances germination.
- **Seed rate:** 12–15 kilograms per hectare.
- Cumin seedlings are sensitive to salinity and emergence from heavy soils is rather difficult so well drained loamy soils with smoth seed bed good for germination
- **Sowing method**: broadcast and line sowing
- The recommended sowing depth is 1–2 cm
- **Important cultivar**: **RZ-19 (Rajasthan jeera-19), RZ-209, RS-1, GC-1 (Gujarat cumin-1), GC-4** (new cultivar good for western Rajasthan and Gujarat)
- Other varieties are UC-89, UC -199 and VC-24 under pre-release multi location trial.
- **Zeeri** (*Plantago pumile*) is an important weed in cumin fields having morphological similarity with cumin plant. As the weed is distinguishable only at the flowering stage; hand weeding is done only to avoid contamination of seeds. Management-Pendimethalin @ 1.0 kg/ha as pre emergence of crop or weeds.

Diseases:

1. **Wilt** (*Fusarium oxysporum* f. sp. cumini)
   - The most important disease is *Fusarium* wilt, resulting in yield losses up to 80%.
   - It is both soil- and seed-borne disease and survives as saprophyte
   - Infected plants show peculiar symptoms of dropping leaves, leading to mortality of the entire plant
   - Disease produces wilting symptoms at seedling and later stage of plant growth
   - Management: seed treatment with carbenazim @ 2g/kg of seed and spray carnendazim 0.2%
2. **Cumin blight** (*Alternaria burnsii*)-
   - Appears in the form of dark brown spots on leaves and stems
   - When the weather is cloudy after flowering, the incidence of the disease is increased.
   - Manamgement: **spray mencozeb (M-45)** @ 2g/litre of water.

3. **Powdery mildew** (*Erysiphe polygoni*)-
   - Incidence of powdery mildew in early development can cause drastic yield losses because no seeds are formed.Later in development, powdery mildew causes discoloured, small seeds.
   - **Management:** Crop should be dusted with Sulphur dust @ 25 kg/ha to control this disease as soon as the symptoms are noticed.

Insect pests:

1. **Aphids**: Cumin can be attacked by aphids (*Myzus persicae*) at the flowering stage. They suck the sap of the plant from tender parts and flowers. The plant becomes yellow, the seed formation is reduced (yield reduction), and the quality of the harvested product decreases.
2. **Mites** (*Petrobia latens*) -which frequently attack the crop. Since the mites mostly feed on young leaves, the infestation is more severe on young inflorescences.

22. Coriandor: *Coriandrum sativum,* F. Apiaceae (Umbelliferae)

- Mainly cultivated for its fruits as well as for the tender green leaves
- In India, highest coriander seed production state is Rajasthan, followed by Madhya Pradesh.
- The odour and taste are due to the essential oil content, which varies from 0.1 to 1.0 % in the dry seeds.
- It is a tropical crop and can be grown throughout the year (except very hot season i.e. March-May) for leaf purpose, but for higher grain yield it has to be grown in Rabi season
- **Seed rate**: 10 -12 kg/ha (irrigated), Seed treatment with *Trichoderma viride* @ 4 g/kg of seed has to be done to control wilt disease
- **Varieties**: CO-1, Gujarat Coriander-1, Rcr-20, Rajendra Swati, Rajendra Sonia, Hisar Sugandh, Hisar Surbhi, Sadhana etc.

Insect-pests and dieases:

(i) **Aphids:** Aphids can be controlled by spraying Methyl demeton 20 EC @ 2 ml/lit or Dimethoate 30 EC @ 2 ml/lit of water

(ii) **Wilt:** Seed treatment with *Pseudomonas fluorescens* @10g /kg followed by soil application of Pf1 @ 5 kg /ha.

(iii) **Powdery mildew** (*Erysiphe polygoni*): Seed treatment with *Pseudomonas fluorescens* (Pf 1) @ 10 g /kg and foliar spray of Pf1 2 g/lit or Spray. Wettable sulphur 1 kg/ha or 0.2 % solution of karathane at the time of initial appearance of the disease and 2nd spray at 10 days interval. Neem seed kernel extracts 5 % spray thrice.

ii.

23. Fenugreek: *Trigonella foenum graecum*, F. Leguminoceae

- **Seed rate**:20-25 kg/ha is required for both season
- Fenugreek is cool season crop.
- Funugreek is used as vegetable as well as pulse. The leaves and young pods are used as vegetables and the seed as condiments.
- Rajasthan state having first position in fenugreek production
- **Seed treatment**: Treated with *Azospirillum* 1.5 kg/ha + *Trichoderma viride* @ 50 g/ha for 12 kg of seeds.
- **Varieties for grain**: RMt- 143, RMt-1, Hisar sonali, Hisar madhavi, Rajendra Kranti, Lam selection-1
- **For leaves (vegetable)** - Rajendra abha, Pusa early bunching, Pant rangini.
- **Powdery mildew**: Dust Sulphur at 25 kg/ha or foliar spray with wettable sulphur 2 g/lit. at the time of appearance of disease.
- **Root rot** (*Rhizoctonia solani*): is a serious disease and can be controlled by drenching carbendazim 0.05% first at the onset of the disease and another after one month.

24. Fennel: *Foeniculum vulgare*, Apiaceae (Umbelliferae)

- Highest production of fennel in Rajasthan, followed by Gujarat
- **Seed rate**: direct sowing: 9-12 kg/ha, for transplanting 3-4 kg/ha (100 m^2 nuresry area).

25. Guar: *Cyamopsis tetragonoloba*

- Also known as **Clusterbean**
- Gum is 30-33%
- **Family**: Leguminaceae
- **Seed rate**: Grain purpose: 15-20 kg/ha
- **Varieties**: RGC-936, RGC-1002, RGC-1066, RGC-1036, HG-563, HG-365, HG-02-20, Goma manjiri.

Other forage crops are-

- **Napier grass**/ Elephant grass: *Penisitum pesrpureum*
- **Anjan grass**/Dhaman grass: *Cenchrus celiaris*
- **Sewan grass**: *Lasiurus scindicus* (found in Western Rajasthan mostly in Jaisalmer).

26. Tea (*Cammelia spp.*)

- ✓ Queen of beverage
- ✓ Family: Theaceae/Camaliaceae

Origin: **China**

Area and distribution:

- India is the **largest producers**, **consumer** and **exporter** of tea in the World.
- In India Assam has first position in production
- Another growing area is Darjeeling (West Bengal), Nilgiree hills (TN) etc.
- The best method of plucking is **two leaves** and a **bud.**
- ***Imperata cylindrica*** (thatch grass) important **weed** of tea.

27. Tobacco

- **Sodic soils** are **unfit** for tobacco production because the plants absorb a lot of **chloride ions** (Cl-) which results in a **poor burning quality** of leaves and a mild **acidic soils** (pH 5-6) are **always better** for the production of **superior quality leaves.**
- Nicotine is produced mainly in **roots** and is carried through stem to leaves where it is accumulated/**stored.**
- First hybrid variety of tobacco in India is**: GTH-1**

Indian tobacco has two cultivated species both are tetraploid (2n=48)

b. ***Nicotiana tabacum***

- ✓ Plant height **150-250** cm.
- ✓ **Large** and narrow leaf
- ✓ **Nicotine** content is **0.5-5.5%**
- ✓ Used for **smoking** and **chewing** purpose
- ✓ Grown on **light soils.**

c. ***Nicotiana rustica***

- ✓ Plant **smaller** than tabacum
- ✓ **Nicotine** content is **3.5-8.0%.**
- ✓ Used for **hookah**, **chewing** and **snuff purpose.**
- ✓ Mostly grown on **heavy soils.**

Indian tobacco (are grouped in to):

a. **Flue Cured Virginia** (FCV):

- **Cigarette** tobacco
- Major **exportable** type
- Covers 30% area and 20% production of total.

b. **Non-Virginia types:**

- **Bidi**, **hookah** and **chewing** tobacco
- **70% area** and **80 % production** of total **tobacco**

Important operations: (TDPC)-

a. **Topping**:
- Removal of **flower heads** either alone or with few upper/top leaves from the plant to **improve** the **size**, and **quality** of **leaves**.

b. **De-suckering**:
- After about 7 days of topping, auxiliary buds grow**; removal** of such lateral branches or suckers (**auxiliary buds**) is called **de-suckering**.
- The main aim of **topping** and **de-suckering** is to **divert energy** and **nutrient** from **flower head** to **leaves**.

d. **Priming**:
- **Removal** of **matured leaves**
- Entire harvest needs **5-6 priming**
- Used in **cigarette** and **wrapper** tobacco.

e. **Curing**:
- It is essentially a **drying process** whereby most of the **moisture** of leaf is **removed** to impart required **colour**, **texture** and **aroma** to the final product.
- E.g. **Flue curing** used for **cigarette process.**

28. Buck wheat: *Phagopyrum esculantum*

- ✓ Family: **Polygonaceae**
- ✓ Known as **pseudocereal**
- ✓ Fruit of buck wheat is known as: **Achene**
- ✓ **Tokyo** is the variety of buck wheat
- ✓ Leaves and flower contains alkaloid: **Rutin**
- ✓ Sowing time of buck wheat: **June –July.**

Crop cultivation terminology:

1. **Arable crops**: Crops which require preparatory tillageeg. Potato, tobacco, rice, sugarcane, maize.
2. **Alley crops**: Crops such as sweet potato,urd, turmeric and ginger are grown in the passages formed by the rows of eucalyptus, subabool and cassia mainly to hasten soil fertility restoration, enhance soil productivity and reduce soil erosion
3. **Augmenting crop**: These crops supplement the yield of the main crops e.g. sowing of mustard with berseem or lucerne to get higher yield in the first cutting.
4. **Avenue crops:** These crops are grown along farm road and fences eg. Arhar, sisal, glyricidia.
5. **Cash crops**: Such crops are grown for sale to earn hard cash e.g. Jute, cotton, tobacco, sugarcane.
6. **Catch crops**: These are grown as substitute for the main crop that has failed on account of unfavorable condition. They are of very short duration, quick growing, early harvestable or usable at any time eg. Moong, urd, cowpea, onion.
7. **Cole crops**: Cole crops are essentially cold weather crops belonging to the family **cruciferae** and capable for withstanding considerable frost eg. Cabbage, cauliflower and sprout.
8. **Cover crops/ Mulch crops**: These crops are grown mainly to cover the soil and to reduce soil moisture and erosion by wind and water e.g. Lobia, groundnut, urd, sweet potato etc.
9. **Complementary crops**: In this cropping each other crop is benefited by intercropping e.g. Jowar + lobia. Jowar receives nitrogen from lobia and lobia requires support from jowar.
10. **Ley crops**: **Such** crops are grown for grazing or harvesting for immediate or future feeding to livestock e.g. Berseem, mustard.
11. **Paired row cropping**: Each third row is removed or growing of crops in paired row is called paired row cropping. It is suitable for dry land and objective is to conserve moisture.
12. **Restorative crops**: Such crops provide a good harvest along with enrichment or restoration or amelioration of soil e.g. Legumes.
13. **Skip cropping**: A line is left unsown in the regular row series of sowing is called skip cropping.
14. **Smother crops**: These crops provide large canopy and quick growing ability e.g. Cowpea, mustard, sorghum.

15. **Trap crop/decoy crops**: Grown on boundary of the field to attract insect pests and soil borne harmful biotic agents such as parasitic weeds e.g. Orobanche (weed) is trapped by solanaceae plants.

16. **Mixed farming**: It is a system of farming on a particular farm which includes crop production, livestock, poultry, fisheries, bee keeping sustaining and satisfying as many needs of the farmer as possible.

17. **Crop rotation**: Refers to the sequence of crops grown on the same land in a recurring succession.

18. **Mono cropping/mono culture**: Refers to growing of only one crop on a piece of land year after year.

19. **Multiple cropping**: Growing two or more crops on the same piece of land in one calendar year is known as multiple crops. It includes inter cropping, mixed cropping and sequence cropping.
    - **Inter cropping**: Growing two or more crops simultaneously on the same piece of land with a define row pattern.
    - **Mixed cropping**: Growing two or more crops simultaneously intermingled without any row pattern.
    - **Sequence cropping**: As growing of two or more crops in sequence on the same piece of land in farming year.

20. **Relay cropping**: Refers to growing a second crop between the rows of the main crop which is ready to harvest shortly.

21. **Ratooning**: The practice of tacking the second crop from previous one is known as ***Ratooning. E.g. ratoon sugarcane.***

Weed science

- Weed as a plant growing where it is not desired or
- **Weeds are u**nwanted plants growing out of time and place.
- The term **'weed'** was coined by **Jethro Tull,** he known as **'father of weed** science' for his contribution of invention of tillage, especially seed drill and horse hoe, which drastically reduced weed competition and prepared soil better, effectively pioneering systematic, mechanical weed management as part of a broader modern agricultural system.
- **NRC for Weed Science** at Jabalpur (M.P.) established in the year 1989.
- From 2014, it was renamed as **Directorate of Weed Research** (DWR)
- **Indian Soceity of Weed Science** (ISWS) established at Hisar (HR), but later shifted to Jabalpur (MP)
- **Indian Journal of Weed Science** published by ISWS at DWR, Jabalpur (M.P).
- In India total losses in agriculture caused by weeds are estimated to be 33%, followed by insects (26%) and diseases (20 %). (DWR, 2015)
- Three families of weeds *viz.*, poaceae, asteraceae and cyperaceae account 44% of the wolrd's worst weeds species.

Pesticides consumption order:

- In world: Herbicides > insecticides > fungicides
- In India: Insecticides > fungicides > herbicides.
- Largest consumption of herbicides among the crops in rice followed by tea and wheat
- The largest manufacturerd and consumed herbicides in India is Isoproturon.
- Per hectare consumption of pesticide in India is less than 0.6 kg/ha
- Highest consumption of herbicides in the world: Japan (5 kg/ha)
- Per ha. herbicides consumption in India is: 40 g/ha/year

List of weeds act as alternative hosts for insects and diseases are:-

Weed	Alternate host for insect-pests/diseases
Echinochloa spp.	Stem borer of Rice (*Tryporyza incertulus*)
Chenopodium album	Stem borer of maize (*Chilo partellus*)
Amaranthus spp.	Gram caterpillar of pigeonpea
Lantana camara	American boll worm (*Helicoverpa* sp.)
Avena fatua	Stem rust of wheat, barley
Parthenium hysterophorus	Tobacco streak virus (cause groundnut stem necrosis)
Sachharum spontaneum	Downy mildew of maize
Cenchrus ciliaris	Ergot disease of Bajra (pearl millet)

Leersia oryzoides	Bacterial leaf blight of Rice
Abutilon spp.	Spotted boll worm & pink boll worm of cotton
Wild mustard (*Sinapis arvensis*)	Club root/black leg of cabbage & cauliflower

Seed production capacity of common weeds:-

Weed	Seeds /plant
Conyza bonariensi	226753
Amaranthus spp.	1,96,000
Lambsquarter *(Chenopodium spp.)*	72,000
Cuscuta spp	16000
Phalaris minor	400

- **Weed seed longevity:** most of weed seeds have longevity of > 20 years.
- **Popular *Proverb for* weed seeds:** One year seeding is seven year weeding.
- ***Weed seed bank:*** it is the reserve of all viable weed seeds and vegetative propagules present in the soil.
- ***Weed seed withdrawal:*** removal of weed seeds from soil. it includes seed germination + fatal germination + seed mortality caused by ants, rodents, birds and other agents and seeds decay or loss of viability due to fungal pathogens *etc.*
- **Weed dormancy: Three type**
- **Weed dormancy**: Harper's classification (given by Harper, 1977).
  1. **Innate/inherent (Primary dormancy):** Due to genetic factors and chemical inhibitors in seed coat.
  2. **Induced (Secondary dormancy):** Dormancy due to environmental factors (Temperature/water logging).
  3. **Enforced**: dormancy in this case enforced (deep placement) & by unfavorable condition.
     - **Wild oat** *(Avena fatua)* exhibited all three type of dormancy **(Reported by Thurston 1959).**

Quiescence: it is resting condition of seed (a type of dormancy).

Harmful effect of weeds:

(i) Dropsy (Blindness in human): Due to *Argemone maxicana*

(ii) Dermal allergies: due to carrot grass (*Parthenium hysterophorus)*

(iii) Prussic acid/HCN/Dhurin poisoning in animal: due to johnson's grass (*Sorghum helepense)*

(iv) Oxalate poisoning in animals: due to pigweed *(Amaranthus spp.)*

Censor mechanism:

- The wind swings the intact plant and forces their mature pods to disperse seeds some distance away from mother plant e.g. *Argemone maxicana.*

Weed dissemination:

- Dispersal of weed seeds from mother plant to other place.
- It includes several mechanisms as follows:

(a) Abiotic dispersal:

1. **Anemochory**: weed seeds dispersal by winds or air: following type structure performes of wind transportation.

   Pappus: a parachute like modification of calyx in to hairs e.g. *Tridax procumbens.*

   Comose: some seeds of weeds are covered with hairs e.g. *Calotropis sp.*

   Baloon: calyx modified that enclosed the seeds loosely along with entrapped air e.g. *Physalis minima*
2. **Hydrochory**: dispersal by means of water
3. **Barachory**: dispersal by gravity

(b) Biotic dispersal:

1. **Myrmecochory:** weed seed dispersal by ants (insects) is known as ***Myrmecochory.***
2. **Autochory:** self dispersal mechanisms of weed seeds is known as ***Autochory***
3. **Zoochory: animals mediated dispersal is following type-**

   (i) **Endozoochory:** viability of weed seeds in animal excreta after digestion is referred as ***Endozoochory***

   (ii) **Epizoochory:** dispersal by adhesion of seeds on out side of animals is known as ***Epizoochory***
4. **Anthropochory:** human mediated dispersal (special kind of zoochory), this is major cause of biological invasion of weeds among the countries.

Classification of weeds:

1. Based on life span or ontogeny:

(a) Annual weed: Complete their life cycle in one year or one season (life span between seed to seed)

(i) Monsoon/Kharif weeds: They germinate and complete their life cycle during mansooon season.

Common Name	Scientific name	Crops
Chaulai, Pigweed	*Amaranthus viridis*	Maize, bajra, sorghum
Spiny or Kanta chaulai	*Amaranthus spinosus*	Maize, bajra, sorghum

Day flower	*Commelina benghalensis*	Kharif crops
Noonia	*Portulaca spp.*	Maize
Small sedge/ motha	*Cyperus irria*	Bajra, sorghum
Yellow nut sedge	*Cyperus esculentus*	Bajra, sorghum
Jungli rice	*Echinochloa colona*	Rice
Barn yard grass	*Echinochloa crusgalli*	Rice
Carpet weed/ Patharchata/ Santhi	*Trianthema spp.*	Kharif crops
Hajardana/stone breaker	*Phyllanthus niruri*	Moong, soybean, urd
Gokharu kanta	*Tribulus terestris*	All kharif crops
Congress grass	*Partheneum hysterophorus*	Non cropped area

(ii) Winter/Rabi weeds:

Common Name	Scientific name	Crops
Maxican poppy (Satyanashi)	*Argemone maxicana*	Mustard
Wild oat	*Avena fatua*	Barley, wheat
Kasni/chicory	*Cichorium intybus*	Barseem
Bathua, common lambsquarter	*Chenopodium album*	Wheat, barley
Khartua	*Chenopodium murale*	Wheat, barley
Krisha neel/scarlet pimpernel	*Anagalis arvensis*	Wheat
Dodder/amar bel	*Cuscuta spp.*	Lucern
Senji/yellow sweet clover/ban methi	*Melilotus indicus*	Gram mustard
Senji/white sweet clover	*Melilotus albus*	Gram mustard
Broom rape	*Orobanche cernua*	Mustard
Gehusa/gulli danda/ gehu ka mama	*Phalaris minor*	Wheat
Pyaji	*Asphodelus tunifolius*	Gram

(iii) Summer weeds: They grow during hot months e.g. *Ageratum conyzoides* and Dudhi *(Euphorbia spp)*.

(iv) Ephemerals: Short living annual weeds, complete life cycle within 2-4 weeks e.g. *Phyllanthus niruri.*

(b) Biennial weeds: Complete their life cycle in two years: they produce vegetative growth during first season/year and produce flowering and seeds in next year.

Examples of biennial weeds: -

1. Wild carrot (*Daucus carota)*
2. Khaki weed *(Alternanthera pungens)*
3. Wild onion *(asphodelus tunifolius)*
4. Musk thistle *(Carduus nutans)*
5. Jangli gobhi *(Launaea procumbens)*
6. Chicory/*Kasni (Cichorium intybus):* it is an annual, binneal or perennial plant.

(c) Perennial weeds: Complete their life cycle in more than two years

1. **Shallow rooted (20-30 cm) perennials:** *Cynodon dactylon, Agropyron repens and Ambrosia sp.*
2. **Deep rooted (1.0 m or more) perennials:** *Cyprus rotandus, Sorgham helepense, Cirsium arvense, Lantana camera* and *Datura sp.* Annual and biennial weeds are reproduced through seeds but perennial weed are reproduced chefly by vegetative parts then seeds.

Examples of perennial weeds:-

Common Name	Scientific name	Crop
Bermuda grass, Doob	*Cynodon dactylon*	all crops
Anjan grass, African foxtail	*Cenchrus ciliaris*	Bajra, sorghum
Water hyacinth/Jal kumbhi	*Eichhornia crassipes*	Aquatic
Purple nut sedge, Motha	*Cyperus rotundus*	bajra, sorghum
Lantana weed/Tiger shrub	*Lantana camara*	non cropped
Field bind weed, Hirankhuri	*Convolvulus arvensis*	Wheat
Gajar ghas/ Carrot grass	*Parthenium hysterophorus*	non cropped
Prickly pear/Nagphani	*Opuntia spp.*	non cropped
Baru / Johnson grass	*Sorghum halepense*	Bajra, sorgham, maize
Kans /wild cane/Tiger grass	*Saccharum spontaneum*	Bajra, sorghum

The perennial weeds propogated by:

S.N.	Vegetative parts	Weeds
1	Rhizomes (stem)	*Sorghum helpense, Saccharum spontaneum*
2	Stolons/runners (horizontal stem)	*Cynodon dactylon*
3	Tubers (underground stem)	*Cyprus spp.*

4	Bulbs (underground swollen leaf bases or scales)	*Allium canadense* (wild garlic or Canada onion) *Asphodelus tenuifolius* (wild onion)
5	Stem fragments	*Cuscuta spp.*
6	Creeping roots piece	*Cirsium arvense*
7	Creeping roots	*Convolvulus arvensis*

2. Classification of weeds based on place of occurrence:-

a. **Obligate weeds** (weeds of cropped land): Weeds occurring under disturbed land or cultivated fields in cropped situations. E.g. *Phalaris minor* in wheat, *Echinochloa spps.*in rice, *Ipomoea triloba* in sugarcane, *Amaranthus spp.* in kharif crops.

b. **Facultative weeds** (weeds of non-cropped area): Weeds of uncropped lands such as waste land, **roadsides** *etc*. ***Examples:*** *Lantana camara, Parthenium hysterophorus, Calotropis gigantean, Tribulus terrestris, Datura metal, Abutilon spp., Alternanthera pungens etc.*

c. Xerophytic weeds: Desert weeds such as *Opuntia spp, Pluchea spp.*

3. Classification of weeds based on origin:-

(a) **Alien / invasive/ Anthrophyte/non-native weeds:** Introduced in India from other countries.

List of common alien or introduced weeds:-

S.N.	Weeds	Introduced from
1	Carrot grass (*Parthenium hysterophorus)*	USA, 1954 with wheat
2	Canary grass (*Phalaris minor*)	Europe/North America
3	*Eichhornia crassipes*	South America
4	*Conyza bonariensis*	North America
5	*Lantana camara*	Tropical America
6	*Xanthium strumarium*	North America
7	*Cynodon dactylon*	Middle East
8	*Digitaria sanguinalis*	Europe
9	*Amaranthus viridis*	South US/Mexico
10	*Sorghum helpense*	Mediterranean region
11	*Cuscutta chinensis*	China
12	*Orobanche spp.*	Europe
13	*Cyprus rotandus*	Europe/Eurasia
14	*Argemone mexicana*	Mexico
15	*Salvinia molesta*	Brazil

(b) **Native weeds:** these are indigenous e.g. *Echinochloa colona, Phyllanthus niururi, Digera arvensis etc.*

4. Classification of weeds based on cotyledons:

(a) **Monocots or narrow leaf weed:** Seeds having single cotyledon. These weeds are tolerant to 2, 4-D application. **Examples-** *Sorghum helepense, Echinochloa colona/crugalli, Phalaris minor, Cynodon dactylon, Commelina bengalensis.*

(b) **Sedges:** Having narrow leaves, thin and triangular cross section stem. Sedges are monocots but susceptible to 2, 4 –D application.eg. *Cyprus spp.*

(c) **Dicots:** These weeds are mostly broad leaved and susceptible to 2, 4-D Application. **Examples:** *Chenopodium album, Amaranthus spp, Parthenium hysterophorus, Lantana camara, Datura etc.*

5. Aquatic weeds:

(a) **Floating weeds:** waterer hyacinth (*Echhornia crassipes*), water fern (*Salvinia molesta*), water lettuce (*Pistia sp.*), water velvet (*Azola pinnata*).

(b) **Rooted weeds with floating leaves:** e.g. lotus (*Nelumbo nucifera*), white water lily (*Nymphaea alba*).

(c) **Submerged weeds:** Weed germinates, grow and reproduce beneath the water surface e.g. BGA (*Anabaena spp.*), Pond weed (*Potamogeton spp*), *Vallisnaria, Hydrilla spp.*

(d) **Emergent weeds /Semi-aquatic weeds:** Grow in shallow water. E.g. *Typha spp., Ipomoea sp.,* pickerel weed (*Pontederia cordata*). Among all aquatic weeds the **water hyacinth** (*E. crassipes*) is the most troublesome aquatic weed.

6. Parasite weeds (crop bound weeds):-

(a) **Stem parasite:** after emergence they get attached to the stem of the host crops. These are two types

1. **Total or holo stem parasite:** They don't have chlorophyll and derieved food from host. E.g. Dodder (*Cuscuta spp.*). **Host plants**: Lucerne, Onion, Carrot, Chilly, Pulses, Soybean, Groundnut, Niger, Fennel etc.
2. **Semi or Partial stem parasite**: Loranthus (*Dendrophthoe 130erseem*): have chlorophyll but depends on hosts for water, minerals, nutrients etc. **Host plants**: Plantation, fruit crop and timber trees (Mango, Tea, and Teak etc.).

(b) **Root parasite:** they attached to roots for derived food or nutrition.

1. **Total or holo root parasite:** Broomrape (*Orobanche spp.*, **Host plants:** Tobacco, Tomato, potato, Brinjal, **Mustard**, Sunflower, Pea, Oak, and Bean*setc.*
2. **Semi or Partial root parasite:** Witch weed (*Striga spp.*).**Host plants:** Millets, Maize, Sorghum, Sugarcane, Rice *etc.*

7. Based on photosynthetic pathway: they categorized as C_3, C_4 and CAM types:-

C_3 weeds	C_4 weeds	CAM weeds
Phalaris minor	*Cypruus rotundus*	*Conyza nutans*
Avena fatua	*Cynodon dactylon*	*Ageratum conyzoids*
Chenopodium album	*Amaranthus spp.*	*Cirsium texanum*
Convolvulus arvensis	*Sorghum helpense*	*Calocasia spp.*
Ageretum conyzoides	*Portulaca oleracea*	
Cammelina bengalensis	*Imperta cylindrical*	
Echhornia crassipe	*Trianthema portulacastrum*	
Phyllanthus niruri	*Echinocloa spp.*	
Xanthium strumarium	*Diditaria sanguinalis*	
Parthenium hysterophorus	*Eleusine indica*	

8. Classification of weeds on basis of relativity:

(a) **Absolute weeds:** These are always weeds. Plant of no economic importance for human being. E.g.*Cyprus spp., Argemone mexicana*

(b) **Relative weeds:** These are other crop plants. Undesirable crop plants in the main crop field e.g. barley in wheat.

(c) **Volunteer plants:** Growing of plant in main crop due to seed shading from the previous sown crop.

(d) **Rogue weeds:** Unwanted plant that grow profusely where it is not desired or any economical plant growing out of its proper place.e.g. Plants of C-306 variety grown in Raj- 4037 variety of wheat crop. **Roguing:** Removal of off type plant from main crop.

(e) **Objectionable weeds:** Weeds which produce seeds that are difficult to seprate once mixed in crop seeds because of resemblance with crop seeds.

S.N.	Objectionable weeds	Main crop
1	Satyanashi, *Argemone maxicana*	Mustard
2	Wild oat, *Avena fatua*	Oat
3	*Kasni*/Chicory, *Cichorium intybus*	Lucerne/Berseem
4	Dodder, *Cuscutaspp.*	Lucerne seeds
5	Wild onion, *Allium canadense*	Onion seeds

6	Senji (*Melilotus spp.*)	Methi (fenugreek)
7	Hiran khuri (*Convolvulus arvensis*)	Wheat

(f) Noxious weeds: Undesirable, troublesome and difficult to control. E.g. *Cyprus spp., Cynodon dactylon, Parthenium sp., Lantana, Cuscuta, Orobanche, Eichornia, Salvinia, Sorghum helpense etc.*

(g) Mimicry weeds: These weeds are morphologically similar with main crop. E.g. *Phalaris minor* in wheat and *Echinochloa spp.* in rice.

(h) Satelite weeds: weeds mature at the same time and height along with crop and seeds are similar to main crop so difficult to separation of seeds from main crop.e.g. *Avena fatua* in Barley

Herbicides:-

1. **Herbicide:** means chemicalsthat kills or suppress the unwanted vegetations.
2. **Weedicide:** it was restricted term, chemicals that kills the weeds only
3. Before 1940, the era of inorganic herbicides like ashes, salts, copper sulphate, sodium arsenate, sodium nitrate, sodium chloride etc.
4. Era of organic herbicides was started with the discovery of 2, 4-dichloro-phenoxy-acetic acid (2, 4-D) in 1940 (used as plant hormone).
5. Hamner and Turkey (1944) firstly observed that 2, 4-D successful controlled weeds in cereal crops.
6. 2, 4 –D is the first selective organic herbicide used in the World.
7. Revolution in weed control by discovery of 2,4-D (Pokorny, 1944)
8. In India 2, 4-D first tested as herbicide in 1946 and introduced at commercially in filed level in 1948.
9. Herbicides are registered in India under the Insecticides Act, 1968, at present (Oct, 2015) 60 technical herbicides and 17 herbicide pre-mixes have been registered for use.
10. In India, the 'Central Insecticide Board and Registration Committee **(CIBRC)**' regulates and manages the use of herbicides and insecticides.
11. The major crops where maximum herbicides are used are rice, wheat and plantation crops.
12. The state where highest herbicides are used is Punjab followed by Andhra Pradesh and U.P.
13. The mostly used herbicides in India are isoproturon, butachlor, 2, 4 –D, glyphosate and atrazine.
14. **Bioremediation:** some weeds remove or accumulate heavy metals from soil or aquatic systems. E.g. *Echhornia crassipes* and Hydrilla serve for bioremediation of Lead, Mercury, and Cadmium.

    ii.

Important herbicide group:

Chemical group	Herbicide	Site of action
Amides Chloro acetamides	Propanil Acetochlor, butachlor, alachlor	Inhibition of photosynthesis at PS-II Inhibition of cell division
Bipyridyliums	Paraquat, diquat	Photosystem-1-electron diversion
Dinitroanilines	Fluchloralin, pendimethalin	Inhibition of microtubule assembly
Imidazolinones	Imazethapyr	Inhibition of ALS (Acetolactase synthase) or AHAS (acetohydroxy acid synthase)
Phenoxy carboxylic acid	2,4-D , 2,4-DB, MCPA,MCPB	Inhibition of auxin synthesis
Sulfonylureas	Sulfosufuron, metsulfuron methyl	Inhibition of ALS (Acetolactate synthase)
Triazines Triazinones	Atrazine, simazine, terbutryne Metribuzin	Inhibition of photosynthesis at PS-II Inhibition of photosynthesis at PS-II
Glycines/phosphon ic acids	Glyphosate, glufosinate	Inhibition of EPSP Synthase
Phenyl ureas	Isoproturon, diuron, monouron	Inhibition of phosynthesis at PS-II
Diphenyl ethers	Nitrofen, oxyfluorfen, lactofen	Inhibition of protoporphyrinogen oxidase (PPO)
Thiocarbomates	EPTC, thiobencarb (benthiocarb)	Inhibition of lipid synthesis

Classification of Herbicides:-

(1) Based on time of application:-

(i) Pre-plant incorporation (PPI): Application of herbicide in finally prepared levelled soil one day or just before sowing of a crop. These herbicides usually have higher vapour pressure or volatile nature. **Examples:** Fluchloralin, EPTC, trifluralin.

(ii) Pre emergence: Application of herbicide 1-2 days after sowing or immediately after sowing of a crop but before the emergence of crop. **Examples:** pendimethalin, atrazine, alachlor, butachlor, metribuzine

(iii) Post emergence: Application of after the emergence of crop and weeds both (generally 15-30 days after sowing). **Examples:** 2, 4-D, isoproturon, sulfosulfuron, metsulfuron, clodinafop-propargyl *etc.*

(2) Based on Selectivity.

(a) Selective herbicide: Herbicide which kills selectively a particular group of weeds in a mixed stands of plants. **Examples:** all the pre emergence and post emergence applied herbicides of field crops.

(b) Non selective herbicide: Herbicide which kills any species of plant irrespective of weeds and crop. They are used in controlling weed in road sides and non cropped area. **Examples:** paraqaut, diquat, glyphosate, glufosinate, acrolein, piclroram, amitrol etc.

(3) **Based on Spectrum of weed control:**

(a) Narrow spectrum herbicide: Herbicide which control particular group of weeds and due to their narrow range usually do not control even all the weeds of particular group. **Example:** 2, 4 D (kills only broad leaved weeds and sedges, not grasses), metsulfuron-methyl etc.

(b) Broad spectrum herbicide: Herbicide which control wide range of weeds (grasses, broadleaves, sedges). **Example:** atrazine, pendimethalin, imezathypr, alachlor, butachlor

(4) **Based on site of application:**

(a) Soil applied/ soil active herbicides: Applied to soil or active through soil and kill germinative weeds and their parts. E.g. all pre plant incorporation and pre emergence herbicides are include in this category.

(b) Foliage applied: Applied to foliage of plant or non cropped weeds. E.g. all post emergence herbicides (2, 4-D, isopruturon, sulfosulfuron, metribuzin etc.) + non cropped herbiceds (paraquat, glyphosate etc.).

(c) Both soil and foliage active herbicides: atrazine, metribuzin have both action.

(5) **Based on residual action in soil:**

(a) Non residual / Zero persistence herbicide: these herbicides easily degraded or metabolized after application and leaves no or less residues in applied soil.e.g. ·Paraquat, diquat, glyphosat these are used in zero/minimum tillage).

(b) Residual herbicide: Herbicide which are more resistant to degradation and maintain their effect in soil for a considerable period of time after application e.g 2, 4-D (3-4 week) Triazines (1-2 Year), Fenec (> 2 Years).

(6) **Based on duration of weed control:**

(a) Temporary soil sterilants / fumigants: Herbicide which are basically fumigants and sterilize the soil for a relatively short period (Usually 15-16 weeks) e.g. methyl-bromide, metham etc.

(b) Permanent soil sterilants: They remain active in soil for a relatively long period e.g. Triazines & Phenyl ureas.

(7) **Based on mode of action:**

(a) Contact herbicide: These herbicide are being less mobility, kills the plant primarily by contact with the plant tissue e.g. paraquat, diquat, propanil, oxyflurfen etc.

(b) Systemic or translocated herbicides: Herbicide moves within the plant from the point of treatment to its other part to variable extent e.g. all herbicide except contact are systemic in nature like 2, 4-D, atrazine, pendimethalin, glyphosate, metribuzin etc. Atrazine is systemic and contact herbicide.

Successful example of bio agents used for biological weed control:

Bio- agents	Kind of bio agent	Weed
Crocidosema lantana	Insect	*Lantana camara*
Ophiomyia lantanae	Insect	*Lanatana camara*
Dactylopius tomentus	Insect	*Opuntia dellini*
Teranychus desertum	Mite	*Opuntia dellini*
Cactoblastis cactrum	Insect	*Opuntia inermis (Australia, Hawaii)*
Zygograma bicolorata	Insect	*Parthenium hysterophorus*
Epiblema strenuana	Insect	*Parthenium hysterophorus (Australia)*
Bectra verutana	Insect	*Cyprus rotundus*
Neochetina eichorniae/bruchi	Insect	*Eichhornia crassipes*
Sameodes albiguttalis	Insect	*Eichornia crassipes*
Orthogalumna terebrantis	Mite	*Eichornia crassipes*
Cryptobagous salviniae	Insect	*Salvinia molesta*

Mycoherbicide or bioherbicide: pathogens cultured artificially & made available in sprayable formulations just like chemical herbicides are known as bio/mycoherbicides.

Product	Content	Target weed
De Vine (it was first mycoherbicide, 1981)	Liquid suspension of fungal spores of *Phytophthora palmivora*	Strangler-vine in citrus *(Morrentia odorata)*
Collego (1982)	Contains fungal sopres of *Colletotrichum gloesporoides*	Jointvetch in rice *(Aschynomene virginica)*
Biomol	Wattable powder containing fungal sopres of *Colletotrichum gloesporoides*	*Malva pusila*
Biapolaris	Fermentation product of *Streptomyces hygroscopicus*	*Sorghum helepense*
Lubao2	*Colletotrichum gloesporoides sub spp cuscuta*	Cuscuta

Herbicide resistance:-

- In India herbicide resistance was first reported in wheat crop in Haryana where some biotype of *Phalaris minor* found resistance to Isoproturon (reported by Malik and Singh, 1993). Highest number of weeds found resistance against atrazine herbicide (66 weed species) and maximum number of weeds resistance has been reported in wheat crop.

Types of resistance:-

1. **Cross resistance:** In this a weed biotype is resistance to two or more herbicide due to presence of single resistance mechanism.
2. **Multiple resistances:** In this a weed biotype is resistance to one herbicide due to the presence of two or more distinct resistance mechanism.

Weed management in common field crops.

Crop	Herbicide	Dose (*a.i.*)	Time of application	Effective for control
Rice (Nursery)	Anilofos	300-450 g/ha	PE (3-4 DAS)	Grasses
	Butachlor	1.5 kg/ha	PE (6-7 DAS)	Grasses &some BLW
Rice (Direct seeded upland rice)	Bispyribac-Na	25 g/ha	POE	Broad spectrum weed
	Benthiocarb	2 kg /ha	PE	Grasses & BLW
	Pyrazosulfuron	20 g/ha	Early POE	Grasses & BLW
	Fenoxaprop-ethyl	60 g/ha	POE	Grasses
	Azimsufuron	17.5 g/ha	POE	BLW & Cyperus
Transplanted rice	2,4-D	0.75 Kg/ha	POE	BLW and sedges
	Pendimethalin	1.0 kg/ha	PE (3-5 DAT)	Grasses
	Propanil	0.84 Kg /ha	(3-4 Leaf Stage)	Grasses & BLW
	Nitrofen	2 kg/ha	PE (2-3 DAT)	Grasses
	Anilofos	300-450 g/ha.	PE (3-5 DAT)	Grasses
Wheat	Isoproturon	1.0 kg/ha	POE (25-30 DAS)	Grasses & non resistant biotype of *Phalaris minor*
	Clodinafop-propargyl	60 g/ha	POE (30-35 DAS)	Wild oat and *Phalaris minor*
	Sulfosulfuron	25 g /ha	POE (30-35 DAS)	Broad spectrum control of weeds
	Carfentrazone	20 g/ha	POE (25-30 DAS)	BLW and *Malva spp.*
	Pinaxaden	40 g /ha	POE (25-30 DAS)	Wild oat & *Phalaris minor*
	Metsufuron methyl	**4 g/ha**	POE (30-35 DAS)	BLW
	Fenoxaflop ethyl	100-120 g/ha	POE (30-35 DAS)	Wild oat & *Phalaris minor*
	2,4 –D	0.50 kg/ha	POE (30-35	BLW

			DAS)	
Maize/ Bajra / Sorghum	Simazine (withdrawl from use in India)	1.0 kg/ha	PE & Early POE	Broad spectrum weed
	Atrazine	0.5-0.75 kg/ha	PE & Early POE	Broad spectrum weed
	2,4-D	0.5-0.75 kg/ha	POE (30-35 DAS)	BLW
Pulses	Pendimethalin	1.0 kg/ha	PE	Grasses & some BLW
	Oxyfluorfen	150-200 g/ha	PE	Broad spectrum weed
	Fluchloralin	1.0 Kg/ha	PPI	Grasses & some BLW
	Imazethapyr	75-100 g/ha	Early POE	BLW, Sedges, Grasses
	Quizalofop-ethyl	40-50 g/ha	POE (15-20 DAS)	Grasses
	Alachlor	1.0 kg /ha	PE	Grasses & Some BLW
Soyabean	Trifluralin	1-1.5 kg/ha	PPI	Grasses & Some BLW
	Pendimethalin	1.0 kg/ha	PE	Grasses & some BLW
	Metribuzine	0.5 kg/ ha	PE	Grasses & some BLW
	Fluchloralin	1.0 Kg/ha	PPI	Grasses & some BLW
	Imazethapyr	75-100 g/ha	POE (15-20 DAS)	BLW, Sedges, Grasses
	Quizalofop-ethyl	40-50 g/ha	POE (15-20 DAS)	Grasses
	Alachlor	1.5 kg /ha	PE	Grasses & Some BLW
Ground nut & Other oil seeds	Fluchloralin	1.0 kg/ha	PPI	Grasses & Some BLW
	Pendimethalin	1.0 kg/ha	PE	Grasses & some BLW
	Metribuzine	0.5-0.7 kg/ ha	PE	Grasses & some BLW
	Fluchloralin	1.0 Kg/ha	PPI	Grasses & some BLW
	Imazethapyr	75-100 g/ha	POE	BLW, Sedges, Grasses
	Quizalofop-ethyl	40-50 g/ha	POE	Grasses
	Alachlor	1.5 kg /ha	PE	Grasses & some BLW
Sugarcane	Atrazine	1.5-2.0 kg/ha	PE	Broad spectrum weed
	2,4 –D	2.5 kg/ha	POE	BLW
	Glyphosate	1.0	POE	All weeds
	Metribuzine	0.88 kg/ ha	PE	Grasses & some BLW
Cotton	Diuron	0.5-0.75 kg/ha	PE (3-4 DAS)	Grasses &some BLW
	Oxadiazon	0.5-0.75	PE (1-3DAS)	Grasses & some BLW
	Butachlor	1.0-1.25	PE	Grasses & some BLW
	Fluchloralin	1.0 kg/ha	PPI	Grasses & some BLW
	Pendimethalin	1.0 kg/ha	PE	Grasses & some BLW
	Quizalofop-ethyl	40-50 g/ha	POE	Grasses

	Glyphosate	1.0 kg/ha	POE (over 15 cm height)	All weeds
	Glufosinate	0.45 kg/ha	POE (over 15 cm height)	All weeds
Vegetables				
Almost all vegetables	Fluchloralin	1.0 kg/ha	PPI	Grasses & some broad leaf weeds
All vegetables	Pendimethalin	1.0 kg/ha	Pre or post transplanting	Grasses & some broad leaf weeds
Onion & garlic	Oxyflurfen	200-250g /ha	Pre or post transplanting	Grasses & some broadleaf weeds
	Oxadiargyl	90 g/ha	0-3 DAT	Grasses & sedges
	Oxadiazon	0.5-0.75 kg/ha	Pre or post transplanting	Grasses & some broad leaf weeds
Potato & onion	Butachlor	1.0 kg/ha	Pre or post transplanting	Grasses & some broad leaf weeds

***Simazine** (withdrawn from use in India), so at present atrazine is the only herbicide most commonly used for weed control in maize, sorghum and pearlmillet.

Note: At the time of application of most of herbicides in the crops, ensure sufficient moisture level in the field.

BLW= Broad leaf weed, **PPI**= Pre plant incorporation, **PE**= Pre emergence, **POE**= Post emergence, **DAS**: day after sowing, **DAT**: day after transplanting.

***a.i.*:** active ingredient of toxicant (all doses of herbicides geven in above table are in a.i. g/ha or a.i. kg/ha)

Apoplastic: herbicide movement through the xylem (non living tissue) is called as *apoplast*. **Symplastic:** herbicide movement through the phloem (living tissues) is called *symplastic*. E.g. glyphosate

Trade name of important herbicide:

Herbicide	Trade name	Time of application
Alachlor	Lasso, crop star	Pre emergence
Atrazine	Atrataf	Pre emergence
Butachlor	Mechete	Pre emergence
Bispyribac-Na	Nominee Gold	Post emergence
Fluchloralin	Basalin	Pre plant incorporation
Glyphosate	Round up	Post emergence
Imazethapyr	Guard	Early Post emergence
Isoproturon	Ronak, Aerolon	Post emergence
Metribuzine	Sencor	Pre emergence
Metsulfuron	Escort	Post emergence
Nitrofen	Tok-E-25	Post emergence
Oxyflorfen	Goal	Pre emergence
Pendimethalin	Stomp	Pre emergence
Paraquat	Sweep, Gramoxone	Non selective

Propanil	Stamp-F-34	Post emergence
Pyrazosulfuron	Sathi	Early post emergence
Sulfosulfuron	Leader	Post emergence
2 ,4-D (Na salt)	Fernoxone	Post emergence (1 kg / ha)
2 ,4-D (amine salt)	Weedar	Post emergence (0.75 kg / ha)
2 ,4-D (esters)	Weedon	Post emergence(0.50 kg / ha)

Important points:-

- Discovery of 2, 4-D herbicide heralds the birth of Weed science in Agriculture.
- The site of action of glofosinate herbicide: Inhibition of Glutamine synthetase.
- Wheat is currently known as account for most number of weed species in world.
- Water hyacinth suitable habitat for insect vector transmitting diseases like Dengue, Chikungunya and Malaria and accumulate heavy metals –lead, cadmium and mercury.
- The selectivity of Triazines due to Glutathions –S- transferase.
- ***DNOC*** (4,6-dinitro-ortho-cresol), developed as the first organic herbicide in 1932.
- Metasulfuron-methy is an exclusive broad leaved killer and good substitute of 2, 4-D.
- The yield losses due to weed infestation in Direct Seeded Rice are higher than transplanted rice.
- Glyphosate is a systematic/ translocated and non selective herbicide.
- Paraquat and diquat are contact and non selective herbicides.
- Continuous use of isoproturn in wheat leads to the development of resistant biotype of *Phalaris minor.*
- Chlorophyll is absent in parasite weeds like *Orobanche* and Dodder and present in *Striga* and loranthus.
- 2, 4-D sodium salt is used to control: **Striga.**
- Drift hazard in to adjoining pulse crop is due to spray of: **2,4-D .**
- Oat is most sensitive to triazines and sorghum is to: **Imazethapyr.**
- In wheat + mustard/gram promising herbicide is: **Isoproturon**

Some important terms:

1. **Weed management:** it is broad term includes prevention, eradication and control of weeds.
2. **Suicidal germination:** Non host crop of striga, like cotton secretes Strigol (chemical release from roots), which stimulate germination of striga seeds and germ tube fails to have attachment to non host crop this is called suicidal germination.
3. **Allelopathy:** The phenomenon of one plant having detrimental effect on another through the production of chemical compounds.
4. **Stale seed bed:** The techniques in which weeds are allowed to germinate (flushing) by application of pre sowing irrigation and killed before sowing by the application of herbicide.
5. **Blind hoeing:** The hoeing is done after sowing of the sugarcane and it control mainly sedges weed in sugarcane.
6. **Selectivity Index:** The ratio of maximum dose of herbicide tolerated by the crop to the minimum dose of herbicide required to control of weeds.
7. Higher the selectivity index means greater the safety to a crop if the Selectivity Index is greater than 2, the herbicide is safe for a crop.

$$\textbf{Selectivity Index (SI)} = \frac{\text{Maximum dose of herbicide tolerated by a crop}}{\text{Minimum dose required to control weeds}}$$

Important points of agronomy:

Angstrom (Å):

- It is the one ten-millionth of a meter.
- Å = **10^{-10}** m **or 10^{-7}** mm or **10^{-8}** cm. ***(RPSC, AO-09)***

A-value: *(JRF-Agronomy)*

- Proposed by **Friend** and **Dean,** 1952 and defined as an available soil nutrient determined in terms of a standard fertilizer used.
- Used for the assessment of **available P** and **S** in soils.
- Expressed as = **B (1-y)**.

Acid forming fertilizer:

- Ammonium sulphate
- Ammonium chloride,
- Anhydrous ammonia
- Urea.

Acid sulphate soil:

- **Very acid soil** (pH < 4) in which sulphuric acid is formed by the oxidation of **S-bearing** ferrous pyrite minerals.
- Found primarily in **coasty**, deltaic estuarine area of the humid tropics. ***(RAS -09)***
- Also referred as **cat clays.**

Adhesion:

- The force of attraction that binds the molecules of **different kind** e.g. **soil** and **water**.

Azolla:

- An **aquatic fern** which fixes atmospheric **N in symbiotic** association with ***Anabaena azollae***, blue green algae which lies in its **cavities**.

Aeroponics:

- It is a system for growing plants with their **roots supplied** with moisture in the **air** e.g. **citrus, olive**.

Anhydrous ammonia:

- Dry and **highly acidic**
- Having **148** equivalent acidity
- **82** % N: **highest** nitrogen % among all **N** fertilizer.

Alumina silicate

- These are most common and principle soil forming minerals i.e. **feldspars**, micas, quarts.

Argillation:

It's a process by which the **dispersed** clay particles are migrated from upper to the lower soil horizons resulting a **textural horizon.**

Agronomic efficiency:

- Unit of crop produced per unit input (nutrient) added.

Antidote:

- A chemical applied to prevent the harmful effect of a specific pesticide on human beings.

Blind cultivation:

- Cultivation before the crop is emerged especially practiced in **potato** and **sugarcane** is done for **weeds** control.

Biostatic effect:

- **Reduced** microbial decomposition of a pesticide in the presence of some chemical called **extender.**
- **Boron** is a biostatic to **2, 4-D**.

Baule unit:

- Amount of a nutrient associated with 50% of the maximum possible yield.

Blotch:

- Disease of barley caused by ***Helminthosporium spp.***

Base temperature:

- Lowest **temperature** at which there is **no growth** is referred as **base** temperature.
- **4** °C for wheat.
- **10** °C for rice/sorghum.

Cupping:

- The upward turning of leaf edge to form a cup-like structure as seen in case of Mo deficiency.

Cohesion:

- The force of attraction that binds the molecules of **same kind** e.g. **water** and **water molecules.**

Ceiling leaf area index:

- Leaf area index at which **mutual shading** of leaves in a **canopy** is complete and net photosynthesis is **zero**.
- Gross photosynthesis **equals** the respiration losses of the whole crop.

Cropping intensity:

- Total sown area/net cultivated area x 100.

Cardinal temperature:

- Concept by **Sachs.** 1860.

- Maximum + minimum + optimum combines called **cardinal temperature.**

For cool season cereals:

- Minimum 0-15 ^{0}C
- Optimum 25-30 ^{0}C
- Maximum 31-37 ^{0}C

For warm season cereals:

- Minimum 15-18 ^{0}C
- Optimum 31-37 ^{0}C
- Maximum 44-50 ^{0}C

Epinasty:

- Increased growth on the upper surface of a plant organ or part (leaves) causing it to **bend downwards.**

Ecesis/ecess:

- Adjustment of an organism to a new habitat where it can complete its **life cycle**.

Ephemerals:

- Plants complete their life cycle **within 2-4 weeks** due to **moisture stress** (drought escaping plants).

Furrow slice:

- Surface soil with depth of **0-15 cm** over one ha is known as **furrow slice.**
- Weight of furrow slice is **2.25 x10^6** kg soil/ha or 1 cm soil over one ha **=150 ton/cm/ha.**

Intensity factors (available **P**):

- The amount of **$H_2PO_4^-$** and **HPO_4^{-2}** present in **soil solution.**

Methaemoglobinaemia:

- A condition associated with **nitrate toxicity** as result of consuming nitrate rich water
- **Infants** suffering from this condition have been referred to as **blue babies** due to development of **grayish-blue** skin and the problem referred as **blue baby syndrome.**

Micron (μ):

- One micron is one millionth of a meter
- 1 μ = $10^{-6\,m}$ or $10^{-4\,cm}$

Nanometer (nm):

- It is the one nine-millionth of a meter
- nm = 10^{-9} m or 10^{-6} mm or 10 Å

Porometer:

- Instrument used to measure **leaf temperature** in addition to **transpiration** and **stomata** resistance measurements.

Physiological maturity:

- When fruit growth is complete and photosynthesis are no longer translocated to fruits, it is known as physiological maturity

Harvesting maturity:

- At physiological maturity, the moisture percentage of grains is about 20 per cent.

Popping:

- In groundnut pod **without kernel** or unfilled due to: **Ca deficiency**.

Light use efficiency:

- The **conversion efficiency** or quantity of dry matter production per unit light interception is known as light use efficiency.

$$LUE = \frac{\text{Amount of dry matter produced (gm}^{-2}\text{)}}{\text{Amount of cumulative light absorbed (MJ m}^{-2}\text{)}}$$

Light compensation point:

- The minimum light intensity at which photosynthesis and respiration rates are equal.
- Rice crop has high compensation point of 1400 ft. candles at 27 ^{0}C.

Light saturation point:

- The light intensity at which there is no further increase in the rate of photosynthesis is called light saturation point.
- Light saturation is high in C_4 plants compared to C_3 plants.

Crop geometry:

- It is a **spatial arrangement** or **plant rectangularity** of crops.
- The way in which the crop plants are arranged in the field is usually referred to as **crop geometry.**

Kor watering:

- The **first irrigation** of crops is known as kor watering and the depth applied in the first watering is **kor depth**.

Row crop:

- Crops which are sowing in uniform spacing e.g. **cotton**.

Selectivity ratio:

- It is the ratio of **maximum** dose **tolerated** by the crop to the minimum dose required to kill weeds.

Saturation light intensity:

- It is the maximum intensity at which the rate of photosynthesis reaches a maximum value.
- Most of crops reach light saturation point at 1/5th of full sunlight.

Sodium absorption ratio: *(JRF exam. many times)*

$$SAR = \frac{Na^{+}}{\sqrt{Ca^{+2} + Mg^{+2}/2}}$$

- The water contains **SAR** value **> 10** is **harmful** to crops.
- Soils having **SAR** value **>13** are considered as **alkali** or **Sodic soil**.

Thermoperiodism:

- Concept given by **Gardner.**
- Response of living organism to regular changes in temperature either day/night or seasonal in called as **thermoperiodism**.

Transpiration ratio:

- Amount of water transpired by a crop to produce 1 kg of dry matter.
- It expressed in **liter/kg** of dry matter.

Transpiration coefficient:

- Amount of water needed by a crop to produce 1 kg of dry matter. i.e. rice crop used **5000** liters of water to produce 1 kg of dry matter.

VAPSA/Vetier:

- Condition of **optimum moisture** to carry out **agricultural** operation **easily** and **efficiently**.
- Generally at **25** to **50** % moisture **depletion.**

3. SOIL SCIENCE

Important points:

✓	Canker nodules are mostly found in:	**Red soils**
✓	Process of moving out of sesquioxide from soil is known as:	**Podzolization.**
✓	Process of mixing of soils is known as:	**Pedoturbation**
✓	Soils having more than 20% organic matter known as:	**Organic soils**
✓	Law of minimum was proposed by:	**J. V. Liebig, 1840**
✓	O' horizon is absent in which soils:	**Arable soils**
✓	Top most mineral horizon is:	**A' horizon**
✓	Diameter of clay particle is:	**Less than 0.002 mm**
✓	Soil structure which is best for cultivation is:	**Crumby structure**
✓	Particle density of most of the soil is:	**2.65 g/cc**
✓	Bulk density of general soil:	**1.33 g/cc**
✓	Total pore space highest in:	**Clay soils**
✓	Kaolinite is a type of mineral:	**1:1 nonexpanding**
✓	Montmorillonite is a type of mineral:	**2:1 expanding type**
✓	Vermiculate is a type of mineral:	**2:1limited expanding**
✓	**Chlorites** are type of mineral:	**2:1:1type mineral**
✓	Recently developed soils formed from soil which order that has no diagnostic horizons:	**Entisols**
✓	Black cotton soils are found in:	**Maharashtra, MP**
✓	The weight of **furrow slice**of soil is:	**2.25×10^6** kg/ha.
✓	One **cm** of surface soil over **one ha** of land weight is:	**150** tones.
✓	Permissible soil loss by water is:	**12 t/ha.**
✓	Chief constitute of sandy fraction is:	**Quartz.**
✓	Most dominant mineral on the earth crust is:	**Feldspars (48%).**
✓	Which pulse crop does not fix N from the atmosphere:	**Rajma**
✓	**First** manufactured fertilizer in India is:	**SSP**
✓	Classification of soils by ICAR in:	**8 groups.**
✓	The adverse effect of salinity on plantsis due to increased:	**Osmotic pressure**
✓	Apical bud dominance is caused by which hormone:	**Auxin**

✓	The crops which absorb the nitrogen in ammonical form directly are:	Paddy & potato
✓	In rice ammonical fertilizers should be applied in:	Reduced zone of soil
✓	While nitrate fertilizer should be applied in:	Oxidized zone of soil
✓	Major source of nitrogen absorption by plants is:	Nitrate (NO_3^-) form
✓	**Pascal** (Pa) is the **SI** unit of:	Pressure.
✓	The **smallest volume** of soil (1to 10 m^2) is called as:	Pedon
✓	Most of the exchange of gases in soil is due to:	Diffusion
✓	In addition of organic matter bulk density:	Decreases
✓	In **red** soils the dominant clay mineral is:	Kaolinite
✓	The red soils are red in colour due to presence of:	Iron oxide
✓	Black soils are best suitable for:	Dry land agriculture.
✓	In black soils the dominant clay mineral is:	Montmorillonite
✓	Black cotton soils are deficient in:	Nitrogen & O.M.
✓	**Azolla** is widely used as a bio fertilizer in:	Rice crop
✓	soil particles < 0.001mm diameter have:	Colloidal property
✓	The term 'Colloids' was given by	Grahm (1861)
✓	Formula of **backing** soda is:	$NaHCO_3$
✓	Formula of**castic**soda is:	NaOH
✓	**USA** is the first country to introduce:	Zero tillage.
✓	**Urea** in the presence of enzyme **urease** converted in to:	Ammonium Carbamate
✓	Electrical conductivity (EC) is used to express:	Salinity of the soil.
✓	Pollen sterility in wheat is due to:	Boron-deficiency
✓	Ballast elements are:	Al and Silicon.
✓	In which state red soils are dominant:	Tamil Nadu
✓	In which states lateritic soils are dominant:	Karnataka and Kerala
✓	Organic matter content in Indian soils is generally:	Less than 0.5%
✓	Most outstanding green manure crop is:	Sunhemp (*Crotalaria juncea*)
✓	Wood is mainly decomposed by:	Actinomycetes.

✓	**Nif**: gene responsible for	N-fixation
✓	Population of **actinomycetes** is higher in:	Alkaline soils.
✓	**Fine texture** soils are more sensitive to:	Water erosion.
✓	The soils which are most suitable for the most of the crops are:	Sandy loam
✓	Rice can tolerate very low level even complete ....of oxygen in the soil.	absence
✓	Desired size of the soil sample for soil testing is:	*0.5-1.0 kg*
✓	Crop having the highest tolerance to boron is:	*Sugar beet*
✓	Insoluble fraction of organic matter is:	*Humin*
✓	Lowering of the bulk density of a soil indicates:	*Better physical properties*
✓	ppm=per cent/10000 or ppm equal to	*0.0001 %*
✓	Or 1 ppm equal to	*1 mg*

- ✓ Red soils are found in ***Tamil nadu***
- ✓ The element required for seed formation: ***Phosphorous***
- ✓ Oxidation: means addition of oxygen or loss of electron
- ✓ Reduction: means addition of H_2 or removal of O_2 or gain of electron
- ✓ On the number basis population of microbes in normal soil is:
  Bacteria >Actinomycetes> Fungi
- ✓ The first antibiotic penicillin was discovered by ***Alexander Flemming, 1929***
- ✓ Oldest method of breeding for improving variety is: ***Selection***
- ✓ Zinc deficiency is produced in: ***Calcarious soil***
- ✓ In weed infested pond mostly fishes are suddenly dieddue to: ***O_2 deficiency***
- ✓ Due to water logging the availability of which nutrient is increased in rice field is: ***Manganese***
- ✓ Bulk density of **sub soil** is **higher** as compared to surface soils.
- ✓ **Active acidity**: it is due to presence of H^+& Al^{+3} in **soil solution.**
- ✓ Soil texture is an **inherent property** of soil it **can't be** changed.
- ✓ Provision of **adequate drainage** is the **basic principle** in **reclamation** of saline and alkali soils.
- ✓ Alluvial soils most **recently** developed soils with no diagnostic horizons
- ✓ Red soils are **best** suitable for agriculture.
- ✓ **Many Pedon** within defined limit is called as **soil series** that is **basic unit** of soil classification.

- ✓ **Chisel ploughs** are also used to break **hard pans** of soil even at **60-70** cm deep.
- ✓ In general **ploughs** are used for **primary** tillage and **harrows** for **secondary** tillage.
- ✓ In **zero** tillage, primary tillage is **completely** avoided and secondary tillage is **restricted** to seed bed preparation in the root zone only.
- ✓ **Soil erosion**: detachment and **transportation** of soil particles by **water/wind** is knwon as soil erosion.
- ✓ Portion of capillary water lying between field capacity (1/3 atm.) to wilting coefficient (15 atm) is known as: **Available water**
- ✓ **Tillage** is the **mechanical manipulation** of soil for loosening the surface crust and brings about favorable condition for the germination of seeds.
- ✓ **Tilth** is the **physical condition** of soil in relation to plant growth.
- ✓ Stages of wind erosion: **Saltation** (**75%** of wind erosion)-suspension-surface creep
- ✓ Stages of water erosion: **Splash**, Sheet, Rill & Gully
- ✓ Total **N content** of soils ranges from less than **0.02%** in **subsoil** to more than **2.5%** in peats.
- ✓ **Warabandi**: rotational supply of water in canal irrigated area.
- ✓ **Cabbage** is highly sensitive to boron toxicity (**< 0.33ppm**)
- ✓ Most common urease inhibitor commercially available is **Agrotain** or **NBPT** (**N-serve** in **India**), recently **ATS** (ammonium thiosulphate) as a urease inhibitor.
- ✓ $H_2PO_4^-$ is **greatest** absorbed at acidic **P^H 6.5** or less. ***(IFFCO-09)***
- ✓ PO_4^{-3} is absorbed at higher value of **p^H** (slightly **alkaline**).
- ✓ The uptake of **HPO_4^{-2}** is much **slower** than $H_2PO_4^-$
- ✓ **Chelate:** organic compound with combine **Fe, Mn, Cu, Zn**. E.g. **EDTA** (slow released micro organic **fertilizer**).
- ✓ **Pedalfer**: an absolute term used for a **group** of a soil with an **accumulation** of sesquioxide (**Fe & Al**) in lower part of **Solum**.
- ✓ **Adsorption**: **gathering** of molecules on a surface.
- ✓ Sun hemp (***Crotalaria juncea***) best suited crop for green manuring, it accumulate higher amount of **N-134 kg/ha** followed by dhaicha (***Sesbania acculata***).
- ✓ ***Sesbania rostrata***: found stem nodules, hence aerial N fixation.
- ✓ **Foliar spraying is** useful in apply small quantity of fertilizers e.g. **micronutrients.**
- ✓ **Organic matter** is the major source for **nitrogen**, **phosphorus** (5-60 %) and **sulphur** (**80%**).

- ✓ The temperature of surface soil is **always higher** than atmosphere.
- ✓ For each **unit increase** in pH the concentration of **Mn^{+2}, Fe^{+2}**decreases by **100 fold**.
- ✓ **Ideal** NPK ratio should be: **4:2:1** for cereals and **1:2:1** for pulses.
- ✓ The **most widely** occurring clay mineral in soil is ***kaolinite***
- ✓ **Underground** reservoirs are usually found in porous rock formation called: ***Aquifer***.
- ✓ The **fertilizer** prepared with Rhizobium culture is known as "***Nitrogin***"Nif gene is responsible.
- ✓ **Alkali soil**: ESP > 15, EC < 4 dsm^{-1}, p^H> 8.5 also called Sodic or Usar soil. ***(RPSC, AO-09)***
- ✓ **Mostly** plants absorbed micronutrients in **reduced form** (Fe^{+2}, Mn^{+2}etc).
- ✓ A unit change in pH represent a 10 fold change in H^+ or OH^- activity.
- ✓ Substances added to soils for the improvement of their condition known as: ***Amendments***
- ✓ The book '**Nature & properties of soil"** is written by N.C. Brady. **(RPSC, AAO-09)**
- ✓ Fertilizer which is most concentrates used for nutrient supply to crops is: ***Anhydrous ammonia (82%N).***
- ✓ Complementary reactions between intercrops is: ***Annidation***
- ✓ **Sandy soils** have **higher** bulk density than clay soils.
- ✓ Green leaf manuring crops e.g. ***Glericidia acculata***, ***Pongamia pinnetta***, *Ipomoea cornea*
- ✓ Green manuring crops e.g. ***Crotolaria juncea*** (Sunhemp), ***Sesbania acculata*** (dhaicha), and cowpea.
- ✓ Sheet stage of water erosion most dangerous because it's not noticed by farmers.
- ✓ **Plant geometry** refers to the shape of plant
- ✓ **Crop geometry** refers to the shape of space available for individual plants.
- ✓ **Muck soil**: are having **highly decomposed** organic matter
- ✓ **Peat soil**: organic matter **partially** decomposed and found under excessive moisture condition. The p^H of peat soils is 3.9 hence suitable for **paddy.**
- ✓ **Secondary clay mineral:** A mineral that has been formed as a result of subsequent changes in rocks is known as secondary mineral eg. Kaolinite, montmorillonite, illite, gypsum.

Unit-1: Soil science

- The word 'Soil' originated from **Latin** word '*Solum*' means **Floor.**

Definition of Soil:

- The soil is a dynamic natural body having three dimensional structures which are composed of minerals and organic materials as well as living forms in which plants grow.

a. Edaphology:

- It is the study of various **properties** of soil in relation to **growth, nutrition & yield of crops.**
- **It** derived from two **Greek** words

Edaphos =	Soil
Logos =	Discourse

- According to Edaphology soil is natural habitat for plants.

b. Pedology:

- To study the **origin**, **classification** and **description** of soil is known as**Pedology**.
- It derived from **two Greek words**:

Pedon =	Soil
Logos =	Study

- Father of **Soil science** and **Pedology** is known as: **V.V. Dokuchaev**.
- According to pedology, soil is a **natural body**.

Types of rocks:

Rocks:	**Examples:**
• Igneous rocks	Granite, basalt
• Sedimentary rocks	Lime stone, sand stone and dolomite
• Metamorphic rocks	Gneiss, marble, quartz and slate

Metamorphic rocks and their parent material:

Metamorphic product:	**Derived**
• Gneiss	From granite
• Marble	From lime stone
• Quartz	From sand stone
• Slate	From shale

Earth minerals:

a. Quartz: silica or S_iO_2:

- Chief constituent of sandy fraction
- Constitute **36** % of earth's crust.
- Rocks containing free silica in abundance, which is not combined with bases, are called **'acid rocks'**.

- In **granite** (acid rock), **quartz** is present in a pure form as a prominent constituent.
- In basic rocks like **basalt** contain only a **little**, if any of free silica.

b. **Feldspars**:

- Formula **K_2O. AlO_3. 6 S_iO_2**
- Such mineral constitute about **48** % earth's crust.
- Feldspar weather easily and give rise to clay on hydrolysis.

c. **Micas**:

- **White mica** is **more resistant** to weathering than black mica.
- Micas are more **resistant** to weathering than feldspar and other silicates.
- Micas constitute **10 %** of earth's crust. Two types:

(i) Potash mica: Potash mica is **white**, clear and transparent and known as **muscovite** mica

(ii) Magnesium mica: It is called ***biotic black*** mica.

Soil formation:

- Soil is formed from weathering of rocks and minerals.
- Soil formation is a slow process; formation of one inch of top soil takes 200-1000 years.

1. **According to Joffe:**

- Joffe identifies two kinds of soil formers, passive and active.
- Formation of soil is determined by the relative influence of these factors:

a. **Passive factors**:

(i) **Parent material**: it is the unconsolidated mass from which the solum develops.

(ii) Relief/topography:

(iii) Time or age of land

b. **Active factors**:

(i) Climate:

(ii) Organisms (biosphere):

2. **According to H. Jenny** -

- The fundamental equation of soil-forming factors is: (tricks as ***clorpt***)

$$s = f\,(cl, o, r, p, t)$$

- *S* = soil forming factors
- f = Independent variables
- cl = Climate
- 0 = Organisms

- r = Relief/topography
- p = Parent material
- t = Time

Soil profile:

- **A vertical section** of the soil through all its horizons and extending into the parent material and the Individual layers are regarded as ***horizons***.
- It is grouped into: **O, A, B & C horizons.**

1. **O' horizon:**
   - **It is organic** horizon is present above mineral soils.
   - It is generally found in **forest** areas.
   - Such 'O' horizon is visible in **virgin** soil and absent in **arable** soils.
2. **A' (Eluvial) horizons:**
   - **Top most** mineral horizon.
   - Containing mixtures of **organic matter** which **tends imparts a darker** color than that of the lower horizon.
   - It is zone of **washing out/maximum leaching**.
   - **Mobilization** of **nutrients** takes place.

A' horizon having sub horizons:

(a) ***A_1 Horizon***: It is uppermost mineral soil horizon rich in organic matter.

(b) ***E' or A_2' Horizon:*** Horizon of maximum eluviations of clay, Fe Al oxides.

(c) ***AB or EB horizon***: Transition horizon between A/E and B (also called **A_3**).

3. **B' (Illuvial) horizon:**
   - It is zone of **washing in** or horizon of **maximum accumulation** (**immobilization**) of materials such as Fe and Al oxides and silicate clays.
   - Deposition of $CaCO_3$, $CaSO_4$ and other salts in **arid zones.**
   - This horizon some time also referred to as **sub soil.**
   - It **can't** be **cultivated** by **tillage** operation.
4. **C' horizon:**
   - It is **unconsolidated** material underlying the ***Solum*** (**A & B horizon**).
   - It is zone of least weathering.
   - Accumulation of Ca, Mg carbonates, and cementation.

Major component of soil:

Mineral	=	45%,	Organic matter	=	5%
Air space	=	25 %,	Water	=	25 %.

Properties of soil:

1. Soil texture:

- Soil texture refers to the relative proportions of various types of soil particles. OR
- It refers to the ***size of the soil particles***.
- It is basic property of a soil, it **can't bechanged.**
- In field condition soil texture is generally determined by **feel method.**

Mechanical analysis of soil:

- It is the process of determining the amount of soil separates below 2 mm in diameter.
- It helps in determining soil textural classes.
- The international pipette and bouyoucos hydrometer method are most common methods for mechanical analysis of soil separates.
- The international pipette method is based on the principle of **sedimentation**.
- The textural class of soil is determined by triangular textural diagram.

Textural classes: Total 12 textural classes but 3 main i.e. sand, silt and clay.

- **Sandy soil**: Soil contains more than **85%** sand.
- **Silt soil**: When more than **80%** silt.
- **Clay soil**: When more than **40%** clay.
- Clay soils have **highest pore space** then other.

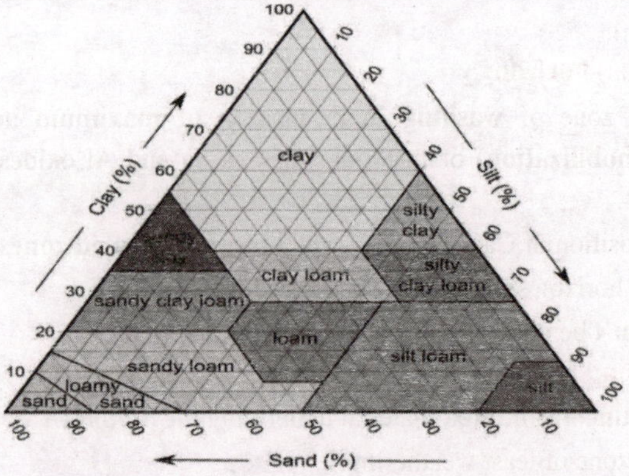

Soil TEXTURE TRIANGLE, showing the 12 major textural classes

Classification of soil particles:

Classification	According to ISSS (dia. in mm)
Gravel	> 2.0
Coarse sand	0.2 - 2.0
Fine sand	0.02 - 0.2
Silt	0.02 - 0.002
Clay	**< 0.002**

Loamy soil:

- These are **best** for **agricultural production** (especially **sandy loam)** because they retain more water and nutrients than sandy soils and have better drainage, aeration and tillage properties than clay soils.
- Suitable for most of crops.

Heavy soils:

- The clay soil imparts cohesion and stickiness and fine texture to the soil and induces **resistance to the passage of implements.**
- In other words it makes the soil **"heavy"**
- These soils are suitable for **cotton,** rice, sorghum.

Light soils:

- In comparison to clay soil sandy **soils** offer **little resistance** to implements and called **light** soil.
- E.g. Light soils suitable for **groundnut**, **potato**, **tobacco** and **legumes**.

Hence;

- The term **heavy** and **light** refers to the **degree of resistance** offered by soil to the passage of **implements.**
- But not to weight in fact the heavy soil weights less than an equal volume of the light soils.

2. Soil structure:

- Refers to the **arrangement** or grouping of primary soil particles viz., sand, silt, clay and their aggregates into certain defined pattern is called soil **structure**.
- It refers to the physical ***shape of the soil aggregates***
- Soil structures can **be changed or altered** by agronomic practices i.e. ploughing, cultivating, drainage, liming and manuring.
- The primary particles sand, silt and clay usually occur grouped together in the form of ***aggregates.***
- Natural aggregates are called ***Peds.***
- ***Clod***: It is used for a coherent mass of soil broken into any shape by artificial means such as by tillage.

Classification of Soil Structure:

- Structure is studied in the field under natural conditions and it is described under three categories

1. **Type**: Shape or form and arrangement pattern of peds
2. **Class**: Size of Peds
3. **Grade:** Degree of distinctness of peds

(A) Types of Soil Structure: There are four principal forms of soil structure.

1. Plate-like (Platy):

- The horizontal axis or dimensions are larger than the vertical axis.
- When the units/ layers are thick they are called **platy**.
- When they are thin then it is **laminar**.
- Platy structure is most noticeable in the surface layers of **virgin soils** but may be present in the subsoil.
- This type is inherited from the parent material, especially by the action of water or ice.

2. Prism-like:

- The vertical axis is more developed than horizontal, giving a pillar like shape.
- Vary in length from 1- 10 cm. commonly occur in sub soil horizons of Arid and Semi arid regions.
- When the tops are rounded, the structure is termed as **columnar** when the tops are flat / plane, level and clear cut **prismatic**.

3. Block like:

- All three dimensions are about the same size.
- Irregularly six faced with their three dimensions more or less equal.
- These types usually are confined to the sub soil and characteristics have much to do with soil drainage, aeration and root penetration.

4. Spheroidal (Sphere like):

- Rounded aggregates or pads not more than **2 cm** in diameter often in a loose condition in the A" horizon.
- This is specific to surface soil particularly high in organic matter/ grass land soils.
- In this structure, infiltration, percolation and aeration are not affected by wetting of soil.
- It is grouped in to granular and crumbs.

(a) Granular:

- The aggregates of this group are usually termed as **granular** which are relatively less porous.
- These are fine aggregates of soil, good for crop cultivation.

(b) **Crumb:**

- When granules are especially **porous** known as crumb.
- It is **most suitable** for cultivation of crops.
- Hence, both granular and crumb structures are most favorable for plant growth.

(B) Classes of Soil Structure: it determined by size of peds

- Each primary structural type of soil is differentiated into 5 size classes, these are:

  1. Very fine or very thin
  2. Fine or thin
  3. Medium
  4. Coarse or thick
  5. Very coarse or very thick

(C) Grades of Soil Structure:

- Grades indicate the degree of distinctness and durability of the individual peds.
- It is determined by the **stability** of the aggregates.
- Grade of structure is influenced by the moisture content of the soil.
- Grade also depends on organic matter, texture etc.

Factor affecting soil structure:

- Addition of organic matter increase the aggregate formation
- Intensive cultivation destroy the soil structures
- For good soil structure tillage operation should be done at optimum moisture condition.
- Sodium ion destroy the soil structure due to deflocculation
- Calcium ion destroy the soil structure due to flocculation
- Mostly grasses and legume improve the granulation of soil
- Roots and root hairs helps to developed crumby soil structure.

3. **Soil density:**

- It is measured by ***Pycnometer***
- Two types:

(a) **Particle density:**

✓ The weight of an individual soil particle per unit volume is called **Particle density or** It is the density of the oven dried soil solids when air is excluded or excluding pore space

$$\text{P. D.} = \frac{\text{Weight of soil solids}}{\text{Volume of soil solids}}$$

- ✓ Particle density of mineral soils is: ***2.66 g/cm³*** *or g/cc.*
- ✓ P.D. of organic soils: 2.1-2.4 g/cm³
- ✓ P.D. of organic matter: 1.1-1.4 g/cm³
- ✓ P.D. sometime also known as ***real/true density*** or *gain density or absolute density.*
- ✓ Particle density is **lower** when the **organic matter** added in soil.
- ✓ In presence of heavy minerals increase the particle density
- ✓ While water has density of : **1g/cm³**

(b) **Bulk density/apparent density**:

- ✓ It is the ratio of the weight of oven dried soil to its bulk volume including pore space.

$$\text{B.D.} = \frac{\text{Weight of } \textbf{oven dry} \text{ soil}}{\text{Volume of oven dry soil}}$$

- ✓ It is also called the volume weight.
- ✓ Itis determined by multiplying particle density by the volume of solids: $2.66\ g/cm^3 \times 0.5\ cm^3 = 1.33\ g/cm^2$
- ✓ Therefore, bulk density is generally **half** of particle density.
- ✓ Bulk density of mineral soils are: **1.33 g/cm³**
- ✓ Bulk density of coarse textural (sandy) soils is: 1.4-1.75 g/cc
- ✓ Bulk density of fine textured soils (clay) is: 1.1- 1.4 g/cc
- ✓ Bulk density of organic matter is: 0.5 g/cc
- ✓ Bulk density of compact sub soil is: 2.0 g/cc
- ✓ Sandy soils have higher bulk density than clay
- ✓ Addition of organic matter decreases bulk density
- ✓ Compactness of soil increase bulk density
- ✓ Soil with large pore space lower the bulk density
- ✓ Tillage operations decrease bulk density.

4. **Thermal conductivity**:

- It is defined as amount of **heat flowing** in unit time through a unit cross section of soil.
- The thermal conductivity is **high** for **sandy** soils compared to clay soils.
- When soil **moisture** content **increases,** thermal conductivity also increases.
- Addition of **organic matter** in soil **decrease** thermal conductivity.
- Thermal conductivity decreases with increase in **porosity**
- **Tillage** decreases thermal conductivity.

5. **Soil temperature**: Diurnal variation in soil temperature
   - Dry soil > Moist soil
   - Loose soil > Compact soil
   - Deep soil > Shallow soil
6. **Redox potential:**oxide reduction potential of a system
   - It is denoted by: **E^h**
   - Measurable unit is: **mv**
   - For well-drained (oxidized) soil: + **400 to + 700mv**
   - Water logged (reduced) soil: **-250 to -350mv.**
7. **Soil plasticity:**
   - Soil plasticity is determined by an apparatus called as **liquid limit device**.
   - Soil containing **> 15 % clay** exhibit plasticity.
   - Range of moisture content in soil which exhibit plasticity is known as ***plasticity index*** *or limit.*
   - **Plasticity index** given by **Alterberg**, 1911 also gave limits of soil.

   (a) ***Lower plastic limit:*** **Lowest moisture** content at which soil can be deformed **without cracking** and it is upper limit of moisture for tillage operations except rice.

   (b) ***Upper limit*** *or* ***Liquid limit***: Higher moisture content which destroys soil aggregates or moisture at which **cease** to be plastic.
   - **Soil plasticity** determined by apparatus: ***liquid limit devices***.
8. **Soil consistency**:
   - Resistance of soil at various moisture content to **mechanical stress**
   - Soil consistency order is: dry soil >moist soil >wet soil(less resistant)
9. **ODR** (oxygen diffusion ratio): movement of O_2 to plant roots.
   - It measured by **platinum weir** electrode in soil, unit is: **$g/cm^2/min$**
     Range of ODR for crops: **22.9×10^{-8} to $39.5 \times 10^{-8} g/cm^2/minute$**
   - **For irrigated pea**: **$39.5 g/cm^2/min$** (highest ODR)
   - **For rice**: **22.1** x 10^{-8} **$g/cm^2/min$** (lowest)
10. **Soil colour:-**
    - Determined by ***Mulsan colour chart***.

    (a) **HUE:** Itdenotes the dominant spectral colour (yellow, red blue, brown).

    (b) **VALUE:** Denotes the lightness and darkness of colour.

    (c) **CHROMA:** Denotes purity of colour.

11. **Specific heat: unit is (Calories/g)-**

- For water: 1
- For soil: **0.2**
- For organic matter: 0.5

Note:

- The heat gain or loss by the soil is five times quicker than by the water because the specific heat of the soil is **0.2** and the specific heat of the water is **1.0 cal/g**

12. **Soil porosity (%) -**

- **Porosity** = 100 - (bulk density/particle density × 100)
- Or porosity = (1-Bd/Pd) x 100
- Porosity range: 30-60%
- Medium to fine textured (clay) soil: 40-60%
- Sandy soils: 35-50%
- Total number of pores (micro pores) is found highest in clay soils
- Large sized pores (macro pores) are found in sandy soils
- Addition of organic matter increases the porosity of soil
- Cropping reduce the pore space than outcropped soil
- Crumby and granular (spheroids) soil structures have more porosity than other soil structure
- Ideal ratio of micro to macro pores is 1:1

Silicates minerals:

a. **Olivine**:

- It is Ferro-magnesium silicate $(Fe\ Mg)_2\ SiO_4$
- Two hydrated forms of olivine are **talk** and **serpentine**, is a hydrated silicate of **Mg**.

b. **Tourmaline**:

- It is boron- alumina silicate.

Some Silicates which are source of plant nutrient:

- Phosphorus: **Apatite**
- Potassium: **Biotite**, Orthoclase
- Magnesium (Mg): **Dolomite, Olivine**
- Calcium: Anorthite, Dolomite
- Boron: **Tourmaline**
- Molybdenum: Olivine
- Manganese (Mn): Pyrolusite

Minerals resistant to weathering (descending sequence):

- **Quartz**> Muscovite > Feldspar >Biotite> Calcite.

Composition of earth's crust:

Element	Amount (%)
Oxygen	**46.6**
Silicon	**27.7**
Aluminum	**8.1**
Ferrous	**5.0**
Calcium	3.6
Sodium	2.8
Potassium	2.6
Magnesium	2.1

Silicate clay mineral:

- The most important silicate clay is known as ***phyllosilicate***.
- The silicate clay unit consists of alternate sheets comprised of one type if sheet is dominated by **silicon** (Si) and other by **Al/ Mg**.
- Silica-dominated sheet is called **tetra hedral** and Al/Mg sheet is called **octahedral** because of eight sided building block.
- Si_4^+ is surrounded by four oxygen ions and Al/Mg by six hydroxyl/oxygen ions.

Classification of silicate clays:

A. On the basis of number and arrangement of tetrahedral and octahedral sheet, silicate clays are classified in to 3-different group:-

1. **1:1 type minerals** (one ***tetrahedral (Si)*** and one *octahedral* ***(Al)*** sheet): e.g. ***Kaolinite***.
2. **2:1 type minerals**: (two **Si** sheet and one **Al** sheet). Means alumina is **sandwiched** between two tetrahedral **silica** sheets. Eg. ***Montmorillonite***, ***illite*** and ***vermiculate.***
3. **2:1:1 or 2:2 types mineral:** e.g. ***Chlorites.***
   - These are basically ferro-magnesium silicates with some amount of aluminum present.
   - It is having an **extra** layer of **Mg-**dominated alumina sheet, called ***Brucite*** ($Mg(OH)_2$).
   - But Mg^{2+} also dominates the alumina sheet of 2:1 type minerals.
   - Thus crystal unit has **two silica** sheets and **two magnesium** dominated alumina sheet.
   - Hence it is also ***called 2:2 type clay minerals.***
   - It is **non- expanding** in nature.

B. On the basis of expending and non-expending nature the silicate minerals are grouped in to following types:-

1. **Expanding type**:
   - It includes **montmorillonite** (Smectite group) and **vermiculate** group.

   a. **Montmorillonite:**
   - **Montmorillonite** is prominent in clay soils.
   - It is most common **Smectite** in soils where Mg^{+2} is substitute for Al^{+3} in alumina sheet (octahedral).
   - It is loosened by "***wonder wall force***".

   b. **Vermiculate**:
   - In this group the degree of swelling is considerably less than Smectite, so also called ***limited expansion clay mineral.***
   - But the **CEC** (cation exchange capacity) of vermiculites **exceeds** that of all other silicate clays due to very high ***negative charges.***

2. **Non-expanding type**:

   a. **Illite:** Found in clay soil

   b. **Micas:** (Muscovite, Biotite): These are found in sand and silt.

Basic properties of major silicate minerals:

SN	Property	Kaolinite	Illite	Montmorillonite	Vermiculate	Humus
1	Type of silicate clays	1:1type non expanding	2:1 type non expanding	2:1 expanding type	2:1 limited expanding	
1	Total surface area m^2/g	37-45	120-170	580-750	**780-900** (highest among **silicates)**	1200
2	External surface area	Very low	Low	Medium	High	-
3	Internal surface area	None	Low	High	High	-
4	Cohesion, plasticity, swelling	Very low	Low	High	medium	Low
5	CEC (meq/100 g soil)	2-15	30-40	60-150	100-150	>200
6	Size (in μ)	0.1-5.0	0.1-2.0	0.01-1.0	-	-
7	P fixing	Highest	Low		Medium	-

Soil taxonomy:

- Father of soil taxonomy: **Dr. Guy D. Smith**

Modern soil classification:-

1. Compressive system of soil cassification:-

- It is **US comprehensive** soil classification system based on 7th **approximation**.
- This system is developed under the leadership of Guy. D. Smith and maintains the natural body concept and two major features adopted were (1975):

a. **Primary basis** for identifying different classes on the soil properties rather than genesis of soil.

b. **Latin** or **Greek** words are the basis for nomenclature.

Six categories were adopted for this system:-

- **Order**: it is highest category, there are 12 soil orders
- Sub-order: 63 sub order in this classification
- Great group: > 240
- Sub-group: > 1000
- Family and
- **Series:** it is most specific and lowest category (> 200 in India)

- The **smallest** unit of soil classification is the ***soil series***.

Soil orders in the 7th **approximation** and their derivation: **12** soil orders.

- Tricks to remember '*A VAGAMI HOUSE*'

SN	Soil order	Features
1	Inceptisols	Name derived from Latin word (inceptum, 'beginning') are soils that exhibit minimal horizon development. **It is largest soil order in world (20%) of total soil**. Young, embryonic soils with few diagnostic horizons. They are more developed than entisols. They have no accumulation of clays, iron oxide, aluminium oxide or organic matter. These soils found in all India except arid and hot regions.
2	Entisols	It defined as soils that do not show any profile development other than an A horizon. An entisol **has no diagnostic horizons**. Entisols are the second most abundant soil order (after inceptisols)
3	Vertisols	A **vertisol** is a soil in which there is a high content of expansive clay known as **montmorillonite** that forms deep cracks in drier seasons or years. Alternate shrinking and swelling causes ***self-mulching***, where the soil material consistently mixes itself, causing vertisols to have an extremely deep A horizon and no B horizon. This heaving of the underlying material to the surface

		often creates a micro relief known as ***gilgai***. Vertisols typically form from highly basic rocks, such as basalt, also known as deep black or "**black cotton**" soils. Found in the Deccan Plateau of India.
4	**Oxisols**	It occurrence in tropicalrain forest, They are classified as feral sols; some oxisols have been previously classified as **laterite soils**. The main processes of soil formation of oxisols are weathering, humification and pedoturbation due to animals. Found in humid areas. They are defined as soils containing *at all depths*. Oxisols are always a red or yellowish color, due to the high concentration of iron and aluminiumoxides and hydroxides.
5	**Alfisols**	Alfisols form in semi arid to humid areas, typically under a hardwood forest cover. They have a clay-enriched subsoil and relatively high native fertility. "Alf" refers to aluminium (Al) and iron (Fe). Because of their productivity and abundance, the Alfisols represent one of the more important soil orders for food and fiber production. They are widely used both in agriculture and forestry, and are still very deficient in nitrogen and available phosphorus. Alfisols have undergone only moderate leaching. By definition, **they have at least 35% base saturation,** meaning calcium, magnesium, and potassium is relatively abundant. Found mostly in laterite and red soil areas.
6	**Histosols**	Greek word 'histos' means **tissue**. These are organic materials rich soils. Typically, histosols have very low bulk density and are poorly drained because the organic matter holds water very well. Most are acidic and many are very deficient in major plant nutrients which are washed away in the consistently moist soil. Histosols are known by various other names in other countries, such as **peat** or **muck**. This occurs because of restricted drainage precluding aerobic decomposition, and the remains of plants and animals remain within the soil. Thus, histosols are very important ecologically because they, and gelisols, store large quantities of organic carbon. If accumulation continues for a long enough period, coal forms. Fossil histosols are known from the earliest extensive land vegetation in the Devonian.

7	Aridosols	**Aridisols** (or **desert soils**) from the Latin *aridus*, for "dry", and *solum* form in an arid or semi-arid climate. Aridisols dominate the deserts of western and north western India. Aridisols have a very low concentration of organic matter, reflecting the paucity of vegetative production on these dry soils. Water deficiency is the major defining characteristic of aridisols. These subsoil horizons can also be cemented by carbonates, gypsum or silica. Accumulation of salts on the surface can result in salinization.
8	Spodosols	Also known as Podzol means "under-ash". The term was introduced in 1875 by Dokuchaev.Mostly found in cool and humid areas
9	Ultisols	**Ultisols**, commonly known as **red clay soils**, are defined as mineral soils which contain no calcareous material anywhere within the soil, have less than 10% weatherable minerals in the extreme top layer of soil, and **have less than 35% base saturation** throughout the soil. Ultisols occur in humid temperate or tropical regions. They are typically quite acidic, often having a pH of less than 5. The red and yellow colors result from the accumulation of iron oxide (rust), which is highly insoluble in water. Ultisols can have a variety of clay minerals, but in many cases the dominant mineral is **kaolinite**. This clay has good bearing capacity and no shrink–swell property.
10	Mollisols	**Mollisols** are formed in semi-arid to semi-humid areas, typically under a grassland cover. Their parent material is typically base-rich (**base saturation > 50%**) and calcareous and include limestone, **loess**, or wind-blown sand. The main processes that lead to the formation of grassland Mollisols are melanisation, decomposition, humification and pedoturbation. Mollisols have deep, high organic matter, nutrient-enriched surface soil (A horizon), typically between 60–80 cm in depth. This fertile surface horizon, known as a **mollic epipedon**, is the defining diagnostic feature of Mollisols. These environments have historically been strongly influenced by fire and abundant pedoturbation from organisms such as ants and earth worms. As the world's most agriculturally productive soil order, the Mollisols represent one of the more economically important soil orders.
11	Andisols	**Andisols** are soils formed in **volcanic ash** and defined as soils containing high proportions of glass and amorphous colloidal materials. These are most fertile soil found in Andoman & Nicobar land. They can usually support intensive cropping, Other andisol areas support crops of fruit, maize, tea, coffee or

		tobacco and forests.
12	Gelisols	'**Churning**' soils. They are soils of very cold climates which are defined as containing permafrost within two meters of the soil surface. The word "gelisol" comes from the Latin *gelare* meaning **'to freeze'**, a reference to the process of **cryoturbation** that occurs from the alternating thawing and freezing characteristic of gelisols. They are known as **cryosols**. These soils are not found in India.

Land capability classification:

- On the basis of capability or limitations, the lands are grouped into **eight** classes by **U.S. Soil conservation service.**
- These classes divided into **two major** groups.

1. Lands suitable for cultivation:
   - Such lands have **four** classes.
   - But for the **regular cultivation**, only **first three** classes viz. I, II, III are **suitable.**
2. Land not suitable for cultivation:
   - **Remaining four** classes are not capable of supporting cultivation of crops.
   - These lands are used for **growing grasses**, **forestry** and supporting **wild life.**

a. Class I: Good cultivable lands without any limitations.
- ✓ **It has no limitation,** good for **intensive** crop cultivation.
- ✓ Such lands need only **crop management practices** like fertilizer, manures, crop rotation etc. to maintain their productivity.
- ✓ Example of class I soils are **alluvials oils** of Indo-gangetic plains.

b. Class II: Good cultivable lands with some limitation
- ✓ **It has some limitations** such as **gentle slope, moderate erosion** hazard, slightly alkali or saline conditions which reduce choice of crops.
- ✓ Hence require moderate **conservation tactics** like **strip** cropping, contour tillage, crop rotation etc.
- ✓ Example: **deep red** and **black soils.**

c. Class III: Moderately good cultivable lands
- ✓ **Have severe limitations** like **steep slope**, high **erosion hazard, slow water permeability** and **restricted root zone**
- ✓ Fertility status of such soils is poor.
- ✓ Requiring **special conservation measures.**
- ✓ Examples; **shallow red soils, slightly saline black** soils.

d. Class IV: fairly good lands
- ✓ **Very severe limitations** on choice of crops.
- ✓ Suitable for occasional cultivation.
- ✓ Mainly use for **pasture/hay.**
- ✓ Example: shallow soils, saline soils, alkaline soils.

e. Class V: Not suitable for arable farming but suitable for grazing
- ✓ **Stony** or **rocky** soils and pond area where drainage is not possible.
- ✓ Mainly used for **grazing** and **forestry.**
- ✓ E.g. **Arid** and rocky soils.

f. Class VI: Suitable for grazing with some limitations
- ✓ Have some limitations on use for grazing/forestry.

g. Class VII: Suitable for wood land and wild life
- ✓ Severe limitation on use for grazing/forestry.

h. Class: VIII: Suitable for **aesthetic purpose** and **watershed protection.**
- ✓ Extremely rough land, **not suitable for any kind of crop production**.
- ✓ E.g. **Sandy beaches**, river washes etc.

The land capability classes are usually shown on the map by standard colours:

SN	Capability Classes	Colour denoted
1	Class – I	Light green
2	Class- II	Yellow
3	Class-III	Red
4	Class-IV	Blue
5	Class-V	Dark green
6	Class-VI	Orange
7	Class-VII	Brown
8	Class-VIII	Purple

Soils of India:

- Broadly **eight** groups but **five major** groups are described as:

1. Alluvial soils:

- These soils are formed by the sediments brought down by the rivers.
- They are also rich in chemical ingredients.
- The rivers deposit very fine particles of soil called **alluvium** in their plains during their long course of journey.
- Alluvial soil is also known as **riverine** soil because it is mainly found in the river basin.
- **It has about 45.6 per cent** of the total land area of the country
- The alluvial soils are yet immature and have weak profiles.
- They differ in consistency from nitrogen and phosphorous are generally low, but rich in potash.

- The widest occurrence of the alluvial soils is in the Great Indo-Gangetic Plain starting from Punjab in the west to West Bengal and Assam in the east.
- They also occur in deltas of the Mahanadi, the Godavari, the Krishna and the Cauvery, where they are called **deltaic alluvium**.
- Along the coast they are known as **coastal alluvium**.
- Geologically, the alluvium of the Great plain of India is divided into newer or younger known as **khadar** and older are **bhangar** soils.
- The **khadar** soils are found in the low areas of valley bottom which are flooded almost every year.
- **Bhangar** is a more clayey composition and is generally dark coloured, and is found on the higher reaches about 30 metres above the flood level.
- A few metres below the surface of the bhangar are beds of lime nodules known as **kankar**.
- Along the Shiwalik foothills, there are alluvial fans having coarse, often pebbly soils. This zone is called **bhabar.**
- To the south of the bhabar is a long narrow strip of swampy lowland with silty soils. It covers an area of 56,600 sq km and is called **tarai.**
- The tarai soils are rich in nitrogen and organic matter but are deficient in phosphate.
- These soils are are suitable for a number of crops such as wheat, rice, sugarcane, jute and soyabean under reclaimed conditions.
- Due to their softness of the strata and fertility the alluvial soils are best suited to irrigation and respond well to canal and well/tube-well irrigation.
- Alluvial soils formed mainly from **Inceptisols** and **Entisols** soil orders.

2. **Black soils**:

- Formed mainly from **vertisols soil** order.
- Rich in **montmorillonite** clay minerals.
- In Maharashtra soil derived from **Deccan** trap is called ***Regur*** or ***black cotton*** soils because cotton is the most important crop grown on these soils. These soils are mostly high in **clay** content.
- These are dark, **fine grained** soils.
- **Poor** in **phosphorus**, **nitrogen** and **organic matter.**
- These soils have been formed due to the solidifaction of lava spread over large areas during volcanic activity in the Deccan Plateau, thousands of years ago.

- Geographically, black soils are spread over 5.46 lakh sq km (i.e. **16.6 per cent** of the total geographical area of the country
- Major crops grown on the black soils are cotton, soyabean, wheat, jowar, linseed, Virginia tobacco, castor, sunflower and millets.
- These soils are mainly found in **Maharashtra, Madhya Pradesh**, parts of Karnataka, Andhra Pradesh, Gujarat and Tamil Nadu. South-east part of Rajasthan.
- The black colour of these soils has been attributed to the presence of a small proportion of **titaniferous magnetite** or even to iron and black constituents of the parent rock.

3. **Red soils**:
   - Formed mainly by **Alfisols** soil order.
   - Red soils have high phosphorous fixation capacity due to presence of **kaolinite** clay mineral.
   - The red colour due to presence of **iron.**
   - These soils are rich in Fe and Mn.
   - Red soils also called as **early soils** because in the light and frequent rains of south west monsoon, red soils permit to sowing being done earlier than other soils.
   - They occupied about 3.5 lakh sq km which is about 10.6 per cent of the total geographical area of the country.
   - These soils are spread on almost the whole of Tamil Nadu, parts of Karnataka, south-east of Maharashtra, eastern parts of Andhra Pradesh and Madhya Pradesh, Chhattisgarh, Orissa and Chota Nagpur in Jharkhand.
   - In the north the red soil area extends in large parts of south Bihar; the Birbhum and Bankura districts of West Bengal; Mirzapur, Jhansi, Banda and Hamirpur districts of Uttar Pradesh; Aravallis and the eastern half of Rajasthan, parts of Assam, Nagaland, Manipur, Mizoram, Tripura and Meghalaya.

4. **Laterite and lateritic soils**:
   - Formed mainly from **ultisols** and **oxisols** soil groups.
   - **Kaolinite** is dominated silicate clay.
   - Both occupied **40 mha** (mainlylateritic about 25 mha)
   - Mainly found on hills of Kerala, Karnataka, Orissa, Assam and coastal area of Goa, Maharashtra.
   - Formed under **heavy rainfall** and **high temperature.**
   - **Maximum leaching** takes place.

- Poor physical condition due to presence of hydrous oxide of Fe & Al and laterite soils are more weathered than lateritic and red soils.
- adopted Shifting cultivation
- Suitable for **plantation crop** and **rice cultivation.**
- Laterite soil is formed under conditions of high temperature and heavy rainfall with alternate wet and dry periods
- These soils mainly contain iron oxide which gives them characteristic pink or red color.
- These are residual soils are mostly composed as calcite depositions.

5. Desert Soils

- These soils are found in **Thar Desert** in the Rajasthan.
- This soil is formed from arid conditions with practically negligible rainfall. This type of soil is highly pervious and has a low density.
- It requires densification to increase its bearing capacity and shearing strength. Commonly recognised plants that grow in these soils are **cacti**.

Soil fertility:

- The **inherent** capacity of the soil to **supply** plant nutrient in adequate amount in suitable proportion for the growth of plants.

Soil productivity

- The capacity of the soil to produces in a unit area, expressed in kg/ha or Rs/ha. A fertile soil may or may not be productive but a productive soil always fertile.

Basic differentiate between soil fertility and productivity:

Soil fertility	Soil productivity
It is the ability of the soil to provide all essential plant nutrients in available forms in a suitable balance.	Capability of soil to produce specified crop yield under well-defined and specified system of management of inputs & environment conditions.
It is an index of available nutrient to plants	It is used to indicate crop yields
Influenced by the physical, chemical and biological factors of the soil	Depends upon fertility and location
All fertile soils are not productive	All productive soils are fertile
It is an inherent property of the soil	It is not the inherent property of the soil

Humus:-

- Humus is a complex and rather resistant mixture of brown or dark brown amorphous and colloidal substances modified from the original tissues or synthesized by the various soil organisms.
- Humus contains **40-45% lignin** and **30-33 % protein**, hence humus is also called **lingo-protein** complex.
- C: N ratio of microorganism: **4:1 to 9:1**
- The **C: N** ratio of humus is: 10:1
- C: N ratio of normal soils: 12:1
- C: N ratio of Legumes: 20:1 to 30:1
- C: N ratio of FYM: 20-30:1
- C: N ratio of Crop straw: 100:1

Organic matter:

- **Humus** and **organic** matter **contains nitrogen** about **5** to **5.5** % while in **humus** & **organic** matter **carbon** ranges from **50** to **58%** that given a **C: N** ratio **9:1** to **12:1**.
- Year after year the quantity of organic matter in soil is **declined**.
- Organic matter is the key of production & maintain of soil health
- Organic matter is found **higher** in **grass** land than forest area.
- Method of organic matter is calculated by multiplying the organic carbon values by a conversion factor of: **1.724.**
- Organic matter = Organic carbon × 1.724
- **1.724** is called ***Bemlen factor***.
- The status of nitrogen in general soil is **0.03-0.05%**, that means presence of **1000** kg of N/ha.

Soil buffering:

- Means resistance to change in soil **p^{H}**.
- Its regulate **availability** of nutrients to plants.

Cation adsorption orderis:

- $Al^{+3} > Ca^{+2} > Mg^{+2} > K^{+}$
- Here, **K^{+}** is less tightly held by colloids hence K^{+} ion **radially available** for plant or subjected to **leaching**.

Anion exchange order:

- $H_2Po_4^{-2} > OH^{-} > SO_4^{-2} > NO_3^{-}$
- NO_3^{-} is less tightly held hence, **easily available** to plants or subjected to **leaching**.

Flocculation:

- The colloidal particles are coagulated by adding an oppositely charged ions process is called flocculation.

Order of flocculation:

- $Al^{+3}>Ca^{+}>H^{+}>Mg^{+2}$

Sequence of availability/absorption of nutrient:-

- **Nitrogen >potassium>phosphorous**

Availability of nutrients with p^H values:-

pH value	pH < 6.0	pH 6.0-6.5	pH > 7.0
Available nutrients	**Al, Fe, Mn, Zn**	**P, B**	**Ca, Mo**

- Availability of **Ca & Mo increases** with **increase** in soil p^H

Soil acidity:-

- **Active acidity:** as the acidity develops due to presence of **hydrogen** (H^+) and **Al^{+3}** ions on the **soil solution**.
- **Exchange acidity:** as the acidity develops due to **H^+** and **Al^{+3}** ions (**adsorbed**) on the **soil colloids**.
- In acidic soils conc. of Fe, Mn Al and H ions increases and the concentration of P, Mo, N, K, S decreases.

Problematic soils of India:

a. Acidic soils:

- Highest acidic soils found in **West Bengal, North-East states, Orissa, TN, Bihar**
- Acidity due to adsorption of H^+, Al^{+3}, Fe^{+2}, Mn^{+2} ions on soil colloids.

Acid soil management:

- Use of agriculture **liming material** such as oxide, hydroxide, carbonates of Ca & Mg and use of **rock phosphate, press mud** etc.

b. Saline soils:

- Found white crust of salt hence called **white alkali**/solon chalk.
- Soluble salt i.e. Cl & SO_4 ions of sodium and calcium in root zones
- EC is used for measurement of soil salinity
- Found in arid and semi-arid zones.

Reclaimation of saline soil:

- **Flooding** or **leaching** of soluble salts by **good irrigation** water.
- Keep soil moist or frequent irrigated the field.
- Use of *FYM* for saline reclamation.

c. Alkaline soils:

- Boron, Mo, and sodium toxicity in alkaline soils
- Black coloured, found in semi-arid & sub humid area
- Developed due to excess **$NaCO_3$** and **$NaHCO_3$** ions in soil
- In alkali soils should be used CAN or DAP instead urea

Reclamation of alkali soils:

1. **Gypsum** @ 8 qt/ha used for reclamation of sodic or alkali or soils having Ph> 8.5.
2. **Iron pyrite (FeS_2)** @ 12 qt/ha can be used for amendments of alkali soil.
3. Green manuring should be adopted
4. Cultivation of salt tolerant crops.

Relatively salt sensitive crops are:

Crop	High tolerant	Medium tolerant	Sensitive crops
Field crops	Barley, dhaicha, sugarbeet, tobacco, cotton, mustard	Rice, sorghum, pearl millet, arhar, wheat, rye, oat	Sunhemp, pea, groundnut, moong, gram, urd
Fodder crops	Rhodes grass, Khushgrass	Sudan grass, berseem, Lucerne, sorghum	Cluster bean
Vegetables	Spinach, *raphanus*, sugar beet, *B. repa*.	Cabbage, cauliflower, tomato, carrot, onion, potato	Sam, English varieties of *raphanus*
Fruit crops	Date palm, Phalsa	Pomegranate, grapes, guava, mango, banana, ber	Apple, lemon, straw bery, almond

✓ In case of **saline** soils **Gypsum shouldn't** be recommended because sulphate ion of gypsum also increases salt concentration of soil.

✓ Substances that influence the plant growth favorably by emulating the soil physical and chemical properties are called ***soil amendments.***

Parameters of different problematic soils:

Parameters	Saline soil	Alkaline soil	Saline-Alkali
pH	< 8.5	> 8.5	> 8.5
EC (ds/m)	> 4	< 4	>4
ESP (%)	<15	>15	>15

Physical classification of soil water:

a. **Hygroscopic water**:

- ✓ It held mostly by soil **colloids.**
- ✓ Tension varies from **31-10,000** atmospheres
- ✓ It held at **hygroscopic coefficient**.
- ✓ Mostly non-liquid
- ✓ Moves in **vapor** form therefore biologically inactive.

b. **Capillary water**:

- ✓ It present in **capillaries** of soil.
- ✓ Held between **field capacity** and **hygroscopic coefficient** in micro pores.
- ✓ At a tension of **0.3** (1/3) to **31** atm.
- ✓ Function as **soil solution.**
- ✓ It is mostly available form of water to crops

c. **Gravitational water**:

- ✓ It held by a negative tension of **less than 0.3** bar.
- ✓ It is free water which drains out
- ✓ It is also called **drainage water**.
- ✓ **Undesirable** and usually present in **macro pores**.

Soil erosion

- The process of the removal of soil particles from the parent body and transportation of such particles by wind and/or water.

1. Water erosion: four stages of water erosion-

a. **Splash/rain drop**:

- It is **first stage** of water erosion.

b. **Sheet erosion**:

- Highest and uniform removal of soil particles.
- It is **unnoticed** by farmer hence **most serious** form of water erosion.

c. **Rill erosion**:

- Can be removed by normal tillage operation.

d. **Gully erosion**:

- It can't be removed by normal tillage operation.

2. Wind erosion: Grouped into three stages:

a. **Saltation**:

- The movements of soil particles through bounces/jumps on ground.
- Saltation is the first stage of movement of soil particles in series of jump.
- Particles **0.1-0.5** mm diameter is moved by saltation.
- Soil loss account for **50-75** % by Saltation.

b. **Surface creep:**
- Rolling or sliding of larger soil particles
- Particle size > **0.5mm**.
- Generally **5-25 %** soil loss by surface creep.

c. **Suspension:**
- Floating of small size particles **< 0.1** mm in size,
- Soil loss accounts 3-4 % by suspension.

3. **Wave erosion:**
- Erosion by combined action of wind and water is known as wave erosion also known as strive bankes
- The permissible limit of soil loss in India is: ***12 ton/ha/year***
- The rate of soil erosion in India per hectare is: ***16.4 t/ha***
- When erosion is caused by excessive grazing, deforestation etc. (human being intervention) is called: ***Anthropogenic erosion (Pre-PG, 1993)***

Deposition of soils:
- The accumulation of **transported** soils at another place is known as **deposition**.
- It may be following types according to transportation source

Transportation source:	Termed as:
Gravity	Colluvial
Water	**Alluvial**
Ice	Glacial
Wind	**Aeolian** (sandy soil)
Wind	**Loess** (silt soil)

Mineralization and immobilization:

a. **Mineralization:**
- It is the conversion of an element from an immobilized form to an available form as a result of microbial decomposition.

b. **Immobilization:**
- It is the reverse of mineralization where available form of an element is fixed as immobile form.

In case of nitrogen:
- When C: N ratio **exceed** 30:1, **immobilization** occurs and
- When C: N ratio is **below** 20:1, **mineralization** takes place.
- C: N ratio > 30:1 = **Net Immobilization**
- C: N ratio < 20:1 = **Net Mineralization**
- C: N ratio = 20-30 **both** immobilization and mineralization takes place.

Aminization:

- The **final stage** is the decomposition of protein and the release of **amines, amino acid** and **urea**.
- This step is termed **aminization** and these produced turned in to **ammonia** (NH_3) is called ***ammonification***.

Nitrification:

✓ The microbial conversion of NH_3 into nitrate is mediated by microbes known as nitrification.

(i) **NH_3^+** ______(Nitrosomonas)______**NO_2^- (Nitrite)**

(ii) **NO_2^{-2}** ______(Nitrobacter)________**NO_3^- (Nitrate)**

- Optimum temperature for nitrifying bacteria is 30-35℃ and optimum pH is 6.5-7.5.

In case of phosphorus:

- **25-80** % of **P** present in soil is in **Organic form**.
- Major source of P in **humus** is present in the form of **phytin** and small amount as 0-10% in **nucleic acid**.
- **C: N: P** ratio = 100:10:1
- C: N: P ratio > **immobilization occurs**.

Unit: Manure and fertilizers:

Manures:

- Manures supply plant nutrients in **small quantities** and provide **organic** matter in large quantities.e.g. green manures, FYM, composts etc.
- Compost that is prepared with the help of earthworms is called ***vermicompost***.
- ***Night soil*** is human excreta, both solid and liquid.

Green manure:

- Green, undecomposed plant material used as manure is called green manure. It is two types;

a. **In situ green manuring**: growing of green manure crops like Sunhemp, dhaicha, clusterbeans and *Sesbania rostrata* and decomposed in the same field. ***Sesbania rostrata*** is a stem nodulating green manure crop which is a native of West Africa.

b. **Ex-Situ or *Green leaf manuring***: Collecting green leaf along with twigs of plants like Glyricidia, Pongamia, Neem, Mahua and buried in another field is known as ex-situ green manuring .

Nutrient contents in some Manure and cakes:

Product	N	P	K
FYM (AAO Exam-2009)	0.5	0.25	0.5
Town compost	1.5	0.4	1.4
Vermi-compost	3.0	1.0	1.5
Night soil (human excreta)	5.5	4.0	2.0
Ground nut cake	7.3	1.5	1.3
Safflower cake (decorticated)	7.9	2.2	1.9

Fertilizers:

- Fertilizers are generally inorganic in origin and they supply one or more essential plant nutrients in large proportions.

Classification of fertilizer:

a. **Straight fertilizer:** Fertilizer which supply only **one major** plant nutrient e.g. **Urea**

b. **Complex fertilizer:** Fertilizer which supplies **two or more** of the primary nutrients they produced by chemical combination e.g. **DAP**

c. **Mixed fertilizers:** Are the products made by **physical mixing** two or more fertilizer e.g. **NPK 20:20:20**

d. **Complete fertilizer**: Fertilizer having all **three primary major** nutrients e.g. **NPK 12:32:16**

e. **Low analysis fertilizers**: Having **less than 25 %** of the primary nutrient e.g. **SSP**

f. **High analysis fertilizer**: Contains **more than 25 %** of the total primary nutrient e.g. **Urea**

Fertilizer grade:

- ✓ Refers to the guaranteed analysis of its plant nutrients.
- ✓ It is minimum guarantee of the plant nutrient content in the terms of available N, P, and K e.g. NPK: **19:19:19.**

Fertilizer ratio:

- ✓ Refers to the relative percentage of N, P_2O_5 and K_2O.
- ✓ E.g.A ratio of **2:1:1** means 80 Kg N, 40 Kg P_2O_5 and 40 Kg K_2O.

Order of hygroscopic nature of fertilizers (decreasing)-

- **Ammonium nitrate**> Urea > ammonium sulphate> CAN

Acidic residual nature of fertilizer:

- Anhydrous ammonia > ammonium chloride > ammonium sulphate> urea > ammonium nitrate (lowest).

Order of basic residual nature of fertilizer:

- **Calcium cyanamide** > sodium nitrate > di calcium phosphate > calcium nitrate.

Equivalent acidity of fertilizers:

- Anhydrous ammonia: 148 meq./100g.
- Ammonium chloride: 128
- Ammonium sulphate: 110
- Urea: 84
- DAP: 77
- Ammonium nitrate: 60

Equivalent basacity of fertilizers:

- Sodium nitrate: 29
- Calcium nitrate: 21

1. Nitrogenous fertilizers:

- The ultimate source of nitrogen is ***atmosphere*** and immediate and important source of nitrogen is ***organic matter***.
- **Per hectare maximum** consumption of fertilizer in the state: **Punjab, while total quantity of fertilizer is highest used in UP among the states.**

A. **Nitrate fertilizers**: properties-

- ✓ Highly **mobile** in soil
- ✓ Suitable for top dressing.
- ✓ Nitrate fertilizers are **basic residual** in nature.
- ✓ **Leaching** and denitrification loss high.

Examples of nitrate fertilizer:

- **Sodium nitrate**: **16% N**, good for **acidic** soils
- **Calcium nitrate**: 15.5% N.
- **Potassium nitrate** (KNO_3): **13** % N and **44%** K_2O.

B. **Ammonical fertilizers**:

- ✓ Less leaching but **readily soluble** in soil.
- ✓ Suitable for water logged soils.
- ✓ **Left acidic** residual nature in soil.
- ✓ **Volatilization** and **denitrification** losses more.

Examples of ammonical fertilizers:-

a. **Ammonium sulphate** (**NH_4SO_4**):

- Contains **20.6 % N** and **24.5 % Sulphur.**
- **Most suitable** for paddy, tea, groundnut & sugarcane

b. **Ammonium chloride** (NH_4Cl^-):

- It contains **25.5 % N**
- Being acidic residual nature it is extensively used in **paddy**.

C. Both NH_4^+ and NO_3^- N fertilizer:

(i) **Ammonium nitrate** ($NH_4^+ \cdot NO_3^-$):

- It contains **33 %** N (half of ammonium form and half nitrate)
- Highly **hygroscopic** nature
- **Explosive** fertilizer

(ii) **Calcium ammonium nitrate** (CAN):

- Commonly known as ***Kishan khad***
- It contains **26%** N
- **Neutral** in nature i.e. neither acidic nor alkaline in soil.

D. Amide fertilizer: two types

1. **Organic form:-**

Urea (NH_2 CO NH_2):-

- It is most commonly used N- fertilizer in **India**
- Cheapest source of nitrogen
- It contains **46% N**
- It is **hygroscopic** therefore produced in **granular/pellet** form.
- **Acidic in residual:** continuous use of urea for several years results reduce soil pH.
- **Neem coated urea:** being slow release properties it increase the nitrogen use efficiency.

Sulphur coated Urea: it is new slow release N fertilizer contains 37% N and 17% sulphur. It is marketed as trade name of **Urea Gold.**

2. **Inorganic form:**

Calcium cyanamide:

- Contains **21 % N**
- Neither import or nor manufactured in India

Analyzing process for determination of available nitrogen is:

- Alkaline permanganate method.

Biuret:

- It is formed by the combination of two molecules of urea during manufacturing of urea.
- **Biuret** toxic to plants hence, conc. in urea should not exceed **1.5%**
- When urea will be applied as foliar spray on crop canopy the biuret concentration should not exceed **0.25%.** (JRF exams many times).

Types of nitrogen losses:

a. **Volatilization** (non biological loss of **NH_3^+**):

- When PH **>8,** nitrogen is lost in the form of NH_3 in **alkaline** medium.
- Volatilization increase in poor drainage e.g. rice field
- About **60 %** of nitrogen loss in India is due to **volatilization**; hence in alkali soil the N application is raised at least by **25%**.
- Gaseous losses of nitrogen (**ammonia)** are carried out by both **volatilization** and **denitrification**.

b. **Denitrification**:

- Formation of N_2 or loss of gaseous form of N by biological reduction of **NO_3^+** and **NO_2^-** are known as ***Denitrification.***
- Denitrifying bacteria: ***Pseudomonas*** and ***Bacillus***
- **Denitrification** and **volatilization** occurs rapid at **higher pH**.

c. **Leaching loss**:

- Leaching or drainage loss mostly occuerd in nitrate fertilizer
- Mostly occurred in humid area

Nitrification inhibitors:

- It is used in **non –nitrate fertilizers**, inhibitors **decrease** the activity of nitrifying bacteria. Examples are following:
- **Thiourea:** **36.8** % N
- Oxamide: 31 % N
- **AM (Pyrimidine)**
- **N-Serve** (Nitropyrin).
- Neem Cake.

Slow release fertilizer:

- Nitrate fertilizers mostly used as slow release due to **overcome** the problem of **leaching.**
- The solubility of nitrogen fertilizer is reduced by two types of compounds:

a. **Synthesizing compound:**

- **Urea formaldehyde** (urea form): **38 - 42** % N
- Isobutylidene diurea (**IBDU**): **32.2** % N.
- Crotonilidene diurea (**CDU**): 32.5 % N

b. **Coating barrier**:

- Sulphur coated urea: new grade- **Urea gold** (37% N &17% S).
- **Neem coated urea:**
- Modified form/**super granules**:

2. Phosphate fertilizers:

a. **Water soluble:**

- **Single Super Phosphate (SSP):** 16% P_2O_5, 19% Ca &12 % S.
- Double Super Phosphate (DSP): 32 % P_2O_5
- Triple Super Phosphate (TSP): 46-48 % P_2O_5
- **Di ammonium phosphate (DAP): 18 % N, 46 % P_2O_5.**
- Mono-ammonium phosphate (**MAP**)**:11 % N, 48 % P_2O_5**

b. **Citrate soluble p fertilizer:**

- Di calcium phosphate: 33-40 % P_2O_5
- Basic slag: **14-18** % P_2O_5

c. **Citrate & water insoluble p fertilizer:**

- Rock phosphate: 20-40 % P_2O_5 (It is used for acid soil reclamation)
- Row bone meal: 20-25% P_2O_5
- Steamed bone meal: 20-30% P_2O_5

Bio-super:

- It is form of **organic fertilizer** mostly in **Australia.**
- **Biosuper**= Rock phosphate + Sulphur + **S** oxidizing bacteria (*Thiobacillus*).

Conversion factor for P:

- % P_2O_5 = % **P** x **2.29**
- % P = % **P_2O_5x 0.43**

Phosphorous **extracted** in soil by following methods:-

(i) **Bray no. 1**: used for acidic soil (pH 5.5 or less).

(ii) **Olsen method**: **0.5 M $NaHCO_3$** adjusted at **P^H 8.5** and is used for alkaline and neutral soils.

3. Potassic fertilizers:

a. **KCL/MOP:**

- It is most common and cheap fertilizer among potassic fertilizers
- It contains highest (60 % K_2O) among all potassic fertilizer
- Also known as murate of potash or potassium chloride
- This fertilizer is suitable for acidic and heavy soils but not for alkaline soils
- It shouldn't be used in sugarcane, sugar beet, tobacco, tomato and potato because in sugar crops, accumulation of sugar is affected due to chloride ion present in the fertilizer.
- Higher content of chloride ion reduce burning quality of tobacco and interfere with starch content in solanaceae crops.
- In these crops should be used K_2SO_4 and KNO_3 instead KCL.

b. K_2SO_4 (Potassium sulphate):

- Contains 48 % K_2O, 17.5 % S.

c. KNO_3 **(potassium nitrate)**:

- Also known as **salt petre** or nitre
- It contains 44 % K_2O, 13% N
- It is an excellent source of potassium and nitrogen and mainly used for fruit trees, tobacco and vegetables

Conversion values for K:-

- **% K =** % $K_2O \times 0.83$
- **% K_2O =** % $K \times 1.20$

Activity ratio:

- Measures of the **intensity** of labile **k** in the soil
- It represents 'k' that **immediately available** to crop roots.

$$Ar_k = \frac{\text{Activity } k^+}{\sqrt{\text{Activity of } Ca^{++}, Mg^{++}}}$$

✓ K & Na is determined by: ***Flame photometer***.

Micronutrients:

1. ***Ferrous sulphate*** **:** is the most commonly used fertilizer which is sprayed @ 0.5 % on the standing crops like groundnut, soybean etc. to control iron chlorosis. It contains 19 % Fe.
2. ***Manganese sulphate*** **:** is the most popular manganese fertilizer which contains 26 % manganese.
3. ***Zinc sulphate*** **:** contains 36 % zinc is the most commonly used zinc fertilizer. It applied as both to soil (at 25 kg/ha) and plant (0.5 % as spray). Zinc oxide (78 %) used as seed treatment.
4. ***Borax:*** is the most commonly used for boron fertilizer which is white compound containing 11 % boron. It is highly soluble in water, hence should be used as boron frits.
5. ***Copper sulphate*** **:** is commonly used for both spraying on plants and applying as soil. It contains 22.5 % copper.
6. ***Ammonium molybdate*** : used as foliar spray. In pulses Mo is more important nutrient because it is constitute of leghaemoglobin hence, can be used @ 3g/kg seed of pulses, it contains 54 % Mo.

Gypsum ($CaSO_4 . 2H_2O$):

- It contains 29.2 % Ca & 18.6 % sulphur
- Agriculture grade gypsum 70 %
- In general case 250 kg/ha gypsum used for sulphur & calcium requirement especially in onion, mustard, groundnut and pulse crops.
- In pulses formed indigestible protein in the deficiency of sulphur
- In groundnut popping is caused by deficiency of Calcium.
- For reclamation of alkali soil, zypsum is used @ 800 kg/ha.

Methods of fertilizer application:

a. **Basal dose**:
- Application of fertilizer at the time of sowing
- Generally potassic and phosphorus fertilizers applied at basal dose due to their **slow release nature**.

b. **Deep placement of fertilizer**:
- In paddy only ammonium form (NH_4^+) incorporated in **reduced zone** to prevent **leaching** and **volatilization** losses.

c. **Starter dose**:
- In seedling **of legume** and **vegetable** at the time of transplanting applied NPK at **1:2:1** ratio.

d. **Split doses of N fertilizer**:
- To **increase N efficiency**, apply **2 split** doses in **most of cereals** having crop duration 4-5 months.
- Applied 3 split doses for crop having duration more than five months.

Suitability of fertilizers:

- ✓ **Potassium sulphate** is used where **quality** of crop is required because **KCL reduce** the quality of crops.
- ✓ For **rice** and groundnut crops: **ammonium sulphate** is the best.
- ✓ In **acidic** soils: **Rock phosphate** and **SSP** is used instead **DAP**.
- ✓ In **saline** & **alkali** soils **ammonical** fertilizers are **not** recommended because alkalization of ammonia takes place.
- ✓ Indian soils are universal **deficient** in **Nitrogen** and **Zinc**.
- ✓ Fertilizer grade indicate the **percentage** of plant nutrients in a fertilizer.
- ✓ Fertilizer ratio is the **relative proportion** of three major (**NPK**) plant nutrients.
- ✓ In water logged soils **ammonium sulphate** is **best** suitable N fertilizer.
- ✓ **Potassium chloride** or Murate of potash is the **most common** and **cheap** fertilizer among potassic fertilizers.

- ✓ **Potassic** fertilizer is suitable for **acidic** and **heavy** soils but not for **alkaline** soils.
- ✓ Denitrifying bacteria ***Pseudomonas*** and ***Bacillus***.
- ✓ Conversion of ammonium in to nitrite in the presence of **nitrosomonas**.
- ✓ Conversion of nitrite to nitrate through: **nitrobacter**.
- ✓ Nitrogen % in Thiourea: **36.8%**.
- ✓ **AM & N-serve** are nitrification inhibitors.
- ✓ In anaerobic soils, sulphates are reduced to hydrogen sulphide by the action of ***Desulfovibro*** bacteria.
- ✓ Elemental sulphur when applied to the soil is converted to sulphate by ***Thiobacillus spp.***

Bio-fertilizers or Bio-inoculants:-

- These are preparations containing micro organisms that supply nutrients especially nitrogen and phosphorus.
- It can be used through application of **seed**, and **soil** or **composting** areas with the objective of **increasing** the number of such micro-organisms and accelerates certain **microbial** process to augment the extent of the **availability of nutrient** in a form which can be assimilated by plants.
- On the basis of nutrient supply these are broadly classified in two types:

A. Nitrogen fixation microbes: These are three types.

1. **Free living:**
   a. Aerobic bacteria: ***Azatobacter chroccum:*** Free fixers commonly used for cereal crops like **wheat, rice, sugarcane & cotton,**
   b. Anaerobic bacteria: ***Clostridium.***
2. **Associative symbiotic:**
   a. **Micro-aerobic**: ***Azospirillum*:** Associated with roots of grasses inside root symbiosis.
   - Used for **sorghum, pearl millet**.
3. **Symbiotic fixation:**
   a. Nodule forming:
   (i) Rhizobium.
   - Rhizobium is an **aerobic** and **heterotrophic** bacteria.
   - It symbiotically fixes atmospheric nitrogen with the presence of leghaemoglobin in nodules of legume roots.

Common bio strains of Rhizobium:	
Rhizobium spp.	**Legumes inoculated**
1. *Rhizobium meliloti*	***Medicago* (*Lucerne*), *Trigonella*, *Melilotus***
2. *R. trifoli*	***Trifolium*** (berseem)
3. *R.leguminosarum*	**Pea**, Lentil
R. phaseoli	***Phaseolus***
R. japonicum	**Soybean**, Cow pea, **Groundnut**, Sunhemp

(ii) **Actinomycetes:** From the genus ***Frankia*** and ***Casuarinas alder.***

b. **Without nodule forming:** The blue green algae, *Anabaena azollae* are having a **symbiotic** association with water fern, ***Azolla pinnata***.

- Anabaena is habits in the cavities in the leaves of floating fern, ***Azolla pinnata.***

B. Phosphorus: these are two types:

- (i) Phosphate solubilizer (PSB) and (ii) phosphate absorber (VAM).
- PSB culture should be used in soils because continuous use of DAP & NPK creates problem like sulphate & phosphate residues in soils so PSB solublises them.

a. **Seed inoculants:**

- One packet of Rhizobium culture contains **200g** mass of culture with lignite.
- **3 packets** are sufficient for one hectare. Required quantity of water is boiled with 300g of jaggery and cooled the solution. Rhizobium inoculums is mixed in the jaggery solution and sprinkled over the seeds and seeds are thoroughly mixed.

b. **Soil inoculants:**

- 2kg/ha carrier based **Azatobacter/Azospirillum/PSB** sculture is mixed with 25 kg ***FYM*** and broadcasted in the field uniformly before transplanting.

Utility:

- BGA can add up to about **20-25 kg N/ha** to rice field.
- Rhizobium increases the yield by **25-30 %** in succeeding crop.

Soil micro-organism:

1. **Micro flora:** includes bacteria, fungi, actinomycets and algae
   - Bacteria grows well within p^H range of 6.5 to 8.0
   - Fungi prefers acidic range 4.5 to 6.5 p^H
   - While actinomycetes prefers slightly alkaline conditions.

Order of micro flora presence in **per gram of soil** (on number basis):

- In general soil:- Bacteria (10^{-8}- 10^{-9}) > actinomycetes (10^{-7}- 10^{-8}) > fungi (10^{-5}- 10^{-6}) > algae (10^{-4}- 10^{-5})
- In acidic soils: fungi > bacteria > actinomycetes> algae
- In alkaline soil: actinomycetes > bacteria > fungi

2. **Soil micro- fauna:**
   - Protozoa, earth worms, nematodes.

Plants used for nutrient deficiency indicator:

Nutrients	Indicator plants
Nitrogen	Cauliflower, cabbage
Phosphorus	**Mustard**
Potassium	Potato
Calcium	Cauliflower, cabbage
Sodium	**Sugar beet, Turnip**
Boron	Sunflower
Ferrous	Sorghum

Unit: - Plants nutrients:

A. General classification:

1. **Major or macro-nutrients**: Those nutrients which are required by the plants in concentrations ***more than 100 ppm*** are termed as major or macro-nutrients. These are grouped in to:-
   a. **Basic nutrients**: Provides basic structure to plants e.g. C, H, O.
   b. **Primary nutrient**: required in large quantity at the time of germination of seedling e.g. N, P, K
   c. **Secondary nutrient**: required in lesser quantity then primary nutrient after germination of crops for proper growth & development e.g. Ca, Mg, S
2. **Minor or micro-nutrients:**
   - The elements which are required by plants in concentration ***less than 100 ppm*** are put in this category.
   - They are also called as ***trace elements*** or ***oligo elements***. These are Fe, Mn, Zn, Cu, Bo, Cl and Molybdenum.
3. Immobile nutrients: **Ca & B**
4. Energy exchange: H & O
5. Cation nutrients: K, Ca, Mg, Fe, Mn, Cu, Zn
6. Anion nutrients: P, B, Mo, Cl, So_4
7. Both anion & cation form: N (NH_4^+, NO_3^-)
8. Non metals nutrients: Boron

B. Other classification of plant nutrients:

1. **Essential nutrient:**

- According to Arnon & Stout **there are 16 essential nutrients** for plants e.g. **C,H,O,N,P,K, Ca, Mg, S, Fe, Mn, Mo, B, Zn, Cu, Cl.**
- Nickel is the **17th essential element** known in 1987 by Brown *et al.*

Criteria of essentiality of plant nutrients:

- Given by **Arnon & Stout in 1939**.
- They stated that plant **can't be** completed their **life cycle** without the deficiency or complete absence of a single nutrient among the essential nutrients.
- **Arnon** (1954) refined the criteria of essentiality of nutrients.

Liebing Law of Minimum:

- According to **Justus Von Liebig (1840)**
- The growth of plants is limited by the plant nutrient present in smaller quantity, all others being in adequate amounts. This has been re-stated as ***barrel concept***.

2. **Beneficial** nutrients: are four in number-

- Vanadium (V), Silicon (Si), Cobalt (Co), Sodium (Na).

3. **Functional nutrients:**

- The term functional nutrient was proposed by **D. J. Nicholas 1963**.
- Functional nutrients are **(20)** = Essential nutrients (16) + Beneficial nutrients Co, V, Si, Na (4).
- Includes all sixteen essential nutrients in addition to four beneficial nutrients (Co, V, Si, Na), collectively known as functional nutrients.

Essential elements and their form of availability:

SN	Element	Conc. in plants	Form of uptake
1.	**Major nutrients**		
	Carbon (C)	45%	CO_2
	Hydrogen (H)	6%	H_2O
	Oxygen	45%	H_2O, O_2
a.	**Primary nutrients**		
	Nitrogen (N)	1.5%	$NO^-{}_3$ (mainly), NH_4^+
	Phosphorus (P)	0.2%	$H_2PO^-{}_4$, (mainly), $HPO^{2-}{}_4$
	Potassium (K)	1.0%	K^+
b.	**Secondary nutrients**		
	Calcium (Ca)	0.5%	Ca^{2+}
	Magnesium (Mg)	0.2%	Mg^{2+}
	Sulphur (S)	0.1%	$SO_2{}^{-4}$, SO_2

2.	**Micronutrients**		
	Iron (Fe)	100 ppm (0.01%)	Fe^{2+} (mainly), Fe^{3+}
	Zinc (Zn)	20 ppm	Zn^{2+}
	Manganese (Mn)	50 ppm	**Mn^{2+}**
	Boron (B)	20 ppm	H_3BO_3 (boric acid)
	Copper (Cu)	6 ppm	Cu^{2+}
	Molybdenum (Mo)	< 1 ppm (0.1ppm)	**MoO_4^{2-}**
	Chlorine (Cl)	100 ppm	Cl^-
	Nickel (Ni)	0.1-1.0ppm	Ni^{2+}

- ✓ In acidic soils phosphorus is available in $H_2PO_4^-$ form and at alkaline (high p^H) phosphorus is available in PO_4^{-3} form.
- ✓ Most of plants absorbed nutrients in reduced form (e.g. Fe^{+2} & Mn^{+2}) rather than **oxidized** form (e.g. Fe^{+3} & Mn^{+3}).

Soil test rating chart for available nutrients:

Nutrients	**Compounds (units)**	**Rating**		
		Low	**Medium**	**High**
Carbon	Organic carbon (%)	<0.5	0.5-0.75	>0.75
Nitrogen	Alkaline $KMnO_4$ (kg/ha)	<250	250-500	>500
Phosphorus	Olsen's method (P kg/ha)	<12.5	12.5-25	>25
Phosphorus	Olsen's method (P_2O_5kg/ha)	< 28	28-56	>56
Potassium	Ammonium acetate (kg/ha)	<140	140-280	>280
Calcium	Ammonium acetate, cmol (p^+)/kg	<1.5	-	>1.5
Sulphur	0.01M $CaCl_2$ (kg/ha)	<22.4	22.4-35	>35
Magnesium	Ammonium acetate, cmol (p^+)/kg	<1.0	-	>1.0
Zinc	DTPA extractable (mg/kg)	<0.6	0.6-1.2	>1.2
Iron (Fe)	DTPA extractable (mg/kg)	<5	5-10	>10
Manganese (Mn)	DTPA extractable (mg/kg)	<5	5-10	>10
Copper	DTPA extractable (mg/kg)	<0.2	0.2-0.4	>0.4
Boron	Hot water soluble (kg/ha)	<0.5	0.5-1.0	>1.0
Molybdenum	Ammonium oxalate extractable (pH 3.3) (mg/kg)	<0.2	0.2-0.4	>0.4

Role of nutrient in plants:

1. Nitrogen:

- **Impart** green colour to plant, encourages **vegetative** growth.
- It is present in most of substances of cells.
- **Nitrogen** is essential constituent of **protein.** ***(IARI, Ph. D-09)***
- Constituent of protoplasm of chlorophyll and coenzyme.
- Play important role in synthesis of auxin

Deficiency:

- Lower leaves become **yellow** or light green in colour.
- **'V' shaped** chlorosis on older leaves or **yellowing at tip**.

Disease:

- **Starvation** disease due to nitrogen deficiency.
- **Buttoning** in cauliflower.

2. Phosphorus:

- Increase the disease **resistance**,
- Enhance new cell formation and necessary for **root development**
- Required for grain formation and maturity of grains
- Phosphorous is essential constituent for **nucleic acid** & **phytin.**
- Most essential functions are **energy storage** and transfer of energy (ADP and ATP), act as **'energy currency'**.

Deficiency:

- Due to **deficiency** of single element (phosphorus), the life cycle of plant can't be completed hence, Phosphorous is called as" **key of life**"
- **Deficiency imparts dark** green colour in leaves
- **Bronzing** appearance on **older** leaves.
- Later develops **red purple** coloration on leaves of maize.

Disease:

- Sickle leaf disease.

3. Potassium:

- Most essential function of **k**$^{+}$ is stomata regulation
- Provides **disease and drought tolerance** in plants
- Cofactors for enzymes
- Responsible for quality products
- Formation and **translocation** of sugars
- Helps in chlorophyll formation.
- In plants K+ also reduced the **transpiration rate** and increase **photosynthetic rate**.
- Because it is stress avoider.
- Potash useful in stress condition because it secretes 60 enzymes

- The term **luxury consumption** is often used to describe **nutrient absorption** by the plant that does not influence yield e.g. **Potassium**.

Deficiency symptoms:

- Spot of dead tissue at tips of leaves.
- **Scorching** & **burning** on **margins** of bottom leaves.

Disease:

- Show rosette or die back disease.

4. Calcium:

- It is constituent of cell wall
- Calcium **immobile** in plants and deficient symptoms appear on **meristem tip** portion.

Deficiency:

- Terminal bud may finally die.

Disease:

- **Tip hooking** and curled margins of leaves.
- **Blossom end rot** (BER) disease of tomato.

5. Magnesium (Mg):

- Essential constituent of **chlorophyll**
- **Magnesium** is a central **atom** of chlorophyll.
- Leaves become mottled & chloratic then reddish or upward cupping of leaf.

Disease:

- Sand-drown disease of tobacco.

6. Sulphur:

- It is essential constituent in some amino acid i.e. cystine, cysteine and methionine.
- Sulphur oxidizing bacteria is: ***Thiobacillus.***
- Sulphur is essential for oil seeds and pulses because it improves oil content and protein content in oil seeds and pulses, respectively.
- Thiourea @ 1g/liter spray on standing crops best for N & S deficiency.

Deficiency:

- Deficiency symptoms resemble to those of N-deficiency but young leaves are pale green or light yellow without any spot.

Disease

- **Akiochi** of rice due to excess of H_2S
- Tea yellow disease of tea.

7. Iron (Fe)

- Required for nitrogen fixation because it catalyst of chlorophyll synthesis.
- During respiration act as O_2 carrier (leghaemoglobin)
- Act as enzyme in oxidation-reduction process.

Deficiency

- **Inter veinal complete chlorosis** but main vein remain characteristic green.
- Scorching of leaf margin
- Young leaves become yellowing or Iron chlorosis in groundnut.

8. Manganese (Mn)

- Formation of chlorophyll
- Co-factor of enzyme
- Mn toxicity causes **crinkled leaf** of cotton

Deficiency

- Dead spot on leaves.
- Leaves become chloratic but smaller vein remain green.

Diseases

- Pahala blight of sugarcane
- Grey spike of barley/oat
- Marsh spot of pea.

9 Copper (Cu):

- Tips and margin of young leaves show chlorosis
- It is compound of plastocyanin
- Essential for photosynthesis/respiration
- Rosetting & excess gumming occurs due to: Copper deficiency.
- Gummosis & xanthomenia diseases in citrus due to Cu deficiency
- Die back and reclamation disease of citrus due to deficiency of copper

10. Boron (B):

- It is necessary for pollen germination
- **Boron** is the only **non-metal** element among the **micronutrient**
- It is necessary for translocation of sugars and is involved in reproduction and germination of pollens. ***(RAS-07)***
- It concerned with water reactions in cells and regulates the intake of water in to cell
- It may act as regulator of potassium ratio and keep calcium in soluble form within the plants
- Taste in cauliflower due to presence of boron

Diseases: due to deficiency of boron

- **Browning** or hollow stem of cauliflower.
- Top sickness of tobacco
- Die back & rosset of apples
- Fruit cracking of tomato and pomegranate.
- Hard fruit of citrus
- Hen & chicken disease of grapes
- Heart rot of sugarbeet

11. Zinc:

- In plants it's required for biosynthesis of **hormones.**

Deficiency:

- **White bud** of maize.
- Khaira disease of **rice**
- Little leaf of cotton and apple
- Mottled leaf of citrus
- Rosette formation

12. Cobalt (CO):

- It is component of vitamin B_{12}
- It is essential for formation of a type of **hemoglobin** in N-**fixing nodule** tissue known as **leghaemoglobin**.
- It is act as O_2 **carrier** in roots
- Also known as animal **protein factor**.

13. Molybdenum (Mo):

- **It** absorbed as **molybdate** (MoO_4^{-2}) forms.
- **Mo** is essential component of enzyme: **Nitrate reductage**.
- Functions as N-fixation in pulses
- Whiptail of cauliflower is due to deficiency of Mo.
- Mo is required for carrot & raphanus for sweetness.

14. Chlorine:

- Absorbed as chloride (Cl^-) form
- High Cl^- sensitivity crop: ***Legume and tobacco.***

Other facts:

- ✓ **Fruit cracking** of tomato is due to deficiency of ***boron***.
- ✓ Reclamation disease also called **white tip** is caused due to deficiency of: ***Copper***.
- ✓ **Sodium** (Na) is important constituent of **CAM** plants.
- ✓ **Silicon** absorbed by plants as **monosilicic acid** and it is **essential** for **rice.** ***(RAS-7)***
- ✓ Pollen grain sterility of wheat is due to: ***Deficiency of boron.***

- ✓ **Sodium is** essential for **osmo-regulation in plants**.
- ✓ Sodium is responsible element for drought tolerance in sugar beet.
- ✓ Chlorosis between the veins and veins remain green shows: ***Deficiency of mg***
- ✓ Complete inter veinal chlorosis occurs due to: ***Mn deficiency***
- ✓ **In *iron deficiency***, the principal veins remain conspicuously green and other portions of the leaf turn yellow tending towards whiteness.
- ✓ In manganese deficiency, the principal veins as well as the smaller veins are green.
- ✓ Silicon is essential for: ***Rice and maize***
- ✓ **Black heart** of potato is due to deficiency of **oxygen in storage**.
- ✓ Metabolic transformation of inorganic plant nutrients into organic plant constituents is known as ***assimilation***.

Soil science terminology:-

Catena (Milne):

- An absolute term for a sequence of soils within a soil zone derived from **similar parent** material, **climatic and age** but having **different soil** characteristics due to variation in **relief** and in **drainage**.

Caliche:

- A pedologic layer near the surface, more or less **cemented** by secondary carbonates of $\mathbf{Ca^{+2}}$ and $\mathbf{Mg^{+2}}$ precipated from the soil solution.

Clod:

- It is an artificial produced compact coherent mass of soil ranging in size from 2-200 mm diameter.

Crotovina:

- It is typical kind of Chanel in the soil horizon produce by action of earth worms and other animals and is filled with organic matter or matter from other horizon.

Cat clays:

- Cat clay is **clay-sulphate** which shines like **cat's eye**.
- Wet clay soils have high amount of sulphur in reduced forms that upon being drained become **extremely acid** due to **oxidation** of **Sulphur** compound and formation of **sulphuric** acid.

Solum:

- The horizon **A + B** combines known as **Solum** or ***true soil***.

Regolith:

- The **unconsolidated** product of rock by weathering
- **Regolith =A + B + C** horizons

Regosol:

- Any soil of the **azonal** order without definite genetic horizons.

Organic soil:

- Soil having at least **20%** organic matter in low clay content soil.

Mineral soil:

- Such soil has usually less than **20 %** organic matter.
- Mineral soils are the biochemically and physically weathered upper portion of the **regolith.**

Soil eradibility factor (K):

- Soil eradibility factor gives an indication of the soil loss from a unit plot of 22 m long with a 9 % slope under continuous fallow. It normally varies from near zero to about 0.6.

Soil eradibility:

- It is the **susceptibility** nature of soil to erosion (mostly **< 0.1 mm** particles).

Soil erosivity:

- Soil erosivity is the **capacity** of agent to cause erosion.

Soil colloids:

- Colloidal particles are generally **smaller than 1 um** (micro meter) in diameter. Colloidal solution: size of particles 10^{-6} to 10^{-9}.

Micelles (micro cells):

- The individual particle of a colloid is called ***micelle***.
- This contains negative charges on their surface; have large surface area per unit weight.

Order of strength of **adsorption/flocculate** soil colloids:

- **Al^{3+}**> H+> Ca^{2+}> Mg^{2+}> K^+ >**Na^+** (least).

4. HORTICULTURE

Important facts:

✓	Edible part of cauliflower is:	**Curd**
✓	Blanching is an important process of:	**Cauliflower**
✓	Whiptail of cauliflower is due to:	**Mo deficiency**
✓	*Pusa ruby* is a variety of:	**Tomato**
✓	*Swaran roopa* is the early variety of:	**Litchi**
✓	***Pajaro***: is the variety of:	**Strawberry**
✓	***Bolero*** is the variety of:	**Marigold**
✓	Richest source of **vitamin C** is:	**Barbados cherry**
✓	Largest importer of cut flower in the world:	**Germany**
✓	The dwarf variety of mango:	**Amrapali**
✓	***California Papershell*** is the variety of	**Almond**
✓	Fruit of rose is known as:	**Hips**
✓	Fruit of okra is a:	**Capsule**
✓	Coconut fat is a rich source of**:**	**Lauric acid**
✓	Coconut is propagated by:	**Seeds**
✓	Mango is mostly propagated through:	**Veneergrafting**
✓	Inflorescence of **cauliflower** is known as:	**Cyme**
✓	Inflorescence of **cabbage** is known as:	**Cat ken**
✓	Black heart of potato is due to:	**O_2 deficiency**
✓	**Pusa snowball** is a variety of:	**Cauliflower**
✓	India's share in the fruit production in the World is:	**10%.**
✓	Development of fruits without fertilization is called:	**Parthenocarpy**
✓	Development of embryo without fertilization is known as:	**Apomixis**
✓	**Oleoresin** is an important product of:	**Chilli seeds or spices**
✓	Kesar (**saffron**) belong to the family of:	**Iridiceae**
✓	Concentration of sugar is used for preservation is :	**60-70 %**
✓	Mango variety suitable for high density planting is:	**Amrapali**
✓	Most salt tolerant fruit crop is:	**Date palm**
✓	Fruit repining hormone is:	**Ethylene.**
✓	A form of low pruning upto 2 m height of	**Pollarding**

	stem is called as:	
✓	Early variety of ber is:	Gola, Seb
✓	Planting season for deciduous plants is:	Jan- Feb
✓	The variey of date- palm is used for dry dates (Chhuhara):	Medjool, Sagai
✓	Generally fruits are in:	Acidic nature
✓	***Baradari*** is an important feature of:	Mughal garden
✓	The fruit of pine apple is known as:	Sorosis
✓	Which crop is propogated by means of ***bulbils*** (clove):	Garlic
✓	Edible banana fruit is seedless because of:	Ambryo abortion
✓	Multistoried cropping is popular in:	Coconut plantation
✓	Which is the late ripening cultivar of ber:	Umran
✓	The exclusion of micro-organism is known as:	Asepsis
✓	Production of vegetables out of their normal season **(off season)** is known as:	Vegetable forcing
✓	**'Alphanso'** variety of mango is grown in:	Maharastra
✓	The chemical preservative used for colorless fruits is:	KMS
✓	Shade loving annual flower plant is:	Salvia
✓	The commonly cultivated variety of table pea in India is:	Bonneville
✓	***Crescent*** is well known for:	Flower arrangement
✓	Saffron is obtained from:	Style and Stigma
✓	Grapes are generally dried in:	Sun
✓	***Sago*** (Sabudana) is prepared from:	Roots of Cassava/Topiaca
✓	***Feni*** (drink) is prepared by:	Cashew apple & Coconut
✓	***Cider*** is prepared by:	Apple & pear
✓	Tea is commonly propagated by:	Soft wood cutting
✓	Origin place of Date palm is:	Iraq
✓	Spacing recommended for Pusa Nanha variety of papaya is:	1.25 x 1.25 m^2
✓	Maximum litchi producing state is:	Bihar
✓	Onion variety which is suitable for export & having yellow colour:	Phule suwarna

✓	The **hard fruits** of citrus are due to the:	**Boron deficiency**.
✓	**Bael** is the richest source of:	**Vit-B_2 (riboflavin)**.
✓	Leaves of **Bael** have religious importance and used to offer:	**'Lord Shiva'.**
✓	Causal organism of **Fig** rust is:	***Cerotolium fici.***
✓	Peach **leaf curl** is caused by fungus:	***Taphrina deformens.***
✓	**Sanjose** scale is the most **serious** pest of:	**Apple.**
✓	**Coffee leaf rust** is caused by:	***Hemelia vestratrix***
✓	**Kent** is the mutant variety of:	**Coffee.**
✓	**Rekha**, Swarna Rekha & **Rajjat rekha** are varieties of :	**Tuberose**
✓	Fruits which have polyembryonic nature are:	**Mango, Citrus, Jamun.**
✓	When pollination is carried out by insects is known as:	**Entomophilous**
✓	Pollination in **mango** is carried by:	House flies
✓	Pollination in **fig** is carried by:	Fig wasp (***Blastophaga pens***)
✓	Pollination in **oil palm** is carried by:	Weevil (***Elaeidobius spp***). NET-2013
✓	Center of **origin of Isabgol** and **Sarpagandha** is:	**India**.
✓	Center of origin of **Bael**, **Phalsa** and **Kagzilime** is:	**India.**
✓	Origin of orchid is:	**India**
✓	Double century is the variety of:	**Coconut.**
✓	Finest fruit of the world:	**Mango Steen.**
✓	Dioecious fruit plants are:	**Papaya, Datepalm**
✓	Gynodioecious fruit:	**Fig, Papaya**
✓	Pollu beetle (***Longitarsus nigripenis***) is the pest of:	**Black pepper (JRF-09)**
✓	Pest of sweet potato weevil:	***Cylas formicarius*** **(JRF-05)**
✓	**Pusa chikni** is a variety of:	**Sponge gourd**
✓	The clove required for planting garlic/ha is:	**500 kg**
✓	Cucumber is native of:	**India**
✓	Water melon is native of:	**T. Africa**
✓	Sponge gourd is:	***Luffa cylindirica.***
✓	Most serious disease of rose:	**Die back**

✓	Tissue culture is common in:	**Orchid**
✓	Annual flower of Indian origin is:	***Gomphrena***
✓	Popular climber of Indian origin is:	***Gloriosa superba***
✓	Annual flower suited for planting in shade is:	*Cineraria.*
✓	**Karonda** is richest source of:	**Iron.**
✓	**Maroon** is the variety of:	**Karonda.**
✓	Powdery mildew disease in mango is caused due to:	*Odium mangiferae.*
✓	Naphthalene acetic acid (**NAA**) is used in mango to control:	**Mango Malformation**
✓	Vegetable which is known as multivitamin greens:	**Chekurmanis**
✓	In which flower arrangement fruit, flower & foliage are used:	**Morimona**
✓	Clove, commonly used spice is obtained from which part of the plant:	**Flower**
✓	Higher temperature induces which type of flower in bottle gourd:	**Male**
✓	Planting system which accomodates maximum number of plants in the orchard:	**Filler system**
✓	**Cricket ball** and **Murrabba** are the popular varieties of:	**Sapota.**
✓	**Scartlet gold** and **Red fuzi** are the high yielding varieties of:	**Apple.**
✓	The chemical used to prevent sprouting during storage of onion is:	**MH** (Maleic hydrazide)
✓	**Rayan** as rootstock is most accepted for commercial propagation of:	**Sapota.**
✓	Most promising rootstock for **mandarins** and **sweet orange** is:	**Rangpur lime.**
✓	**Tea mosquito bug** (***Helopeltis antonii***) is a pest of:	**Cashew.**
✓	**Coffee berry borer** (***Hypothenemus hampei***) is a serious pest of:	**Coffee.**
✓	*Trimming* (removal of **undesired** plant parts)is generally done in	**Rose and marigold**
✓	***Bromalin*****:** enzyme is obtained from:	***Pineapple***
✓	Aroma in fruits is due to presence of:	***Esters***

✓	Acid content is a very important consideration for the preparation of:	*Jam*
✓	Domestic quarantine in India exist for:	*Sanjose scale*
✓	Upper part of stem for grafting is called:	*Scion*
✓	In onion bolting takes place due to:	***Low temp.*** *(except N-53)*
✓	Hand pollination is most useful in:	*Date palm*
✓	Highest source of energy:	*Walnut (contains maximum fat)*
✓	Hogarth curve is also known as:	*Line of beauty*
✓	Fruit of new world:	*Avocado*
✓	Orchid is propagated by:	*Micropropogation*
✓	Most important fruit used in processing industries in India:	*Pine apple*
✓	Pagoda is a name of:	*Japanese monument*
✓	Red colour in jelly is due to:	*Charring of sugar*
✓	Vista vision is a theme for:	*Land scape*
✓	Vegetable crop whose all parts are edible is:	*Winged bean*
✓	Sex form of pointed gourd is:	*Dioecious*
✓	Primitive sex form of cucumber is:	*Monoecious*
✓	The best site for the seed production of temperate type varieties of vegetables is:	*Hills*
✓	Edible part of lettuce is:	*Leaf*
✓	Edible part of parsley is:	*Leaves*
✓	Edible portion of celery is:	*Leaf stalk*

- ✓ ***Asparagus*** is a perennial crop
- ✓ Edible parts of asparagus are: ***Underground young shoots*** (known as ***spears***)
- ✓ Sex ratio in cucurbits can be modified by: ***Silver nitrate***
- ✓ Bolting (emergence of seed stalk prior to their time of formation) is a disorder of: ***Onion and cabbage*** (bolting in onion due to cold temp. fluctuations and in cabbage due to raising temperature above 75^0 F)
- ✓ Blindness (absence of curd or head in Cole crops) is a disorder of: ***Cole crops due to frost***
- ✓ Buttoning (Formation of small curds in cauliflower) is due to: ***Deficiency of nitrogen***
- ✓ Blanching in vegetables is done: ***To inactivate enzyme activities***
- ✓ Seed of carrotis known as: ***Schizocarp.***

- ✓ Forking of carrot is due to: ***Compact soil & application of undecomposed FYM***
- ✓ Harvesting stage for long distance transport of muskmelon is: ***Half-slip stage***
- ✓ Chief pollinating agent in onion is: ***Honey bee***
- ✓ Sex form of water melon is: ***Monoecious***
- ✓ Seed rate of hybrid tomato is: ***100-150 g/ha***
- ✓ Origin of cucumber: ***India***
- ✓ Physiological process responsible for rapid deterioration of fruits and vegetables is: ***Respiration***
- ✓ Serious disease of cumin is: ***blight***
- ✓ Sugary disease is common in fennel is caused by: ***Virus***
- ✓ Vitamin C act as: ***Anti-oxidants***
- ✓ Tuberization of potato requires: ***Short day length (temp.< 21 ⁰C)***
- ✓ The nutrient which improve quality of fruits is: ***Boron***
- ✓ Edible part of knoll khol is: ***Swollen stem (knob)***
- ✓ Onion varieties suitable for dehydration: ***White coloured***
- ✓ Anti-fungal factor in onion is: ***Catechol***
- ✓ Vitamin lost during processing is: ***Vitamin C***
- ✓ Rutabaga is a man-made vegetable is across between: ***Brassica oleracea × B. napus***
- ✓ Self-incompatibility exists in mango, aonla, apple, almond and cherry.
- ✓ India is the largest producer, processor, consumer and exporter of cashew in the world.
- ✓ Major tea growing pockets are located in **Assam (53%),** West Bengal (23%), Tamil Nadu (11%) and Kerala (8%).

Horticultural institutes:

Abbr.	Name of Institute	Place	State
CISTH	Central Institute for Subtropical Horticulture	Lukhnow	UP
CITH	Central Institute of Temperate Horticulture	Srinagar	J&K
CTCRI	Central Tuber Crops Research Institute	Thiruvantha puram,	Kerala
CPCRI	Central Plantation Crops Research Institute	Kessorgod	Kerala
CPRI	Central Potato Research Institute	Kufri	H.P.
IIHR	Indian Institute of Horticultural Research	Bangalore	Karnataka
IISR	Indian Institute of Spices Research	Calicut	Kerala
CIAH	Central Institute for Arid Horticulture	Bikaner	Rajasthan
CRIC	Central Research Institute for Chikoo	Muzaffarpur	Bihar
NRCB	National Research Center for Banana	Tiruchirapali	TN
NRCC	National Research Center for Cashew	Puttur	Kerala

NRCC	National Research Center for Citrus	Nagpur	MH
NRCG	National Research Center for Grapes	Pune	MH
DMR	Directorate of Mushroom Research	Solan	H.P.
NRCMAP	National Research Center for Medicinal & Aromatic Plants,	Anand,	Gujarat
NRCOG	National Research Center for Onion & Garlic	Pune	MH
NRCO	National Research Center for Orchids	Gangtok	Sikkim
NRCSS	National Research Center for Seed Spices	Ajmer	Rajasthan
IIVR	Indian Institute of Vegetable Research	Varanasi	U P
NHB	National Horticultural Board	Gurgaon	Haryana

Suitable plants for intercrops in orchard:

Main crop	Intercrop
▪ Mango	Papaya, Garlic
▪ Coconut	Banana, Phalsa

Fruits leading in area:

- Mango > Citrus > Banana

Leading in production:

- Banana > Mango > Citrus

Vegetables leading in area:

- Potato > Onion >Tomato

Leading in production:

- **Potato > Onion > Tomato**
- Highest onion producing state is: **Maharashtra.**
- Total vegetable production is maximumin **West Bengal** followed by **Uttar Pradesh.**

Classification of fruits:

A. According to **Fruit ripening**:

1. **Climacteric fruits**: (Fruit can ripe after harvesting/picking from plants)

Mango	Banana
Sapota	Guava
Papaya	Apple
Fig	Plum
Peach	Pear

2. **Non-climacteric: (Don't ripe after picking).**

Citrus	Grape
Pineapple	Pomegranate
Litchi	Ber

B. According to Climate:

1. **Tropical:**

Mango	Banana
Papaya	Pine apple
Cashew	Coconut

2. **Temperate:**

Apple	Almond
Strawberry	Starberry

3. **Sub-tropical:**

Phalsa	Ber
Litchi	Grape
Bael	Date palm
Citrus	Pomegranate
Guava, Fig, Aonla	

C. According to Fruit type:

1. **Berry:**

Banana	Papaya
Grape	Arecanut
Date palm	Guava

2. **Drupe/ Stone:**

Mango	Ber
Coffee	Coconut
Almond	Jamun

3. Balausta:	Pomegranate
4. Amphisarca:	Bael
5. Pepo:	Water melon (cucurbits)
6. Pome:	Apple
7. Hesperidium:	Citrus
8. Syconus:	Fig
9. Sorosis:	Pineapple, Mulberry, Jack fruit
10. Nut:	Litchi, Cashew
11. Schizocarp:	Carrot
12. Capsule:	Okra, Aonla
13. Lomentum	Tamarind

D. According to Edible parts of fruits:

✓ Pericarp:	Ber
✓ Mesocarp:	Papaya, Mango
✓ Endosperm:	Coconut
✓ Thalamus:	Apple, Strawberry

- ✓ Placenta: Bael
- ✓ Fleshy aril: Litchi
- ✓ Bracts: Pineapple
- ✓ Juicy seed coat: Pomegranate
- ✓ Juicy placental hairs: Citrus
- ✓ Fleshy receptacle: Fig

Daily requirement: (per day per capita):

- Cereals: **475 g**
- Pulses: **80 g**
- Fruits: **92 g**
- Vegetables: **300 g**
- Milk: **240 g**
- Sugar **40 g**

Nutritive value of fruits and vegetables:

Nutrients	Fruits	Vegetables
Vitamin **A/Retinol**	**Mango 4800 IU/100g,** Papaya **2020** IU/100g	**Bathua leaves11,300 IU/100g**
Vitamin- **B1/Thymine**	**Cashew nut**	Chillies 0.55 mg/100 g
Vitamin **B2/Riboflavin**	**Bael** > papaya	Fenugreek leaves
Vitamin **C/Ascorbic acid**	**Barbados Cherry-2000 mg, Aonla 600** mg/100 g	Coriander leaves 135 mg/100 g
Carbohydrates	**Raisins** (dehydratcd **grape**)	**Topica** (38.1%)
Protein	**Cashew nut (21.2 %)**	**Peas (7.2 g/100g)**
Fat	**Walnut (64.5%)**	Potato 11.8g/100g
Fiber	**Fig**	Potato
Calcium	Litchi (0.21%)	Agathi 1130 mg/100g
Phosphorus	**Almond** > Cashew nut	Amaranthus 800 mg/100g
Iron	**Dry Karonda (39.1 %)** Date palm (pind) 10.6 %	Amaranthus (22.9%)
Calorific value	**Walnut** 687calorie/100g	**Tapioca** 338 calorie/100g

Major Nutritional Deficiency diseases:

Disease	Deficiency of nutrient:
❖ **Fruit cracking** in apple	**Boron**
❖ Internal **necrosis** of Aonla	Boron
❖ **Hen &chicken** disease of grape	Boron
❖ **Hollow heart/** heart rot of sugar beet	Boron
❖ **Fruit cracking** of pomegranate/bael	**Boron**
❖ **Necrosis/black tip** of mango	Boron
❖ **Die back**, little leaf &**gummosis** of citrus	**Copper**
❖ Whiptail of cauliflower	**Molybdenum**
❖ **Buttoning** of cauliflower	Nitrogen
❖ **Grey spike** of oat/**Pahala blight** of sugarcane	Manganese
❖ **Khaira** disease of rice & **white bud** of maize	**Zinc**

Causes of colour/bitterness in fruits/vegetables:		
❖	Yellow colour in papaya is due to:	Caricaxanthin
❖	Redness of apple is due to:	Anthocyanin.
❖	Red colour in tomato is due to:	Lycopene.
❖	Red colour in chilli is due to:	Capcyanthin.
❖	Pungency in chilli is due to:	Capsaicin/capsicutin
❖	Orange colour in carrot is due to:	Carotene.
❖	Red colour in carrot is due to:	Anthocyanin
❖	Yellow colour in turmeric is due to:	Curcumin
❖	Bitterness in bitter gourd is due to:	Memordicocite.
❖	Bitterness in cucumber is due to:	Cucurbitacin
❖	Yellow colour of onion is due to:	Quercetin
❖	Red colour in onion is due to:	Anthocyanin
❖	Pungency in onion is due to:	Allyl propyl di-sulphide
❖	Pungency in raphanus is due to:	Isocyanate
❖	Pungency in mustard is due to:	Glucosilates
❖	Pungency in garlic is due to:	Di allyl di sulphide
❖	Green colour in potato tuber due to:	Solanin

❖	Sour taste of gram leaves is due to:	Malic/oxalic acid
❖	Pungency in pepper is due to:	Oleoresins
❖	Pungency in cabbage leaves due to:	Sinigrin

Epithets for fruits/crops and others:-

- ✓ King of fruits — Mango
- ✓ Queen of fruits — Litchi
- ✓ King of pulses: — Gram
- ✓ Queen of pulses: — Pea
- ✓ Queen of spices: — Cardamom
- ✓ Queen of flowers: — Gladiolus
- ✓ Food of god: — Cocoa
- ✓ Queen of beverage: — Tea
- ✓ King of temperate fruits: — Apple
- ✓ Butter fruits: — Avocado
- ✓ King of arid fruits: — Ber
- ✓ Poor man's fruits: — Ber
- ✓ Kalp vrkasha: — Coconut
- ✓ Miracle fruit: — China kiwi fruit
- ✓ White gold: — Cotton
- ✓ Century plant: — Date palm
- ✓ Adam's fig: — Banana
- ✓ King of forest: — Teak

Location and site selection of orchard:

a. **Location:**

- Refers to **geographical area** of the field in **relation to city, highway, town** etc. **(Pre P.G-05)**

b. **Site of orchard:**

- Place in **relation to topography**, **altitude**, **water body** and other factor which affect tree **behaviaring.**

Factors affecting selection of site:

1. **Land slope:**
   - Should not be **more than 5%**. If slope more than **15% unsuitable** for orchard, but suitable for contour cropping.
2. **Direction of slope:**
   - **Southern** slope receive **more sunshine** hence produce **early flowering** so, early crop can be taken.
3. **Soil fertility:**
   - **High fertility** requiring crops: Mango, citrus, grape.

- **Poor fertile** crops are: Phalsa, aonla.

4. **Soil texture:**
   - **Light texture** for **papaya** and **Ber**.

Training of orchard:

- Refers to judicious removal of plant parts to develop a proper shape and strong frame work of plant **to make capable of bearing heavy crop load**.
- While pruning is defined as the judicious removal of parts like root, leaf, flower, fruit etc. **to obtain good and qualitative yield**.

Method of training:

a. **Center leader system** (this method is not suitable for high altitude and hot arid places where wind velocity is high)

b. **Open center system** (not suitable for frost prone area)

c. **Modified leadership system** (this is most acceptable for commercial fruit production.

Training system for grape vines:

a. **Head system**: (suitable for less vigorous cultivar & it is very simple and inexpensive)

b. **Kniffin:** (also known as 4-canes system)

c. **Telephone system** (6-canes system, suitable for vigorous cultivar)

d. **Bower system:** also known as pergola or pandal, vines are spread over a structure (pandal) raised above ground on poles. Suitable for high yielding cultivers. This system is widely used for 80% of commercially grown grapes area in India because this method gives highest yield.

Pruning:

- Refers to removal of plant part especially shoots, roots, buds etc. to obtain better and qualitative yield is termed as pruning
- Pruning is started in later part of plant life when becomes produce fruits and flowers.

Method of pruning:

a. Thinning:

- Selective or complete removal of plant parts is termed as thinning

b. **Ringing or girdling** **(NET-08)**

- In this method a circular ring measuring about 3 cm in length is removed to hasten bearing by allowing greater accumulation of photosynthesis in upward portion of the plant.
- This is important practicedone in **grapes**.

c. Notching:

- Making a notch above a bud by removing a wedge shaped piece of bark is termed as notching
- It checks the influence of hormone and encourages growth
- Notching is commonly practiced in **Fig.**

Lay out of orchard:

- It includes **system of planting** and **planting distance**.

System of planting:

a. **Square system:**
- **Simplest** and **widely** adopted /popular method
- In this method R x R and P x P distances are kept similar.
- The plants are planted exactly at right angle at each corner.
- Number of plants/ha = 10,000 m^2/x^2
- Where, x^2 = R x P.

b. **Rectangular system:**
- **More plants** can be planted as compare to square system.
- Keeping more space between rows to row whereas plant to plant distance is kept comparatively less.
- Number of plants = 10,000 m^2/X × Y.
- Where, X = R x R, Y= P x P.

c. **Hexagonal** or equilateral **triangle** system:
- It accommodated **15% more plants** than square system.
- The plants are planted at the corner of equilateral triangle.
- This is very intense method of planting and **difficult** to layout.
- Requires fertile land.

d. **Quincunx** system/diagonal/**filler** system:
- Most suitable in **highly fertile** soils.
- This method is similar to square system but an additional '**filler**' plant is planted in the center of square.
- Almost **double number** of plants can be planted than square method.
- The common filler plants are **papaya, banana, phalsa** etc.
- **Filler plants** give **additional income** for non bearing orchard.

e. **Contour system:**
- Adopted in **hilly areas**
- Planting distance is **not uniform.**

f. **Triangular system:**
- It is mostly used in **HDP** (high density planting).
- **E.g. Amrapali** variety of mango.

Modes of reproduction of horticultural crops:

1. **Asexual reproduction**:

- Asexual reproduction in plants means a single parent produces genetically identical offspring (clones) from vegetative parts like roots, stems, or leaves, without seeds or flowers, through methods like runners, tubers, or cuttings.

(a) **Underground stems**: these are modified stems beneath the soil for food storage, survival or propagation. Examples are-

Rhizome: ginger, turmeric and banana

Tuber: potato

Bulb: onion, garlic and tuberose

Corm: gladiolus and taro/colocasia

Suckers: mint, chrysanthemum, banana

(b) **Sub-areial stems:**

Offshoots: datepalm, water hyacinth, pistia

Suckers (crawns): pine apple

Runners (creepers): oxalis and doob grass (Cynadon dactylon)

Stolones: mint, jasmine and straw berry

(c) **Aerial stems:**

Phylloclade: opuntia (prinkly pear)

Cladodes: asparagus

Thorn: duranta

Bulbil: agave and lilies

Commercial propagation methods of major fruit crops:

Fruit crop	Propagation method
Lemon/lime	**Seeds**
Sweet orange/Mandarin	T-budding or shield budding
Grape	**Hardwood cutting**
Pomegranate	**Cutting**/Air layering
Tea	Softwood cutting
Litchi	**Air layering**/Gootee
Guava	**Stooling/Mound layering**
Banana	**Sword sucker**
Date palm	**Offshoot**/sucker
Pineapple	**Suckers/ slips**
Strawberry	**Runners**
Aonla	**T-budding/Patch**
Bael	**Patch budding/Seed**
Ber	**Ring & T-budding**

- Custard apple — **T-budding**
- Sapota — **Inarching**
- Rose/Apple — **Shield budding**
- Papaya, Phalsa, Coffee and Jamun are propagated by: ***Seeds***

Important families of fruit crops:

Family	Common name	Botanical name
Bromeliaceae	Pineapple	***Annanas comosus***
Anacardiaceae	Cashew	***Anacardium occidentale***
Annonaceae	Custard apple	***A. squamosa***
Apocyanaceae	Karonda	***Carrisa carandas***
Lauraceae	Avocado	***Persia americana***
Moraceae	Fig	***Ficus carica***
Rosaceae	Strawberry	***Fragaria ananasa***
Rutaceae	Bael	***Aegle marmelos***
Sapindaceae	Litchi	***Litchi chinensis***
Tilliaceae	Phalsa	***Grewia subinequalis***
Rubiaceae	Coffee	***Coffea robusta***

UNIT-1

Fruit production

1. Mango: *Mangifera indica*, Anacardiaceae

- ✓ Known as **national fruit** of India/king of fruits.
- ✓ Fruit type of mango:**Drupe/stone**
- ✓ Origin of mango was: **Indo-Burma**
- ✓ Edible part of mango is known as: **Mesocarp**.
- ✓ Highly cross pollinated mediated by: **House fly**
- ✓ **India's** rankin mango production: **First** (50% of world)
- ✓ State first in mango production is: **Uttar Pradesh**
- ✓ Mango is the richest source of: **Vitamin A**
- ✓ **Rumani** used for dwarfing effect in: **Dasheri**
- ✓ Major pest of mango: ***Mango hopper*** (*Amritodus atkinsoni*)
- ✓ Major disease of mango: ***Powdery mildew***
- ✓ Mango malformation is due to: ***Low temperature & fungi.***
- ✓ Spongy tissue is due to: ***Convection heats.***
- ✓ Internal fruit necrosis is due to: ***Boron deficiency.***
- ✓ Deblossoming is done for: ***Control of malformation.***

- ✓ **Clustering** or **jhumka** in mango is due to**:** ***Low temperature, improper pollination and poor fertilization***
- ✓ Mango is a **climacteric and tropical** fruit
- ✓ In mango, **only 0.1** % flower develops fruits to maturity.
- ✓ **VHT** (vapour heat treatment) is recommended for **disinfection** of mango against **fruit flies** and **stone weevil**.
- ✓ Kanyakumari and Madurai districts of Tamil Nadu produce two crops of mango in a year.
- ✓ Alphonso: it is commercial variety of mango in Maharashtra
- ✓ Propagation of mango is done by: ***Veneer grafting***
- ✓ **Sai sugandha:** It is regular bearing variety and free from **malformation.**

- Soil: **Well drained** loamy soil.
- Planting time: ***June -July.***
- Distance: Desi- **10-12 m**, grafted-8 x 8, **Amrapali**: **2.5 x 2.5** m^2 (1600 plants/ha).

Propagation:

- Commercial method is: **Veneer grafting** (April-Oct.).
- Recent method: **Epicotyl stone** grafting (in Konkan, MH.)
- Other method: **Inarching** & Soft wood grafting.

Varieties:

SN	Variety	Character
1	**Alphonso**	Best **export variety** but *susceptible to **spongy tissue,*** **also** known as ***Happus***
2	**Chausa**	**Sweetest** but late maturing variety, grown in Northern India.
3	**Banganapalli**	Most commercial variety of **Andhra Pradesh**
4	**Bombay Green**	Early maturity variety of **north** India, also known as **Saroli**
5	**Kesar**	It has good **processing quality,** popular in Gujarat
6	**Langra**	North Indian variety, has characteristic turpentine flavor
7	**Niranjan**	**Off season** variety (bear fruits in October)
8	**Lal sindhuri**	Powdery mildew resistant **new variety** of mango
9	**Neelam**	Ideal for **long transport**, **two** crops are taken in a year
10	**Edward**	Resistant to **anthracnose**
11	**Fazli**	Commonly grown in Bihar & West Bengal
12	**Banglaura**	Known as totapuri, south Indian variety
13	**Dashahari**	Grown in UP, best medium ripening variety.
14	**Gulab Khas**	Grown in UP. It has rose like aroma,
15	**Imam pasand**	Sweet and rich in flavor

Hybrid varieties of mango:

SN	Hybrid	Cross	Characters
1	Arka anmol	Alphonso × Janardan pasand (AJA)	Free from spongy tissue
2	Arka puneet	Alphonso × Banganpalli (ABA)	Free from spongy tissue
3	Arka aruna	Banganpalli × Alphonso (BAA)	Free from spongy tissue
4	Ratna	Neelam × Alphanso, (NAR)*	Sponge tissue disease resistant, developed from Kokan Krishividyapeeth, Dhapoli, MH.
5	Amrapali	Dasheri × Neelam (DNA)	Dwarf, Suitable for **HDP** (2.5 x 2.5 m^2), good for kitchen garden, developed from IARI, New Delhi.
6	Sindhu	Ratna × Alphanso (RAS)	First **seed less** variety in world, developed from Kokan Krishividyapeeth, Dhapoli, MH.
7	Mallika	Neelam × Dasheri (NDM)	Highest **vitamin** content, developed from IARI, New Delhi.
8	Sai sugandha	Totapuri × Kesar (TKS)	**New hybrid** variety of mango
9	Ambika	Amrapali x Janardan pasand (AJA)	Developed from CISH, Lukhnow, UP. Skin colour dark red, Yield high
10	Newly released hybrids are **Pusa Deepshikha, Pusa Mahohari, Pusa lalima, Pusa pratibha, Pusa pitambar and Pusa shresth**.		

Types of mango varieties:

a. **North Indian** cultivar: These are **alternate bearer**, monoembryonic, self incompatible.

b. **South Indian** varieties: These are **regular bearing**, **polyembryonic varieties**.

c. **Regular bearing** varieties:
- Neelam, Totapuri, Amrapali, Banglora and most of hybrids.

d. Alternate bearer:
- Langra, Deshehari, Bombay green, malika.

e. **Off season** bearing:
- Niranjan and Madhulica

f. **Poly embryonic** varieties:
- Olour Goa
- Chandrakaran Bellary
- Mulgoa.

Insect -pests:

a. Mango mealy bug: ***Drasicha mangiferae* (Family: Margodidae)**
- It is **major** pest of mango.
- Nymph and adult suck the cell sap from leaves and tender trunks.

Management:
- **Alkathene banding** 30-45 cm **wide** & **30** cm **above** ground.
- Spray **carbaryl** @ 0.25%.

b. Mango Hoppers: ***Amritodes atkinsoni***
- Nymph and adult suck cell sap.

c. Stem borer: ***Bactocera rufomaculata***.

d. Fruit fly: ***Bactocera dorsalis***
- Use vapour heat treatment (**VHT**) for export of mango

e. Mango stone weevil: ***Sternochetus mangiferae***:
- Mango stone weevil and fruit fly lay eggs on fruit skin of developing fruits
- The larvae move in pulp & finally in stone of mango.
- Use of **irradiation**.
- This is **quarantine** importance pest. **(ARS-07)**

Diseases:

a. Powdery mildew: ***Oidium mangiferae***
- Severe during **Feb-March**.
- Spraying of **Karathane** @ **0.1%**.

b. **Anthracnose**: ***Colletotrichum gloesporoides***
- Severe in **humid** and high rainfall area.

c. Malformation: ***Fusarium monilliformae*** var. ***subglutinana***
- Vegetative and floral malformation; bunchy top appearance
- It is a serious problem in north India (in PB, Delhi & UP).
- **Bombay green** and **Chausa** are more **susceptible**.

Management:
- Pruning
- Spaying of **Mancozeb** @ 2g/liter.
- Spraying of NAA (Planofix) @ **200 ppm** by deblossoming at bud burst stage in the month of October.

Physiological disorders:

a. **Black tip**: Due to **deficiency of boron.**
- Near brick kilns **distal end** of affected fruits get **turned black** and become hardened.
- **Smoke** of **brick-kilns** containing gases like **CO, CO_2, SO_2** & **Acetylenes** are responsible for this.

Management:

- Establish the orchard **away** from brick kilns at **least 1 km** distance.
- Three spray of **borax** (0.6%) (before flowering, during flowering and at fruit set)
- Borax contains: **11% boron**.

b. **Spongy tissue**: Due to soil **heat convection** (high temp.)

- Major problem in **Alphonso**
- Fruit from outside look normal, but a patch of flesh becomes spongy, yellowish inside the fruits.
- It is also known as **internal break down**.

Resistant varieties:

- **Arka aruna**, **Ratna**, **Arka puneet**, **Arka anmol** *etc*. hybrids are free from spongy tissue.

c. **Leaf scorching**: Due to **Potassium** deficiency.

- Spraying of **potassium sulphate @ 5%.**

d. **Internal necrosis**: **Boron deficiency**.

- **Management**: Application of boron /borax @ 0.6 %

e. **Alternate bearing:**

- **On year**: Bears flowering.
- **Off year**: Bears no flowering.
- It is problem especially in **Dashahari, Langra** and **Chausa.**

Management:

- Application of **Cultar** (Paclobutrazol) @ 5-10g/plant, it induces flower.

2. Banana/kalpatharu: *Musa sp.*, Musaceae

✓ India is the largest producer of banana in the world.
✓ Propagated by sword suckers
✓ Banana gives one time fruits in life
✓ Weight of sword suckers generally used: *500-750 g*
✓ Most of cultivated banana is: ***Triploid in nature***
✓ In banana commonly found: ***Vegetative parthenocarpy***
✓ Diploid banana variety is: ***Lady finger***
✓ Cut rhizomes are used for propagation is known as: ***Peepers***
✓ Seedlessness in banana is controlled by spray of: *2, 4-D @ 25 ppm.*
✓ High density 1.2 x 1.2m to 1.5 x 1.5m is common now days.
✓ For the ripening of banana low concentration of ethylene at 15-18°C in controlled chamber is used. Or ethrel @ 1000 ppm solution spray on fruits.

- ✓ Banana fruits can be stored at 13 °C temperature & 85-95% humidity for three weeks. **(RPSC-AO 2012)**

Varieties:

▪ Dwarf cavandish	Robusta
▪ Rasthali	Poovan
▪ Honda	Champa
▪ Nendran (Kerala and TN)	

Diseases:

a. Panama wilt: ***Fusarium oxysporium*** p.v. ***cubensis.***

b. Sigatoka leaf spot: ***Cercospora musicola***

c. Bunchy top: ***Viral (BABU virus)***, transmitted by banana aphids.

d. Moko disease: ***Ralstonia solanacearum*** (bacteria)

e. Kokkan disease: ***Virus***, transmitted by aphids

Insect-pests:

a. Rhizome weevil: ***Cosmopolitus sordisus***.

- It is monophagus pest of banana **(JRF-09)**

b. Aphids: ***Pentalonia nigronervosa***, vector of **Bunchy top** virus.

3. Citrus fruits:

(a) Mandarin: *Citrus reticulate,* known as santra orange

- ✓ Mandarin is most common among citrus fruits grown in India.
- ✓ Mandarins are highly susceptible to: ***Water logging***.
- ✓ Mandarin, sweet orange, acid lime and grape fruit are: ***Highly polyembryonic***.
- ✓ Citrus fruits have a special kind of fruit skin referred as: ***Leathery rind***.
- ✓ Spacing of lime, lemon and sweet orange is: **6 m x 6m**
- ✓ Kinnow can be grown in HDP by using "***Troyer citrange***" as a **rootstock** by spacing the plants at **1.8 x 1.8** m^2.

Varieties of mandarin:

- Laddu.
- Coorg: commercial variety in south India.
- **Kinnow**: Cross between ***King mandarin*** × ***Willow Leaf mandarin*** **d**eveloped by **H.B. Frost,** USA -1935). It was first introduced in Punjab.

(b) Sweet Orange: *C. sinensis.*

- ✓ Preharvest **fruit drop** is common in **citrus**.
- ✓ It can be checked by spray of 2, 4-D @ 20 ppm.
- ✓ Ranpur lime is the best rootstock for mosambi.

- ✓ T' budding or Patch budding are most common methods for propagation of sweet orange.

Varieties:

- Satgudi
- Blood red
- Pine apple
- Mosambi
- Jaffa
- Valencia

(c) Kagzi lime: *C. aurantifolia*

- ✓ Known as ***Acid lime/sour lime***.
- ✓ Acid lime is commonly propagated by seeds.
- ✓ **Kagzi lime** *(C.aurantifolia)* is the indicator plant for **Tristeza** diesase. (Highly **susceptible** to this disease).
- ✓ Citrus **canker** is most serious disease of **acid lime**.

Varieties of kagzi lime:

- Chakradhar: **Seedless** variety of **acid lime**.
- Pramalini: Good yielder
- Mithachikna: sweet, juicy variety.
- Sai sarbati: **Tolerant** to **tristeza** and **canker.**

Diseases of citrus:

a. Gummosis: ***Phytophthora spp***.

b. Bacterial canker: ***Xanthomonas compestris*** p.v ***citri***. (transmitted by **leaf minor)**

c. Tristeza: **Virus** (vector is aphids: ***Toxoptera sp***.)

d. Citrus greening: Earlier **Mycoplasma** (now causal organism is ***Ca librobacter***, a gram negative, walled, bacteria).

- Transmitted by citrus **psylla**, ***Diaphorina citri***

e. Exocortis: **Viroids.**

Physiological disorder:

a. Granulation: Due to high **temperature** & **R.H.** during **ripening.** It can be managed by sprayingof **lime** & application of 2,4-D (@ 12 ppm)

b. Exanthema: **Copper deficiency.**

c. Yellow leaf of citrus: **Mo** deficiency.

d. Die back due to: **Cu deficiency**

e. Little leaf: Cu deficiency

Insect-pest of citrus:

a. Psylla: ***Diaphorina citri*** (vector of **greening** disease)

b. Leaf minor: ***Phyllocnistis citrella*** (vector of citrus **canker**)

c. Aphids: ***Toxoptera auranti*** (vector of **Tristeza** disease)

d. Lemon butterfly: ***Papilo demoleus***. (Controlled by **bagging** of fruits).

e. Fruit sucking moth: ***Othreis fullonica*** (only **adult** suck the juice from ripening fruits)

4. Papaya: *Carica papaya* (Caricaceae)

- ✓ Papaya is a: ***Polygamous** plant*
- ✓ It produces fruits throughout the year.
- ✓ It is a tropical fruit and more sensitive to frost.
- ✓ Yellow pigment in papaya: ***Carica xanthin*** **(RPSC, AO-09)**
- ✓ Milk (lactate) obtained from unripe fruit of papaya: ***Papain***
- ✓ In dioecious varieties of papaya: ***10% male** plants are planted*
- ✓ Papaya is commercially propagated by: ***Seed***
- ✓ Seed rate of gynodioecious papaya:***250-300 g/ha***
- ✓ Seed rate for dioecious varieties of papaya***: 400-500g/ha***
- ✓ Enzyme present in dried latex of papaya is: ***Pepsin***
- ✓ Carpine obtained from papaya is utilized as: ***Diuretic** and **heart stimulant.***
- ✓ Recommended spacing for Pusa nanha is: ***1.25 x 1.25 m²*** *(6000 plants/ha).*
- ✓ Frost is the most limiting factor in papaya in: ***North India***
- ✓ Papaya is susceptible to: ***Water logging***
- ✓ Most serious disease of papaya seedlings is: ***Damping off***

Gynodioecious varieties: tricks**'DMCCTSS'**

- Pusa delicious
- Pusa majesty (nematode resistance)
- Coorg honey dew
- CO-3
- Taiwan red lady (**blood red coloured**)
- Surya **(RPSC, AAO-09)**
- Sunrise solo **(pink flesh)**

Dioecious varieties:

- Pusa giant (used for canning & Tooty fruity)
- Pusa dwarf
- CO-5 (papain rich variety)
- Pant C-1
- Betty
- Hatras gold
- Pusa nanha (**extremely dwarf,** suitable for **HDP/pot** garden).

Diseases:

a. Damping off/stem rot is due to: ***Pythium aphanidermatum***.

b. Ring spot**: Viral**

- Upper leaves of papaya is **Mottled.**
- Transmitted by **aphids** (***Aphis gossipi***).

c. **Papaya leaf curl**: **Viral**

- Transmitted by **white flies.**

Management:

- Both **white flies** and **aphids** can be managed by spraying of **Imidacloprid @ 0.3 ml**/liter of water with Streptocyclin @ **0.1 g /liter** of water is also effective.

5. Pomegranate: *Punica granatum* (Punicaceae).

✓ It is highly **drought tolerant** among fruit crops.

✓ Origin of pomegranate probably: ***Iran (Persia)***

✓ Pigment responsible for the red colour in pomegranate fruits: ***Anthocyanin.***

✓ India has first position in the world with respect to area and production of pomegranate.

✓ In India Maharashtra is the leading state in area and production followed by KR, AP, GJ, TN and RAJ.

✓ Juice of pomegranate is useful for patient suffering from ***leprosy***.

✓ **July-August** is ideal time of planting in tropics.

✓ ***Ambe bahar*** is most commonly preferred by the growers because of high yield as compared to other flowering season.

✓ Wild type Anar is known: ***Daru***

✓ Presently ***Bhagawa*** is the leading variety of pomegranate cultivation in India especially in Maharashtra.

✓ Propagated by ***stem cutting*** and ***air layering*** *(Gootee).*

Bahar treatment (Crop regulation):

SN	Bahar	Flowering time	Harvesting	Remark
1	**Ambe**	Jan- Feb.	Jun-August	More flowering and high yield.
2	**Mrig**	June-July	Nov –Jan.	Fruit quality is not very good, more prone to insect-pest and diseases.
3	**Hasta**	Sept-Oct.	Feb- April	High flowering, Goodfruit quality & colour, Preferred for export.

Varieties:

SN	Variety	Features
1	**Jalore seedless**	Commonly grown in Rajasthan
2	**Ganesh**	Selection from Alandi, popular in MH
3	**Kandhari**	Medium fruit
4	**Arakta**	Suitable for processing
5	**Mridula**	Cross from Ganesh x Gul a-Shah red
6	**Ruby**	It is hybrid from a 3 way cross from IIHR
7	**Jyoti**	It is cross between Bassein seedless x Dholka
8	**Dolka**	Grow in Gujarat
9	**Bhagawa**	Bright red coloured, popular in MH, also known as **Sinduri,** Mastani, Jai maharashtra
10	**Amlidana**	Ganesh x Nanha(new pomegranate hybrid)
11	**Wonderful**	Originating from USA, fruit size large.

Insect-Pests:

a. **Anar butterfly**: ***Virachola (Deudorix) isocrates.*** **(JRF-05)**

- **Serious** pest of pomegranate
- Pencil size bored holes can be seen on fruits from which larval excreta comes out continuously.

- It is managed by **covering** of fruits with **butter paper**.
- Spray **Methomyl** 40 SP @ 1.0 ml/l or malathio 50EC @ 1ml/liter spray on plants.

b. **Fruit fly**: ***Bactrocera zonata*** (Tephritidae). **(JRF-05)**

- Minor pest of pomegranate

c. **Fruit sucking moth: *Othreis spp.***

Diseases:

a. **Bacterial blight: ** Caused by *Xanthomonas axonopodis* pv. ***punicae.***

- Also known as **nodal blight** or blak spot
- In Maharashtra commonly knownasoily spot or***Telya.***

Physiological disorder:

a. **Fruit cracking**:

- Due to **Ca**, **B & K** deficiency and irregular rrigation
- More in Rajasthan (dry areas).
- Managed by spray of **Borax** @ 0.5 %.

b. **Internal breakdown:**

- Disintegration of arils in matured pomegranate.
- Arils become brown and blackening.
- The incidence is more in ***ambe bahar*** *(Jan-June).*

6. Ber: *Zizyphus mauritiana.*

✓ Best time for pruning in ber is: April- May (in Rajasthan)

✓ Ber crop is heavily/drastically pruned

Varieties:

- **Dodhia:** Resistant to **fruit fly**.

Early varieties (suitable for **dry area**):

- Gola
- Kaithali
- **Seb** (**RAS-09**)

Mid maturity:

- Mundia
- Banarasi
- Jogia
- Meharun

Late maturity:

- Umran: Originating from **Rajasthan** (**RPSC, AO-09**)
- **Illaichi:** Triplets variety, Found pollen sterility
- **Katha:**
- Pathani:

Diseases:

- Powdery mildew: ***Odium jujube*** var. ***indica.***

Insect-Pest:

- Fruit fly, ***Carpomia vesuviana***

7. Guava: *Psidium guajava* (Myrtaceae)

✓ **Practice** of tacking **winter crop** instead of **rainy season** crop is known as **crop regulation** (to **escapes** the attack of **fruit flies**).

✓ **Stooling** is the most common & cheapest method of guava propagation.

✓ Allahabad region of UP is known for best quality guava production.

✓ Guava is susceptible to soil acidity and alkalinity

✓ **Rainy season** crop can be removed by spraying of **urea** at the time of **peak flowering** period.

✓ *Wilt* is serious problem in guava (common in **Alkali soil**)

Varieties:

- Allahabad safeda
- Lucknow-49 (**Sardar**): It is selection from Allahabad safeda
- Chittidar (**numerous red dots** on skin).
- Harijha (popular in Bihar)
- Hafsi (**Red fleshed** guava)
- Behat coconut (**seedless** guava)
- Lalit (**pink flesh** fruit, dual purpose fruit)

- Kohir safed (Kohir x Allahabad safeda)
- Safed jam (Allahabad safeda x Kohir)
- Arka mridula (soft seeded, **dwarf variety**)
- Allahabad Safeda: famous variety of Allahabad
- Allahabad Surkha (fruit with **deep pink flesh**)
- Apple colour: (pink coloured)

Hybrids:

- **Hybrid-45** = Allahabad safeda x Sardar (L-49).
- G-Vilas Pasand- dwarf variety bear fruits throughout the year
- Pusa Aarushi & Pusa Pratiksha – new varieties developed by IARI.
- Megha Supreme & Megha seedless- for rainfed northeast region

Diseases:

a. Guava wilt: ***Fusarium oxysporium*** p.v. **Psidii**.

Symptoms:

- **Wilting** appearance of plants and leaves of young plants is **dries up**.

Management:

- Soil drench with **Carbendazim** (Bavistin) @ 2g/liter water.
- Grow resistant varieties i.e. **Allahabad Safeda.**(**RAS -09**)

Insect pest:

a. Stripedmealy bug: ***Ferrisa virgata,*** serious pest in South India

b. Guava tea mosquito bug: ***Helopeltis antonii.***(IARI, Ph. D- 07) (Symptoms- corky out growth/scab formation in guava)

Physiological disorder:

a. Bronzing: *Due to* ***Zn deficiency***

8. Aonla: *Emblica officinalis (Euphorbiaceae)*

✓ It is also known as Indian gooseberry is an indigenous fruit

✓ Propagation by patch budding in north India.

✓ Aonla fruits contains vitamin C: ***600 mg/100g***

✓ **Large sized**, fruits are mostly utilized for: ***preservation as candy***

Varieties:

Varieties	Features
Banarasi	Early maturity, best for **Murrumba**
Kanchan	Regular bearing, known as **NA-4**
Chakiya	Alternate bearing, free from fruit necrosis
Hathijhool	Suffers from **fruit necrosis**, also named as **Francis**

Diseases:

a. Ring rust: ***Ravenelia Emblica,*** it is serious disease in aonla.

b. Fruit necrosis: ***B-deficiency.***Francis variety is highly suffered

Insect -Pests:

a. Bark eating caterpillar: ***Indarbella quadrinotata***.

9. Grape: *Vitis vinifera, Vitaceae*

- ✓ Ideal time of **planting** unrooted grapecuttings is: ***October.***
- ✓ Rooted cuttings are planted in January-February.
- ✓ Grape is a deciduous crop.
- ✓ Propagated by hard wood cuttings.
- ✓ **Thompson seedless** with its clone occupies **55% area** under **grape** cultivation.
- ✓ **Training system** of **grapes** mostly adopted in India is: ***Bower system.***
- ✓ Element that is universal deficient in grape is: ***Mg***
- ✓ **Tartaric** acid is present in grape fruits
- ✓ In northern India grapes are pruned in: ***December-January***
- ✓ The **chemical** used to reduce post harvest **fruit drop** in grape is**:** ***NAA (50 ppm).***
- ✓ Outstanding cultivar for **raisin (dehydrated** grape) is: ***Kismis Bely***

Varieties of grape:

Varieties	Features
Arka Trishna & Soma	**Hybrid varieties, for wine purpose**
Dilkhush	-
Arka Hans	**Hybrid variety, White wine**
Arka Kanchan	**Hybrid variety, Late maturity**
Thomson Seedless	**Mostly grown** variety
Muscat	**Coloured seeded variety**
Perlette	Mostly grown in Rajasthan
Anab-e-Shahi	-
Sonaka Super	Large sized, sweetest variety

Diseases:

a. Downy mildew: ***Plasmophora viticola*** (major disease)

b. Powdery mildew: ***Erysiphe viti***

Physiological disorder:

a. **Hen** & **Chicken** disease is due to: ***Bo** deficiency.*

b. **Berry** or Blossom **drop** in Grape: ***Improper** pollination & fertilization.*

c. **Pink berry** formation is due to: ***High temp,*** common in **Thompson seedless.**

d. **Calyx end rot**: ***Ca** deficiency.*

Insect-pests:

a. Thrips: ***Rhiphiphothrips cruntatus***

- It can be managed by spray of malathion 50 EC @ 1ml/liter

b. Flea beetle: ***Scelodenta strigicollis***

- It is managed by spray malathion 50 EC@ 1.5 l./ha in orchard.

10. Coconut: *Cocus nucifera*, Aracaceae/palmae

✓ **India's** rank **3rd** in **coconut** production after **Indonesia** and **Philippines**.

✓ **Kerala** share in **coconut** production is: **42%**

✓ Chowgghat green dwarf and Pratap are cocout varietiees

Insect-Pest of coconut:

a. Rhinoceros beetle: ***Oryctus rhinoceros***

- Fan like appearance of leaves.

b. Red palm weevil: ***Rhyncophorus ferrugineus***

- Gummosis (oozing of brown liquid)

c. Black headed caterpillar: ***Opisinia arenosella***

- Monophagus pest of coconut.

d. Coconut mite: ***Aceria guerreonis***. **(IARI, Ph. D-05)**

Diseases of Apple:

a. Crown gall: ***Agrobacterium tumefaciens***

b. Fire blight of apple/pear: ***Erwinia amylovora***

c. Apple scab: ***Venturia subinaequalis*** (fungi)

d. Water core: ***Bo deficiency***

e. Bitter pit and cracking: ***Ca deficiency***

Insect-pests of apple/pear/peach:

a. **Sanjose** scale: *Quadraspidiotus perniciosus*, Diaspideae ***(BHU-2015)***

- Suck the cell sap from leaves, twigs and fruits, the infested fruits also have a scaly appearance & each spot is surrounded by a scarlet or red area.
- Important parasitoid of sanjose scale is ***Encarsia perniciosi***

b. **Wooly** apple aphid: ***Eriosoma lanigerum***

- The infested plants have pale-green leaves and **whitish cotton patches** on stem and branches, **characteristic galls/knots** are **formed on roots** & underground portion of plants.
- Its main parasitoid is: ***Aphelinus mali*** (attacks on 4th and 5th nymphal stage).

UNIT-2
Vegetable production

Important points:

- ✓ Vegetables are rich source of **vitamins** and **minerals.**
- ✓ **India** 2nd largest producer of **vegetable** after **China**.
- ✓ Vegetable crops in India occupy only **2.8 %** of the **total cropped** area.
- ✓ India accounts for **13.38 %** of World **production** of vegetables.
- ✓ State having largest **area** and **production** of vegetables: ***West Bengal***
- ✓ State having maximum **productivity** of vegetable: ***Tamil Nadu***
- ✓ **Market** garden is very intensive method of vegetable cultivation.
- ✓ **Truck** garden is very **extensive** method of **vegetable** cultivation.
- ✓ Green leafy vegetables are rich source of **folic acid**.
- ✓ Major **mineral** present in **fruit** and **vegetable** is **potassium**.
- ✓ **Cucumber** is the **non climacteric** vegetable fruits.
- ✓ **Quecertin** compound presence in onion and garlic is capable to protect against **cancer** and **heart** disease.
- ✓ Vitamin present in chilli is: ***Vitamin P***.
- ✓ The **flavor in garlic** is due to sulphur compound: ***Di allyl di sulphide***
- ✓ Flower colour of **pointed gourd**, **bottle gourd** and **chilli** is: ***White** (ARS-09)*
- ✓ Flower colour of **cruciferous** is: ***Yellow***
- ✓ Popular variety of **pumpkin** is: ***Pusa vishwas and Arka chandan***
- ✓ **Arka harit** and **Pusa visesh** are varieties of bitter gourd
- ✓ Pointed gourd is propagated by vine cutting @ 2000-2500 cuttings/ha.
- ✓ Ridge gourd commercially trained on kniffin system.
- ✓ **Pusa nasdar** and **Satputia** (hermaphrodite) are the varieties of ridge gourd.
- ✓ **Pusa supria** and **Pusa chikni** are varieties of **sponge gourd**.
- ✓ **PBOG-1**, **Pusa manjari** and **Pusa maghdoot** are the varieties of: ***Bottle gourd.***
- ✓ **Poinsett** and **Pusa sanyog** are varieties of: **Cucumber**
- ✓ **Pusa parvati** is mutant variety of **French** bean.
- ✓ **Pusa suffaida** and **Pusa lal** are the variety of: **Sweet potato**.
- ✓ **RS-1** and **RZ-19** are varieties of cumin.
- ✓ **Rcr-41** and **Acr-2, Sadhna** are the varieties of coriander.
- ✓ **Rmt-141, Hisar Sonali** are varieties of: **Fenugreek**

Chapter -1

Classification of vegetables:

1. Botanical classification of vegetables:-

Family	Common name	Botanical name	Origin	Edible part
Alliaceae	Onion	*Allium cepa*	Central	Bulb
	Garlic	*Allium sativum*	Asia	**Cloves**
Lilliaceae	Asparagus	*A officinalis*	Europe	**Spwears**
Araceae	Taro	*Colocasia esculenta*	Srilanka	**Corms**
Chenopodiaceae	Beet root	*Beta vulgaris*	Mediterrian	**Roots**
	Palak	*Beta vulgaris varbengalensis*	Indo-China	Leaves
	Spinach	*Spinacea oleracea*	Iran	**Leaves**
Compositae	Lettuce	*Lactuca sativa*		**Leafy heads**
Amaranthaceae	Amaranthus	*Amaranthus sp.*	**India**	Leaves, stem
Convolvulaceae	Sweet Potato	*Ipomea batata*	S. America	**Tuber**
Cruciferae /Brassicaceae	Cabbage	***Boleracea*** var. ***capitata***	Mediterrian	**Head**
	Cauliflower	***B oleracea*** var. ***botrytis***	Mediterrian	**Curd**
	Broccoli	***B oleracea*** var. ***italic***	Mediterrian	**Flower bud**
	Turnip	*Brassica rapa*	China	Root
	Radish	*Raphanus sativus*	Europe	Root &leaves
Cucurbitaceae	Cucumber	*Cucumis sativus*	**India**	Fruit
	Musk Melon	*Cucumis melo*	T. Africa	Fruit
	Water Melon	*Citrullus lanatus*	T. Africa	Fruit
	Pumpkin	*Cucurbita moschata*	Mexico	Fruit
	Bottle gourd	*Lagenaria siceraria*	**S. Africa**	Fruit
	Bitter gourd	*Momordica charantia*	**Indo-Burma**	Fruit
	Pointed gourd	*Trichosanthus dioca*	**India**	Fruit
	Snake gourd	*Trichosanthus anguina*	**India**	Fruit
Malvaceae	Okra	*Ablmoschus esculantus*	Africa	Fruit

Leguminaceae	Peas	***Pisum sativum***	**Asia**	Pod
	French bean	***Phaseolus vulgaris***	Mexico	Pod
	Cluster bean	***Cyamopsis tetragonolobus***	**India**	Pod
	Cow pea	***Vigna unguiculata***	**Africa**	Pod
Umbeliferae	Carrot	***Daccus carota***	Afghanistan	Root
	Coriander	***Coriandrum sativum***	-	leaves
Solanaceae	Potato	***Solanum tuberosum***	**S. America (Peru)**	**Tuber**
	Tomato	***Lycoperisicon esculentum***	**S. America**	Fruit
	Brinjal	***Solanum melongena***	**India**	Fruit
	Chilli	***Capsicum annum***	**Mexico**	Fruit
Euphorbiaceae	Cassava	***Manihot esculanta***	**Brazil**	**Tuber**

2. Classification of vegetables based on life cycle:

a. Annual vegetables:

- Okra — Tomato
- Cow pea — Brinjal

b. Biennial vegetables:

- Radish — Carrot
- Cabbage — Cauliflower
- Onion

c. Perennial vegetables:

- Sweet potato — Parval (pointed gourd)

3. Classification of vegetables based on photoperiod

a. **Long day vegetables:**

- Onion — Cabbage
- Cauliflower — Radish
- Palak — Carrot
- Turnip — Potato (Flowering point of view)

b. **Short day vegetables:**

- Sweet potato — Cluster ban

c. **Day neutral vegetables:**

- Tomato — Brinjal
- Chilli — Okra
- Cucurbits — Cowpea

4. Classification of vegetables based on presence of sex forms

a. **Dioecious vegetables**: male & female flowers are present on separate plant. Examples are:-

- Pointed gourd   Beet root
- Spinach

b. **Monoecious**: both flowers are present on same plant.

- Cucurbits   Cassava
- Amaranths

1. Tomato: *Lycopersicon esculentum, Solanaceae*

- ✓ Botanical name of tomato: ***Lycopresicon esculentum***
- ✓ Origin of tomato: **Peru**
- ✓ Fruit of tomato is: **Berry**
- ✓ Red colour in tomato due to: **Lycopine**
- ✓ Popular variety of tomato: **Pusa ruby, Arka vikas**
- ✓ Seed rate of tomato: **350-400 g/ha**
- ✓ Major pest of tomato: Fruit borer (***Helicoverpa sp***.)
- ✓ Tomato susceptible for: **Frost**
- ✓ Tomatofruits generally regarded as: **Protective food**
- ✓ In adverse conditions tomato crop is sprayed with: ***PCPA*** for higher fruit setting.
- ✓ ***Punjab chhuhara*** is a determinate variety of tomato
- ✓ A period of drought followed by **sudden heavy irrigation** or rain during fruiting may cause **cracking of fruits**
- ✓ Chemical used for preservation of tomato sauce: **Sodium benzoate**
- ✓ BER (blossom end rot) disease of tomato is due to: **Ca deficiency**
- ✓ Tomato is day neutral and **self-pollinated** plant
- ✓ It is basically a **warm season crop** but may be grown in all seasons.

Seed rate:

- For normal: 350-400 g/ha
- For hybrid: **100-150 g/ha**

Varieties:

- Pusa Rakshit: Suitable for protected cultivation
- Pusa ruby: Sioux x Improved Meeruti.
- Pusa-120: Nematode resistant
- Arka sourabh:
- Pusa gaurav:
- Arka vikash:
- Hissar lalit: Nematode resistant **(RPSC, AO-09)**

- Pusa sheetal: Winter season
- SL-120: Nematode resistant
- Roma: Suitable for transport and processing
- Parker: Fruit borer (*H. armigera*) resistant
- Punjab chhuhara: Suitable for **processing.**
- Arka Rakshak & Arka Abhed- new hybrid from IIHR
- **Flavr savr:** Delay ripening a genetically modifiedvariety of tomato developed through Biotechnology in 1994 by Calgene Company.

1. Blight of tomato: **bacterial** disease.
   - It can be managed by use of **Copper fungicides** i.e. **Copper oxychloride** @ 3g/liter of water with **Bacteromycin @ 0.1** g/liter of water or Mencozeb @ 3g/litre.
2. Leaf curl: **virus**
   - Transmitted by **white fly.**

Insect-pests:

1. Fruit borer: ***Helicoverpa armigera***.
2. White fly: ***Bemisia tabaci*** (transmitted **leaf curl** disease)
3. Serpentine leaf minor: ***Liriomyza trifoli***.

Physiological disorders:

SN	Disorder	Symptoms	Causes
1	**Fruit cracking**	Cracking in tomato fruits is common, radial and concentric, cracking are severe	Effect of soil moisture and **Boron deficiency**
2	**Blossom end rot (BER)**	BER is severe disorder of tomato cultivated in green houses or protected structures	Due to **Ca deficiency** and high temperature
3	**Puffiness**	Fruit surface is generally flattened and locules are unfilled with pulp and seeds	Poor pollination, Low or high temperature
4	**Sun scald**	Whitish, sunken and papery lesion develops due to sun heat	Expose of fruits to high temperature > 40 ^{0}C
5	**Cat face**	Due to abnormal growing condition	Primary Causes-due to environmental stress during flowering, especially cold nights.
6	**Gold Fleck/Fruit Pox**	Gold-colored specks on the fruit surface; chlorophyll development is affected.	High calcium/potassium (K) ratio. Excess calcium (Ca) uptake compared to potassium (K) leads to calcium oxalate crystal formation

2. Brinjal: *Solanum melongena, Solanaceae*

- ✓ Pigment present in brinjal is: **Anthocyanin.**
- ✓ **Brinjal** fruits are good source of: **Vitamin B.**

Seed rate:

- **200 g/ha** for nursery sowing

Varieties:

- Pusa purple long (extra early maturity)
- Pant samrat (**Phomopsis** blight and bacterial **wilt** resistant).
- Azad kranti
- Black beauty (nematode resistant)
- Arka sheel, Arka nidhi

Hybrids:

- Pant rituraj:
- Pusa bindu:
- Pusa upkar:
- **Pusa bhairav** (Phomopsis blight resistant)
- Arka navneet (Highest yielding):
- Arka neelkantha (Nematode resistant).
- Annamalai: Aphid resistant, recommended for TN
- Arka **keshav**
- Pusa Oishiki & Pusa Vaibhav- new varieties from IARI, New Delhi

Insect-pests:

a. Fruit and shoot borer: ***Leucinodes orbonalis, Pyralidae, Lepidoptera*** (it is major pest of brinjal)

- ✓ Cause **'dead heart'** in young plants, tender growing ***shoot*** is killed by larvae in early stage and later bore into ***fruits.***
- ✓ It managed by application of Flubendiamide 48 SC @ 60 g a.i. /ha.

b. Hadda beetle: ***Epilachna vignitictopunctata (F. Coccinelidae)***

c. Brinjal brown leaf hopper: ***Cestius phycitis*** (vector of little leaf of brinjal).

d. The white flies and jassids/leaf hopper in brinjal can be controlled by application of Imidacloprid 200 SL @ 75 g a.i /ha

Diseases:

- Little leaf of brinjal is due to: Phytoplasma
- Transmitted by leaf hoppers: ***Cestius phycitis.***

3. Chilli: *Capsicum spp.*(Solanacae)

- ✓ Red colour of chilli is due to: ***Capcyanthin***
- ✓ Pungency of chilli is due to: ***Capsaicin or capsicutin***
- ✓ Major pest of chilli is: ***Thrips (Scirtothrips dorsalis).***
- ✓ Anthracnose/die back due to fungus: ***Colletotrichum capsici.***
- ✓ NAA (Planofix) @ **10 ppm** and Triacontanol are used to control: ***Fruit drop***
- ✓ India is a major producer, consumer and exporter of chilli in the world
- ✓ Andhra Pradesh has first position in chilli production.
- ✓ *Capsicum annum* is also known as sweet pepper or bell paper
- ✓ Leaf curl of chilies: Viral, transmitted by **thrips/whiteflies.**
- ✓ To prevent fruit rot spraying calcium nitrate (CN) @ 1 litre/ha or broadcasting about 25 kg/ha.

Seed rate:

- **For chilli: 1.0-1.5 kg/ha**
- For capsicum annum (hybrid): 250g seeds/ha
- Sown in nursery and transplanted after 35-45 days of seedling

Varieties:

- California wonder
- Arka mohini
- Andhra jyoti
- Pusa jwala (thrips resistance)
- Jwalamukhi (suitable for HDP)
- Arka lohit (powdery mildew tolerant)
- Bhaskar
- Bhagyalakshmi
- Indra
- Punjab lal (resistant to fruit rot, die back, curl)
- Mathania long
- Bharat

Pests management:

1. Chilli thrips: ***Scirtothrips dorsalis***
   - **Inward curling** of leaves is a typical symptom
2. White flies: ***Bemisia tabaci***
   - It is responsible for transmission of **chilli mosaic virus**

- Both **white flies** and **thrips** can be managed by spraying of any **systemic** insecticides i.e. **Acephate @ 1.0 ml**/liter of water.

- **Imidacloprid** @ **0.5 ml**/liter of water with Streptocyclin @ **0.1 g /liter** of water.

4. Okra: *Ablmoschus esculantus.*

- ✓ Major pest of okra is: Fruit borer ***(Earias vitella)***
- ✓ Blister beetle: ***Mylabris pustulata*** *is a pest of okra.*
- ✓ Yellow vein mosaic is due to: **Virus**, transmitted by white fly.

Seed rate:

- For kharif season: ***12-15kg/ha is optimum***
- For spring-summer the seed rate is: ***18-20 kg/ha***

Varieties:

- Pusa sawani: Day neutral and spineless variety
- Varsha uphar: Resistant to white flies
- Punjab padmini
- Pusa mukhmali
- Prabhani kranti: White fly resistant
- Arka anamica
- Arka abhay
- Hissar unnat
- Kashi Chaman- developed from ICAR-IIVR, resistant to YVMV.

Cole crops:

- ✓ Ancestor of Cole crops is Colewort (wild cabbage) here from the word Cole is derived
- ✓ All Cole crops are: ***Protogyny***.
- ✓ Major pest of Cole crops is: ***DBM*** *(Plutella xylostella)*
- ✓ Club rot of cauliflower is due to: ***Plasmodiophora brassicae***
- ✓ Curd rot/soft rot of Cole crops due to: ***Erwinia crotovera.***
- ✓ White rust/white blister of cruciferous: ***Albugo candida***
- ✓ Self-incompatibility is most common in: ***Cruciferae family***
- ✓ The characteristic flavor of Cole crops is due to presence of: ***Di methyl tri sulphite.***

Insect-pest of cole crops:

1. **Diamond back moth** (DBM): ***Plutella xylostella***
   - It is **serious** pest of **cabbage**
   - Larvae feed on leaves and later make galleries in head/cabbage fruits

- DBM is most notorious among all insect-pests because has developed the resistance against most of insecticides, followed by *Helicoverpa armigera*.

Management:

- Indian mustard which attract 80-90 % DBM used as **trap crop.** **(ARS-07)**
- **Alternative sprays of following pesticides:**
- Spray of Quinolphos 35 EC @ 250 ml/ha
- Spray Cartap hydrochloride @ 500 g a.i/ha
- Spray Fenvalerate 20 EC @ 45 g a.i. /ha

2. Mustard saw fly: ***Athalia proxima lugens*** (only one pest from order: **Hymenoptera, F:** Tenthredinidae)
3. Cabbage semi-looper: ***Plusia orichalcea*** (Noctuidae)

5. Cabbage: *Brassica oleracea* var. *capitata*

✓ India **rank 3rd** in cabbage production.

✓ Cabbage has **anticancer** property due to presence of **Indole-3-carbinol**.

✓ The incidence of club root of cabbage is 100% at pH 5.7

✓ Fruit of cabbage is known as: **Head**

✓ Fruit type of cabbage: **Siliqua.**

Seed rate:

- **500-750g** of seeds is required for one ha.

Varieties:

- Golden Acre
- Copenhagen Market (early variety).
- Pride of India (early and round variety)

Hybrids:

- Pusa drumhead: Black leg resistant
- ***Pusa mukta:*** Black rot resistant
- Pusa sambandh
- Diamond express-.
- Arka sugandhi- it is candidate variety of cabbage under DUS (distinctness, uniformity and stability) test.

6. Cauliflower: *B. oleracea* var. *botrytis*

✓ **Blanching** is mostly done in early varieties of cauliflower to **protect curd** from **attaining yellow** color after their direct expose to sun and to **arrest** enzymatic activity.

✓ Fruit of cauliflower is known as: **Curd**

- ✓ Three season varieties are available in: **Cauliflower**
- ✓ **Scooping**: removal of central portion of curd for earlier initiation of flower stalk in cauliflower
- ✓ **Pusa himjyoti** and **Pusa snowball** have self-blanched habit.

Seed rate: **450-700g** seeds are sufficient for one ha.

Early variety:

- Pusa early synthetic
- Early kunwari
- Pusa deepali
- Pusa kartiki
- Pusa sharad

Late variety:

- **Pusa snowball**
- **Pusa himjyoti**: Can be grown from **April to July** in the hills.

New released varieties-

- Arka Spoorthi- from ICAR-IIHR (early maturity variety)

Disorders of cauliflower:

S N	Disorder	Character	Causes
1	**Blindness**	Plants having no terminal buds and fail to form any curd	Damage in early stage by insects, low temp. & **frost**
2	**Buttoning**	Development of small curds in young plants	**Nitrogen deficiency**
3	**Hollow stem (RAS pre-2012)**	Excessive growing plants of cauliflower develops hollow stem and curd	Excessive use of fertilizers especially **'nitrogen'**and deficiency of **boron**
4	**Whiptail**	Only midrib of leaf blade is develops so the name'***whiptail***' is given	**Mo** deficiency in acidic soils
5	**Chlorosis**	Inter-veinal, yellow mottling of lower leaves	**Mg** deficiency
6	**Riceyness**	The surface of curd is loose and has velvet appearance due to formation of small white flower buds at the curding stage	Fluctuation of temp., Heavy application of N & high humidity
7	**Browning**	Light brown to dark spots on stem, brown color of curd	**Boron** deficiency

7. Onion: *Allium cepa*

- ✓ Rich source of: **Vitamin B.**
- ✓ Pollination of onion is chiefly mediated by: **Honey bees**
- ✓ Pungency in onion is due to: **Allyl propyl di sulphide**.
- ✓ Onion is a cool season crop
- ✓ The **colour** of outer skin of onion is due to: **Quecertin.**
- ✓ Onion is very useful in: **Sun stroke.**
- ✓ **Bolting**/stem elongation in onion is due to: **Low temperature**
- ✓ In **onion & garlic**: Purple blotch is due to: ***Alternaria porii.***
- ✓ Lassalgoan in Maharashtra is the biggest onion market in India.
- ✓ **Cate chol** is a phenolic factor present in onion has **antifungal property**.
- ✓ To check **sprouting** or germination of onion during **storage** is spraying of **MH @ 2500 ppm**15-20 days after storage is beneficially
- ✓ Important pest of onion is **thrips**. It can be controlled by sprayingof **Acephate @ 1 ml/liter** of water.

Seed rate:

- **10-12 kg Seeds/ha** for nursery or 750 to 1000 kg bulbs for direct planting
- Nursery area required for 1ha. Onion crop is: 500m^2

Varieties:

Rabi season:

- Pusa red
- Nasik red
- Patna red
- Udaipur 101, 102, 103
- Pusa ratnar
- N-53
- Kalyanpur red round
- Pusa madhavi
- Arka bindu

Kharif season:

- Arka kalyan
- Agrifound dark red
- N-53
- Arka pragati (grow in both season).

8. Root crops:

- ✓ **Allicin** is the **antibacteria**l substance of: **Garlic**
- ✓ Allin is a water soluble **amino acid** present in: **Garlic**

a. Radish: *Raphanus sativus*

- ✓ **Mustard saw fly** and **painted bug** are the pests of radish.
- ✓ Roots of radish prepared for seed production is known as: **Stackling**
- ✓ Brown heart in radish is due to: **Boron** deficiency
- ✓ **Seed rate**: 5.5-11 kg/ha

Varieties:

- Arka nishant (multiple diseases resistant)
- Pusa chetki (suitable for summer)
- White icicle
- Japanese white.
- Pusa safed, Punjab safed, Pusa desi
- Scarlet globe
- Pusa rashmi
- Pusa himani (can be grown throughout the year)
- Round Cherry Belle: Red coloured variety of radish

b. Carrot: *Daccus carota*

- ✓ Carrot is an annual crop for root production and biennial for flowering and fruit set.
- ✓ **Kanji** (beverage is prepared from black carrot)
- ✓ **Forking** of carrot is due to: Hard pan of soil, application of undecomposed FYM, overcrowding, over watering and infestation of *Meloidogyne sp.*

Seed rate:

- **5-6 kg/ha** of seeds is sufficient.

Varieties:

- Pusa kesar: red coloured, suitable for early sowing
- Pusa rudhira: red coloured variety
- Pusa meghali: orange coloured, highest vitamin-A
- Pusa yamdagni: suitable for cool climate
- Punjab black beauty: purple black root
- Chantenay: for canning and storage
- Imperator: late maturity
- Ooty-1

- Nantes
- Amsterdam

9. Cucurbits:

✓ **Pusa vishwas** and **Arka chandan** are varieties of: **Pumpkin.**

✓ **Pusa chikni** is variety of: ***Turai*** (Sponge gourd)

✓ Pusa raunak, Biakner green, Dilpasand, Hisar selection-1, are the varieties of: ***Tinda/Round gourd***

✓ Powdery mildew in cucurbits is due to: ***Erysiphe cichoracearum***

✓ Downey mildew in cucurbits is due to: ***Pseudoperanospora cubensis*** **(RAS-09).**

✓ Yellow vine of cucurbits is due to: ***Serratia marcescens***

✓ Bud necrosis: mostly in water melon, viral, transmitted by aphids.

Insect- pests of cucurbits:

1. **Red pumpkin beetle**: *Raphidopalpa foeicollis*
    - Both grub and adult beetle cause damage, the grub damage the plants by boring in to roots and underground stem.
    - But beetles are very destructive to vegetable particularly during March-April when creepers are very young.

Management:

- Spray Malathion 50 EC @ 0.05 %.

2. **Cucurbits fruit flies:** *Bactrocera cucurbitae*
    - Maggots feed pulp and seeds and infested fruits may fall prematurely & fruit become soft and rotten.
    - It can be controlled by **bait spray** containing 50 ml of Malathion 35 EC and 0.5 kg molasses in 50 liters of water for one ha.
3. **Cucurbit mite**: *Tetranychus cucurbitae*
    - Known as **red spider mite**
    - Suck cell sap by making webbing on **underside of leaves**
    - Damage is confined to lower surface of leaves
    - It can be controlled by spraying **Dicofol/ Ethion 50 EC** @ 1.5 ml/liter.

10. Cucumber: *Cucumis sativus*

✓ Known **as '*kheera*'.**

✓ Primitive sex form of cucumber is: **Monoecious**

✓ **Pillow** disease of cucumber is due to: **Calcium** deficiency.

✓ Vector of cucumber mosaic virus is: **Aphid**

✓ Seed rate: **2.5 kg/ha**

Varieties:

- Pusa sanyog (gynoecious variety)
- Japanese long green
- Poinesett
- Balaam kheera
- Pusa barkha, Pusa Uday, Arka Veera

11. Water melon: *Citrullus lanatus*

- ✓ Water melon fruit contains: ***92-95 % water. (RAS pre-09)***
- ✓ Metallic sound at the time of harvesting shows: ***Immaturity.***
- ✓ Heavy **dull sound** at the time of harvesting shows: ***Maturity.***
- ✓ The drying of tendril at the base of the fruit is also a sign of maturity
- ✓ Bud necrosis in watermelon is due to: virus transmitted by aphids (*Aphis craccivora*)
- ✓ Seed rate: ***4-4.5 kg/ha.***

Varieties:

- Sugar baby (very sweet, 11-13% TSS)
- Durgapura meetha
- Durgapura kesar (yellow fleshed)
- Pusa rasal
- Asahi yamato
- Arka jyoti, Arka muthu
- Improved shipper: an american introduction

Hybrids:

- Arka Manik (Powdery mildew resistant)
- Pusa bedana (**seedless** hybrid, **triploid** variety).

12. Musk melon: *Cucumis melo*

- ✓ High temperature at the time of fruit maturity and ripening increases the sweetness.
- ✓ Musk melon is harvested at **full slip stage**, except **Hara madhu** at **netting**.
- ✓ Seed rate: **1.5-2.0 kg/ha**.
- ✓ Seed rate: 1 kg/ha by dibbling method

Varieties:

- Durgapura madhu: Early variety **(RPSC, AAO-09).**
- Hara madhu: Late variety
- Arka jeet: Excellent flavor, very sweet

- Pusa rasraj: Early variety
- Punjab sunhari:
- Arka rajhans: Excellent transport **quality**
- Other are- Pusa madhuras, Pusa madhurima, DMDR-2

13. Bottle gourd: *Lagenaria siceraria*

- Known as ***lauki***
- **Seed rate : 3-6 kg/ha**

Varieties:

- Pusa summer prolific long (PSPL)
- Pusa summer prolific round (PSPR)
- PBOG-1 (Pant sanker)
- Samrat (highest yield)
- Pusa manjari: grow in spring season
- Punjab komal
- Arka bahar
- Pusa meghdoot (F_1hybrid by public sector)

14. Pea: *Pisum sativum, Leguminacae*

- ✓ Garden pea is: ***Pisum sativum var. hortense.*** It is used in canning &as green vegetable, also known as table pea
- ✓ Field pea is***: Pisum sativum var. arbens.*** It is used as *Dal*, fodder, green manure etc.
- ✓ Pea is rich source of protein: **about 25%**(dried seeds) however, as tender pod contains 7.2 % protein
- ✓ State having maximum area and production of pea: ***UP***
- ✓ In India pea is the most popular canned vegetable
- ✓ Wrinkled seeded varieties are sweeter than smooth seeded.
- ✓ Maturity of pea is measured by: ***Tendrometer***
- ✓ Stem borer of pea: ***Euzophera particella.***
- ✓ Powdery mildew of pea: ***Erysiphe polygoni***.

Seed rate:

- **60-80 kg/ha**

Varieties:

- Azad P-1
- Jawahar mator-1 (susceptible to powdery mildew)
- Boneville (suitable for **dehydration,** introduced from USA)
- Arkel (sickle shaped pods, wrinkled seeds, dwarf variety)
- Sylvia (whole pod is edible)
- Arka Ajit (resistant to **powdery mildew**).

- Harbhajan (extra early variety)
- Hissar harit-1
- Aparna/HFP-4:- first dwarf variety (IARI, Ph.D-09)
- Type -163 (field pea variety)
- Rachna (powdery mildew resistant)
- Swarna rekha (early maturity)
- Meteor: smooth seeded, dwrf vaiety

15. Cow pea: *Vigna sinensis* (Lobia): Leguminacae

- ✓ Native place: **Africa**
- ✓ Seed rate for rainy season crop is: 12-15 kg/ha and for summer vegetable: 20-25 kg/ha
- ✓ For **fodder purpose: 34-40 kg/ha**
- ✓ Pest of cow pea is pod fly: ***Ophiomyia phaseoli***

Varieties:

- Pusa barsati
- Pusa rituraj
- Kansi gouri
- Pusa dofasali
- Pusa komal
- Pusa phalguni (dwarf variety)

17. Cluster bean: *Cyamopsis tetragonoloba (Leguminaceae)*

- Rajasthan is first in production and area of cluster bean
- It is used as cattle feeds and fodder, making gum, resins and explosive.
- But immature pods are also used as green vegetable.
- Seeds rate: 15-20kg/ha

Varieties:

- Pusa sadabahar: For vegetable purpose
- Pusa mausmi: For vegetable purpose
- Pusa navbahar:For vegetable purpose
- Durgapura safed: For grain purpose, early variety

16. Palak (Spinach): *Beta vulgaris* var *bengalensis*

- ✓ Seed rate: **20-25 kg/ha**

Varieties of palak:

- Punjab green
- All green
- Pusa harit
- Pusa jyoti
- Jobner green (mutant variety).
- Pusa bharti: Polyploidy variety)

UNIT-3
Floriculture

- ✓ Leading flower and flower product **exporting** country: ***Netherlands***
- ✓ Leading flower product **importing** country: ***Germany***.
- ✓ Country having largest market of **cut flowers**: ***Germany***.
- ✓ State having maximum loose flower production in India: ***Tamilnadu***.
- ✓ Largest importer of floriculture products from India: ***USA*** *(27%)*.
- ✓ Maximum cut flower production in India: ***West Bengal***
- ✓ Flower crop covering maximum area in India: ***Jasmine***
- ✓ **Hogarth course** is also known as**:** ***Line of beauty***.
- ✓ Book '**Beautiful Gardens**' is written by: ***M.S Randhawa***.
- ✓ **Dr. B.P. Pal** and **Priyadrashani** are varieties of: **Rose**
- ✓ Japanese flower arrangement is known as**:** ***Ikebana.*** **(RAS-06)**
- ✓ India is the largest producer of loose flower in the world

Famous gardens:

Garden	Place
▪ Mughal garden:	**Pinjore**, Haryana, built by Fadai khan
▪ Rose garden:	**Chandigarh**
▪ Mandhor garden:	**Jodhpur**.
▪ Charbag garden:	New Delhi (at Humau tomb)
▪ Buddha Jayanti Park:	**New Delhi**
▪ Brindavan garden:	**Mysore** (Biggest formal garden)
▪ National botanical garden:	**Lukhnow**

Style of gardening:

1. **Formal style:**
   - ✓ Plan is symmetrical
   - ✓ E.g. **Mughal gardens**, **Persian gardens, Pinjore garden**
2. **Informal style:**
   - ✓ Plan is **asymmetrical**
   - ✓ It reflects naturalistic effect of total view and represents natural beauty
   - ✓ E.g. **Japanese garden**
3. **Free style:**
   - ✓ Combination of both formal and informal style.
   - ✓ E.g. **Rose garden** of Ludhiana/Chandigarh**, English** garden.

Main features of garden

1. **Mughal garden**: (formal)
   - ✓ **Terraces**
   - ✓ Running water (life of garden)
   - ✓ **Baradari**
   - ✓ High protecting wall
   - ✓ **Terminal building**
2. **Japanese garden:** features
   - ✓ **Garden lanterns**
   - ✓ Garden pagoda
   - ✓ **Dry landscape**
   - ✓ Wells
   - ✓ E.g. **Buddha jayanti park, New Delhi**
3. **English garden:**
   - ✓ Herbaceous border
   - ✓ Cottage garden
   - ✓ **Lawn**
   - ✓ **Rockery (RAS-09)**
   - ✓ **E.g. Rastrapati garden, Tea garden**

Lawn:

- ✓ Regarded as **heart of garden**
- ✓ No garden is complete without lawn
- ✓ Planting of grass in lawn by **turf method** is best and quickest method for planting of lawn and in this method use turf-piece of earth with grasses.

Flower crops:

1. Rose: *Rosa spp.*(Rosaceae)

- ✓ Origin place of rose: **India.**
- ✓ **Rose known as:** **Queen** of flowers
- ✓ Rose is: **Symbol of beauty**
- ✓ National flower of: **England**
- ✓ Best time of planting: **September-October**.
- ✓ Commercial method of propagation: **T-budding.**
- ✓ Best time of budding: **Nov. to Feb.**
- ✓ Fruit of rose: **Rose hips.**
- ✓ **Dr. BP Pal**is related to: **Rose breeding.**
- ✓ **Gulkand** is prepared by mixing: **Petals** and **sugar** in 1:1 ratio **(RAS-09).**
- ✓ **White** flower variety of rose is**:** **Tushar. (RAS Exam-09)**

- ✓ **Red** flower variety of rose is: **Crimson Glory.**
- ✓ **Thorn less**variety of rose: **Pusa mohit**, **Suchitra**
- ✓ **New varieties: Pusa alpana, Pusa mahak**
- ✓ **For protected cultivation:** Taj mahal, First red
- ✓ To increase shelf-life of flower: **Pulsing treatment is done.**

2. Gladiolus: *Gladiolus hybridus*

- ✓ Family: **Iridaceae**
- ✓ Propagation by: **Corms or tissue culture**

Varieties:

- Priyadrashani Happy end
- Suchitra Friendship

New released varieties:

- Arka Ranjini Arka Manorama
- Pusa Sindoori

3. Chrysanthemum: *Dendranthema gradiflora* (Compositeae/asteraceae)

- ✓ Common name **'Guldaudi'**
- ✓ **Origin: China.**
- ✓ **Propagation by:** root **suckers**.

Varieties:

- **Diana** Basanti
- Rakhi Birbal sahani
- Indira Red gold

New released varieties:

- Pusa Anmol Pusa Guldasta
- Pusa Shatabdi Kirti
- Punjab Shingar Punjab Mohini

4. Mary gold: *Tagetes sp.*(Compositeae or asteracae)

Two types:

1. African, large flower gainda: ***Tagetus erecta***
   - **Varieties**: Climax, Glitters, Snowbird, Sugar, Beauty gold, Pusa bahar, tannis ball, Pusa Deep, Arka Shubha, Buddy purple.
2. French, small flower: ***Tagetus patula***
   - **Varieties**: star of India, Lemon drops, Arka bhanu, Pusa arpita, Pusa deep, Happy orange, Golden boy, Star dust.
   - **Jafrine** alkaloid present in French marigold (*T. Patula*) which is used as anti-inflammatory, antimicrobial effects, insecticidal activity and in various medicines.

Seed rate: **1.0-1.5 kg/ha.**

Fruit preservation

- The **art** and **science** of keeping fruits for longer time without **detoriation** in quality.
- In fruit preservation sodium benzoate works on the principle: **To check the growth of yeast,** whereas KMS/SO_2 works on the principle: **To check growth of bacteria & fungi.**

Principle of preservation:-

1. **Canning**:
   - **Fruits** can be processed at a temperature of **100 ºC**.
   - **Vegetable** can be processed at temperature of **115-121ºC**.
2. **Pasteurization**:
   - Heating of fruits and vegetable juice at **85-90 ºC** for **30** minutes.
   - It kills **only harmful** microbes.
3. **Sterilization:**
   - **Heating** of fruits and vegetables above **100 ºC**
   - It kills both **beneficial** and **harmful** microbes.
4. **Freezing:**
   - Cooler storage: 15 ºC
   - Refrigeration or chilling: 0-5 ºC
5. **Cryo preservation**: Preserve in liquid nitrogen at **-196** ºC.
6. **Drying**: Removal of **moisture** by applying **heat** e.g. **Raisins** (*Kismis*).
7. **Preservation through osmosis:**
   - High concentration of **sugar: e.g. Jam** (68 % sugar), **Agra ka petha**.
   - **Salt preservation**: Salt conc. **15-25** % act as preservatives in **pickles.**
8. **Preservation by chemicals**:

a. **KMS** (Potassium Meta bi Sulphite): @ 350-500 ppm or **0.05**%.
   - **Sulphur di oxide** (SO_2) is responsible for preservation.
   - It used against **colorless** fruit juices/pulp.

b. **Sodium benzoate:** @ 700 ppm/**0.07**%.
   - **Benzoic** acid is responsible for preservation.
   - Used in **colored fruit product** (e.g. Tomato products).

9. **Fermentation**: grape wine (**alcohol 7-20**%) is oldest example of fermented beverage.
10. **Asepsis**: prevent entry of microbes.
11. Oxidation can be checked by **antioxidants** (Ascorbic acid /vit.C)
12. Checking of **enzymatic** spoiling: **Blanching** in vegetables
13. **Blanching:** keeping vegetables in boiling water for 2-5 minutes followed by cooling.

Difference between **Jam** and **Jelly**:

SN	Jam	Jelly
1	**Ripe** and **thick** flaccid fruits are **used.**	Used firmed and ¾th mature fruits
2	Pectin is **not** important than acid	**Pectin is** necessary (0.5-1.0%) Pectin tested by: Alcohol, Jelmeter test and spectrophotometry or colorimetric
3	It is opaque	This is transparent
4	**Sugar** is added according to **acid content** (68.5% sugar)	**Sugar** is added according to **pectin content** (generally 65-70%)
5	Prepared from **mixing** of fruits e.g. apple, papaya, tomato, strawberry	Fruit don't mixed E.g. **Guava, apple**
6	Temperature at end point: 219-221°F	End point temp. 221-222°F or 105.5°C
	TSS: **68.5%**	**TSS: > 65%**
7	Acid should be: 0.5-0.6%	Fruit acid should be **:1.0%** **For judging the End Point of jelly:** The **'Sheet Test'** is commonly used

Difference between **squash** and **cordial:**

SN	Squash	Cordial
1	**Strained juice** containing moderate quantity of **fruit pulp** with concentrate **sugar** and **water**	Cordial is made by Clear sweetest juice of **fruit pulp**
2	Contains 25 % juice and 40 % TSS. E.g. Mango, pine apple, lemon squash.	It contains **25%** fruit juice

Difference between **Squash** and **Sarbat**:

SN	Squash	Sarbat
1	It contains **fruit juice**, sugar and water	Clear **sugar syrup with artificial flavor**
2	**TSS:** 40%	**TSS:** >70%
3	E.g. Mango, pine apple Squash	E.g. Almond, Khus-khus Sarbat

Difference between **preserve** and **candy:**

SN	Preserve	Candy
1	Preserve made from **natural fruits** by **cocking** whole or cut.	A fruit cover with **sugar** and **glucose,** subsequently grind and dried.
2	In heavy sugar syrup it becomes tender transparent e.g. ***Aonla murrabba.***	E.g. ***Agra ka petha.***

Common Jelly Preparation Disorders-

1. **Failure to Set (Runny Jelly):** This occurs when cooking is stopped **before** reaching the jellying point (65% Total Soluble Solids).
2. **Tough or Stiff Jelly:** Results from **overcooking** the mixture beyond the setting point or using insufficient sugar, which requires prolonged boiling to thicken the juice.
3. **Crystallization:** Often caused by cooking the mixture **too slowly or for too long**, which leads to over-concentration of sugar. It can also occur if undissolved sugar crystals on the sides of the pan fall back into the finished product.
4. **Syneresis (Weeping):** The spontaneous release of liquid from the gel, typically caused by **excess acid** which breaks down the pectin structure over time, or storage in a place that is too warm.
5. **Cloudy or Foggy Jelly:** This can be caused by **over-cooling** the jelly before pouring it into jars, or by pouring it too slowly. Other non-time related factors include imperfect straining of juice or using immature (starchy) fruit.
6. **Darker Color:** Primarily caused by **long, slow boiling** of sugar and juice, which leads to caramelization and loss of fresh fruit flavor.

Terminology:

Bonsai:

- Japanese art of growing miniature trees and shrubs by extreme dwarfing. Best example is *Ficus sp.*

Topiary:

- It is an **art** of **training** plants in to **different shape** i.e. shape of animals, birds etc. Plants used in topiary are evergreen. E.g. ***Duranta***, ***Murrya panniculata***, *Clerodendron innerme* (the best one).

Edge:

- When low growing perennial, annual plants are grown on the border of plots, beds are called edge. **E.g. Alternanthera**, **Sunrose**

Hedges:

- When shrubs are planted on boundary for fencing. It may be **ornamental** or **protective.** Karonda, **Bougainvillea**, **Mehndi**, Hibiscus, **Clerodendron** and **Lantana** are the best example of **hedge**.

5. ENTOMOLOGY

Important points:

- ✓ NPV is mostly used to control: ***Lepidopteran larvae***
- ✓ Bud necrosis in groundnut is transmitted by: ***Thrips***
- ✓ **Binomial nomenclature** first given by: *C.* ***Linnaeus*** (1758)
- ✓ The scientific name of maize stem borer is: ***Chilo partellus***
- ✓ Termite control can also be done by : ***Frequent Irrigation***
- ✓ The chief excretory organ in insect is: ***Malpighian tubules (MT)***
- ✓ **Juvenile hormone (JH)** is secreted by: ***Corpora allata (CA) gland***
- ✓ **Tree banding** is useful for control of: ***Mango mealy bug.*** **(NET-2013)**
- ✓ The **main characteristic** feature of an insect is: ***Three pairs of legs***
- ✓ Which is the best suitable species for bee culture: ***Apis mellifera***
- ✓ In lac production which country has monopoly: ***India***
- ✓ What is optimum plant population for Bt – cotton: **10,000 plants/ha.**
- ✓ Largest size of cultivated honey bee is: ***Apis dorsata***
- ✓ **Bromadialone** (0.005%) is an effetive: Single dose anticoagulant ***Rodenticide***
- ✓ Which bacteria is produced at commerical level for insect control: ***Bacillus thuringiensis*** **(Bt.)**
- ✓ House fly spreads/transmits at least 65 diseases to humans, including ***typhoid fever***, ***dysentery***, ***cholera***, ***poliomyelitis***, ***anthrax***, ***leprosy*** and ***tuberculosis***.
- ✓ The safe insecticide for vegetable is: ***Malathion***
- ✓ The most widely used insect pheromone is: ***Female sex pheromone***
- ✓ Major pest of cucurbits is: ***Red Pumpkin beetle***
- ✓ Diamond back moth (DBM) is a pest of: ***Crucifers vegetables (specially cabbage)***
- ✓ The bollgard is: ***Transgenic cotton (Bt. Cotton)***
- ✓ The most suitable fumigant for quarantine purpose is: ***Methyle bromide***
- ✓ The most effective moulting hormone (MH) in insects is: ***Ecdysone***
- ✓ Moulting hormone (ecdysone) is secrted by: ***Prothoracic glands***
- ✓ The major excretory product of terrestial insects is:***Uric acid***
- ✓ Nosema disease of honeybees is caused by a protozoa: ***Nosema apis***
- ✓ **'Silver shoot and onion leaf'** like symptomsin rice is caused by gall midge: ***Orseolia oryzae***
- ✓ Bud necrosis in watermelon and muskmelon is transmitted by: ***Aphids***
- ✓ The storage insect which can be survive at higher temperature and low humidity is: ***Khapra beetle (Trogoderma granarium)***
- ✓ Protein hydrolysate is used as an attractant for the control of: ***Fruit fly***

- ✓ ***Aphelinus mali*** has been a successful parasitoid incontrolling: ***Apple wooly aphis*** **(NET-2013)**
- ✓ ***Rodolia cardinalis*** has been successful predator in controlling: ***Cottony cushion scale*** *(Icerya purchasi)*
- ✓ The infestation of which insects starts from the field: ***Angoumois grain moth, potato tuber moth, pulse beetle and sweet potato beetle***
- ✓ If a pest is confined to a particular area and occurs regularly, its infestation is termed as: ***Endemic***
- ✓ The toxin produced by ***Bacillus thuringiensis*** interferes with the insect's: ***Digestive system***
- ✓ After ingestion of both Bt. and NPV in insect body act as: ***Stomach poison***
- ✓ While fungi and nematodes (EPN) act as***: Contact poison***
- ✓ Phase theory of locust given by: Uvarov, 1921 **(Pre PG, Exam 2016)**
- ✓ The headquarters of Directorate of Plant Protection, Quarantine and Storage (DPPQ&S) is located at: ***Faridabad, Haryana (1946)***
- ✓ ***Epiricania melanoleuca*** is a lepidopteron parasitoid effective against: ***Sugarcane pyrilla (Pyrilla perpusilla)***
- ✓ Enforcement of laws for the control of insect pests in India is governed by: ***Destructive Insects and Pests Act (1914)***
- ✓ Rice stem borer (*Scirpophaga incertulus*) is absolute or **monophagus** pest of rice
- ✓ Vector of rice 'Tungro' viral disease is Green Leaf Hopper (***Nephotettix virscence***)
- ✓ Vector of rice **'Grassy stunt'** disease is Brown Plant Hopper (***Nilaparvata lugens***).
- ✓ **Chaffy grains** with **black spot** in rice are due to Gundhi bug (***Leptocorisa acuta***)
- ✓ In rice, gundhi bug causes damage on the stage: ***Milky stage***
- ✓ Serious pest of wheat is: ***Termite***
- ✓ Wheat stem borer (***Sesamia inferens***) attack in: ***Night***
- ✓ Specific pest of wheat nursery is: Ghujhia weevil (***Tenymecus indicus***)
- ✓ Control of tundu disease: ***Hot water treatment 50 ºC for 2 hrs***.
- ✓ 'Hopper burn' symptoms in rice due to: *BPH* (***Nilaparvata lugens***)
- ✓ Vector of leaf curl viral diseases in plants is: *White fly* ***(Bemisia tabaci)***
- ✓ Flaring of squares in cotton is due to: *Spotted bollworm* (***Earias vitella***)
- ✓ The formation of rosette flower in cotton is due to: *Pink boll worm* (***Pectinophora gossypiella***)

- ✓ Double seed formation in cotton bolls is due to: pink boll worm (***Pectinophora gossypiella***).
- ✓ Large circular bore holes with faecal pellets on bolls is the symptoms of: American boll worm (***Helicoverpa armigera***)
- ✓ Pest population should be kept below: Economic Threshold Level ***(ETL)***
- ✓ The appropriate time for insecticides application to control damaging pests: ***ETL.*** **(NET-2013)**
- ✓ The pest which attacks all the parts of the plant is: ***Termite***.
- ✓ White grubs prefer to lay eggs on: ***Sandy or sandy loam soils***.
- ✓ Pest which has a status of 'International pest' is: *Desert locust* ***(Schistacerca gregarea)***
- ✓ Recently method of control of bollworm is: ***Bt. Transgenic plants***.
- ✓ Number of eggs /second lay by termite: ***One egg/sec.***
- ✓ The insects are ***poikilothermic*** i.e. they have no precise mechanism for regulating the body temperature so also called ***cold blooded*** organisms.
- ✓ The total heat required for the completion of physiological processes in the life-history of a species is considered as constant a ***thermal constant (ARS-04)***.
- ✓ Control of weeds by biological agent known as: ***Parabiological*** *control.* **(IARI, Ph.D-09)**
- ✓ **IIBC**: International Institute of Biological Control in West Indies, 1927
- ✓ **NCIPM**: National Centre of Integrated Pest Management, New Delhi, 1988
- ✓ **PDBC**: Project Directorate of Biological Control, Bangalorc, 1993 (PDBC renamed as **NBAII:** National Bureau of Agricultural Important Insects, in 2009)
- ✓ **IOBC**: International Organization for Biological Control: *Zuriec,* ***Switerzerland***.
- ✓ ***Cryptobiosis:*** a phenomenon when insect become quiescent due to **adverse climatic** condition and show **no visible** sign of metabolic activity
- ✓ ***Limnetic zone***: The open water zone to the depth of effective light penetration.
- ✓ Wax moth (***Galleria melonella)*** most serious pest of **apiculture (JRF-03)**
- ✓ **Curbicula**: in hind legs of worker honey bees a pollen collecting structure is found on tibia which is known as curbicula orpollen basket.

The word entomology derived from **Greek** words:

- ***Entomon*** = Insect
- ***Logos*** = Discourse

Insect Morphology:-

Definition: The branch of science which deals with study of structure and function of insect is known as insect morphology.

Body wall **(Integument)** of Insect:-

- It is ***chitinous exoskeleton***
- It is the outer most covering of insect and it has one of most distinctive feature is to serve as exoskeleton as a frame work for attachment of muscles and barrier for retarding loss of water from outer surface of insect body.

The structure of integument: -

- It is divided in to three layers i.e. cuticle, epidermis (hypodermis) and basement membrane.

1. **Cuticle**:
   - It is outer most layer of integument
   - It is made up of **chitin** (25-60%) and **protein** (25-40%)
   - It is non-cellular, non-living layer and secreted by hypodermis
   - Cuticle is further divided in to ***epicuticle*** and ***procuticle***
     (a) The epicuticle is made of **non-chitinous protein** layer
     (b) The procuticle is having **chitin** and **protein** layer
   - **Chitin**: It is nitrogenous polysaccharides (carbohydrates) consist of several chains of acetyl glucosamine and glucosamine residue.
   - **Arthropodin**: The main structural proteins of insect are known as ***arthropodin.*** It is untanned protein and soluble in water.
   - **Sclerotin:** It is tanned protein that is water insoluble.
   - **Resilin:** Elastic protein responsible for the flexibility of wings sclerites.
2. **Epidermis (Hypodermis):**
   - It is second layer of body wall beneath cuticle
   - It is unicellular & active layer of body wall
   - The main function of epidermis is to secrete cuticle
3. **Basement membrane:**
   - It is innermost non-cellular layer of insect integument to which muscles are attached.

Moulting:

- The casting (shedding) of old cuticle is known ***moulting.***
- It is necessary to growth of insect.
- It has to two process the ***apolysis*** and ***ecdysis.***

- *Apolysis:* It is beginning process with separation of old cuticle from underlying epidermal cells.
- *Ecdysis:* It is ending process of moulting with separation of old remain cuticle.

The insect body is divided in to three segments:

1. **Head 2. Thorax and 3. Abdomen**

1. **The insect head:**
   - On the basis of classical theory (embryonic studies) insect head is made from **six segments**.
   - Insect head bears one pair of antenna, a pair of compound eyes and mouth parts.
2. **Thorax**:-
   - It consists of 3 segments i.e. prothorax, mesothorax and metathorax, usually two pair of wings are present on meso and metathorax which is collectively called **'pterothorax**. Each segment of thorax carries one pair of legs hence; three pairs of legs are present.
3. **Abdomen of insects**:-
   - It has 10-11 rings like segments except collembolla which has 6 abdominal segments, while protura has 12 segments abdomen.
   - Abdominal segments of pterygote insects are divided in to pregenital segments (first 1-7 segments).
   - The 8th and 9th segements are regarded as genital segments because on 9th abdominal segment, the male genitalia is found while 8th and 9th segments are collectively form female genitalia e.g. in grasshopper.
   - Remaining 10th and 11th segments have post genital appendages like **cerci** (a pair of sensory organ located on tail region of insects) and dorsally **one epiproct** and laterally **two paraproct** (sclerotized plate like structure of ovipositor is found on 10 or 11th abdominal segment of female grasshoper).
   - The larvae of pterygote insecs have locomotory structure which is known as **prolegs**. In lepidopteran larvae 1 to 5 pairs of prologs (with crochets) and in mustard saw fly 8 pairs of **prolegs** (without crochets) are present on abdomen.
   - In aphids, on dorsal side of 5th or 6th abdominal segment have **cornicles** which release **honey dew substances**.

The appendages of insect head:

1. The insect antenna:-

- One pair of antenna is present in all insects, except **protura**
- Antenna is consists of three parts i.e. scape (basal part), pedicel and flagella.
- Major functions of antenna is act as organ of **smell, taste** and in some insects **hearing** (tympanal/Jhonston's organ) and **sound producing** (chordotonal organ) are found in second segment of antenna.

The insect antenna:-

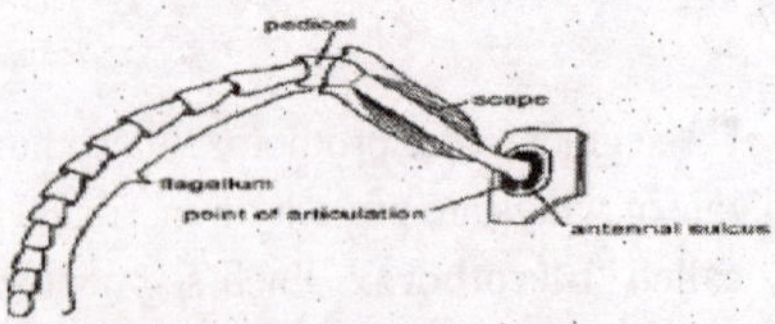

The structure of typical insect antenna

Types of insect antenna:-

SN	Type	Features	Example
1	**Setaceous**	Each segments from base to apex becomes smaller and tapering in to a point	Cockroach, leaf hoppers
2	**Filiform**	Thread like antenna in which all segments of antenna are uniform in thickness	Grasshopper
3	**Moniliform**	All segments are globular in shape and form a prominent construction between the joints	Termites
4	**Serrate**	In this type antenna form saw teeth like projections on side of each segments	Pulse beetle
5	**Pectinate**	In which each segment of antenna bears long comb like projection in one direction	Saw fly
6	**Bipectinate**	Each segment of antenna has long projection on both side	Silk worm moth
7	**Plumose**	The whorls of hairs arises from the joints of each segments which looks like ***plume***	Male mosquito
8	**Whorled**	The whorl of bristles arises at every joint of segments	Mango mealy bug
9	**Clavate**	Terminal segments of antenna form club shaped structure	Butter flies, blister beetle
10	**Capitate**	The terminal segment of antenna form a long knob or cap	Khapra beetle
11	**Geniculate**	The first segment is long and apex form with small segments	Bees, ants, weevils

12	Aristate	The first segment is smaller and broad than second while flagellum are more longer and bear heavy bristles	House fly

Mouth parts of insects:

S N	Types of Mouth parts	Main organs	Food chanal formed by	Examples
1	Biting and Chewing type (These are typical type of mouth parts)	They consist of: (a) **Labrum** (upper lip) (b) **Mandible** (one pair) (c) **Maxillae** (parts-cardo, stipes, galea, lacinia and maxillary palpi (d) **Labium** (lower lip) divided in to submentum, mentum and prementum (e) **Hypopharynx** (tongue like, formed salivarium) (f) **Epipharynx** (taste bunds are located)	-	This types of mouth parts are found in orthoptera, isoptera, and coleopteran e.g. Grasshoppers, locust, crickets, termites, beetles and slightly modified in larval lepidoptera
2	Piercing and sucking type	These are two types- (a) **Bug type**: They suck the cell sap from plant (found 4 stylets: 2 maxillae + 2 mandibles) (b) **Mosquito type**: suck blood from animals (bears 6 stylets, additional two are labrum-epipharynx and hypopharynx)	Food chanal is formed by maxillae Food chanal is formed by labrum-epipharynx with the help of hypopharynx	Found in bugs, aphids, jassids, whiteflies, and leafhoppers (hemiptera) e.g. Mosquito and fleas
3	Sponging type	Mandibles are absent Labium is long proboscis having three parts (rostrum, haustellum and labellum) distal end form '***pseudotracheae***'.	-	House fly
4	Siphoning type	Maxillae modified into ***galeae*** Labrum and mandibles both are rudimentary	-	Butterfly, moths
5	Chewing and lapping type	-	-	Bees and wasp (adult)
6	Rasping and sucking type	Only 3 stylets are present (2 maxillae and 1 left mandible, while right mandible is absent which known as asymmetrical mouth parts)	-	Thrips

The thoracic appendages of insect:

- Thorax carries usually two pairs of wings and three pairs of legs.

The insect leg:

- The insect leg has five segments -coxa, trochanter, femur, tibia, and tarsus (claws).
- In worker honey bees the tibia of hind legs has pollen baskets is called '**corbicula**' that carries pollens.
- In grasshoppers and head louse between the claws, a lobe like structure is known as '**aroleum**'.
- In house fly paired adhesive pads located on the ventral sode of tarsus, between the claws which is known as pulvilli.

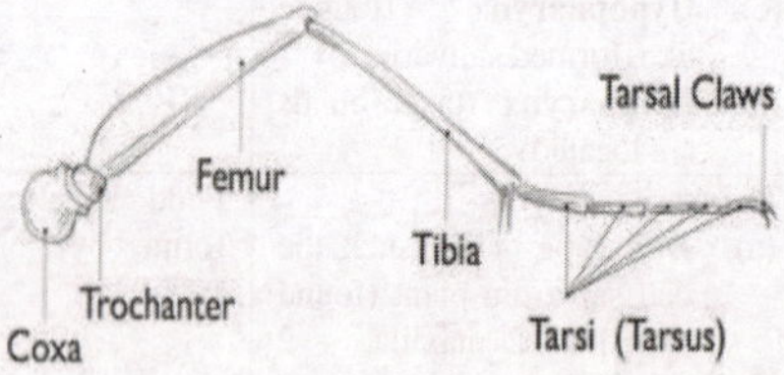

The structure of typical insect leg

Modification of insect legs:-

SN	Modification of legs	Purpose	Example
1	Ambulatorial	Walking	Cockroach
2	Saltorial	Jumping	Hind legs of grasshopper
3	Scansorial	Clinging	Head louse
4	Cursorial	Running	Ants and wasp
5	Stridulatorial	Sound production	Crickets
6	Fossorial	Digging	Fore legs of mole crrickets
7	Natatorial	Swimming	Hind leg of water bug
8	Raptatorial	Prying	Fore legs of mantis
9	Foragial	Pollen collection	Hind legs of worker bee

The insect wings:

- Pterygotes insect posses usually two pair of wings on meso and metathorax which is collectively known as **pterothorax.**
- The wings probably arose as lateral expansion of thoracic body wall.
- A typical insect wing is traigular with three margins and three angles
- Arrangement of veins on wings is called **wing venation.**
- Wings have two types of veins longitudinal and cross veins.

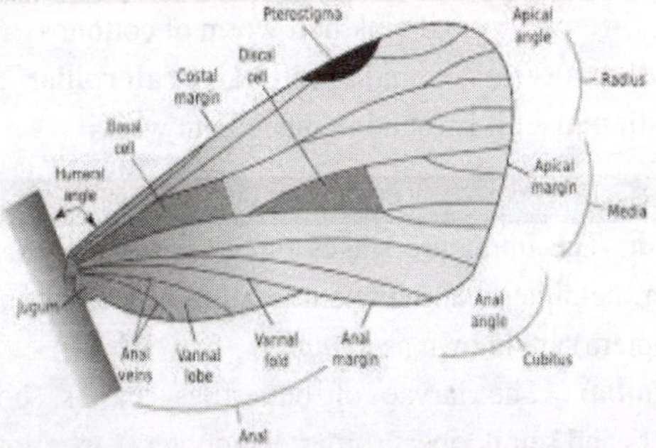

Hypothetical wing venation by Comstok and Needham

Chapter-6. Metamorphosis and diapauses of insects:-

Metamorphosis:

- Series of changes that takes place during the development of insects from eggs to adult are collectively known as metamorphosis.

Types of metamorphosis:

1. **Ametamorphosis**: insects do not undergo any metamorphosis e.g. apterygotes.
2. **Incomplete or direct metamorphosis**: it is also known as simple or hemimetabolous metamorphosis. In this type life cycle has three stages i.e. eggs, nymphs and adult and insects wings develop externally so these are called **'exopterogota'** e.g. Hemiptera (grasshopper), Isoptera, Orthoptera and thysanoptera (it is link between indirect and direct metamorphosis because in some families of thysanoptera pupa may or may not be present).
3. **Complete of indirect metamorphosis**: this is also known as holometamorphosis because it has four stages i.e. egg, larva, pupa and adult in their life cycle. Larvae have both thoracic and abdominal legs which is known as prologs except diptera (in dipteral larva is legless or apodous and known as maggots). In these insects wings develop internally during pupal stage and are called **endopterygota** e.g. Lepidoptera, coleoptera, hymenoptera, diptera, neuropteran, strepsiptera, trichoptera etc.,
4. **Hypermetamorphosis**: some insects have two or more types of larvae in the life cycle e.g. some coleopteran insects (blister beetle).

Insect diapause:

- In insects some stages undergo to diapause or dormancy by arresting growth or development during the physiological process of life cycle to avaoid adverse environmental condtions. It may occur in any stage of life cycle.

(a) **Egg diapause:** it is found in mango mealy bug and silk moth

(b) **Larval diapause:** it is found in maize stem borer, rice stem borer and pink boll worm of cotton.

(c) **Pupal diapause:** it is found in **red hairy caterpillar**

(d) **Adult diapause:** it is found in mango nut weevil

Types of larva:

a. **Nymph**: The immature stages of bugs, white fly, aphids and jassids (order: hemiptera) and immature stages of grasshopper, locusts (order: orthoptera) are known as ***nymphs***.

b. **Caterpillar**: The larvae of butterflies, moths, bollworms, borers, loopers and semiloopers (order: lepidoptera) are known as ***caterpillar***.

c. **Grub**: The larvae beetles, weevils (order: coleoptera) and larvae of bees, wasp, braconids (order: hymenoptera) are known as ***grub***

d. **Maggot**: The larvae of house flies, shoot flies, midges (order: diptera) are legless and known as ***maggots***.

Insect physiology:

- It includes physiology and *biochemistry* of insect organ systems
- Father of insect physiology: ***V. B. Wigglesworth***
- The book **'Principles of Insect Physiology'** is written by: V. B. Wigglesworth and he worked on **kissing bug, *Rhodnius prolixus*.**

1. Digestive system of insect:

Alimentary canal:

- It is a simple hollow tube like structure which extends from the buccal cavity (mouth parts) to annus.
- It is divided into three regions:

1. Fore gut (stomodaeum):

- It is anterior region of alimentary canal and ***ectodermal* in origin**
- It is primary organ of ingestion and storage of food
- It consists following parts:

(a) **Pharynx**:

- It is the region between mouth and oesophagus.
- It is form well developed sucking pump in fluid feeders

(b) **Oesophagus**:

- It is second part of fore gut and undifferentiated tube just after pharynx
- Its function to receive food from pharynx and pass to crop

(c) **Crop**:

- It is bag like structure and act as storage of food

- In honey bees crop is known as ***honey stomach*** (where nectar conversion occurs (honey formation takes place)

(d) Gizzard (pro-ventriculus):

- It is posterior part of fore gut joined to mid gut
- Function as grind the food and pass to mid gut

1. Mid gut/stomach/ventriculus:

- It is ***endodermal*** in origin
- It'sprimarily concerned with secretion of enzymes and digestion and absorption of food.
- It extends from end of gizzard (hepatic caeca) to anterior part of hind gut (Malpighian tubules).
- In mid gut ***intima*** i.e. cuticular linings are absent and form peritrophic membrane to protect mid gut epithelium from rough food particles especially in solid feeders.

2. **Hind gut/Proctodeum**:

- It is ***ectodermal*** in origin
- Extends from posterior end of mid gut to anus
- It differentiates in to three regions, Ileum, Colon and posterior part near to anus is rectum.

Peritrophic membrane:

- Fore gut and hind gut have cuticular linings (initima) in the epithelial layer while mid gut is being endodermal in origin do not have these cuticular linings and a thin membrane is formed to protect from abrasions of rough food particles which is known as peritrophic membrane.
- It is present in solid feeder i.e. larva of lepidoptera, gurb and adults of coleopteran insects.
- Peritrophic membrane is absent in fluid (liquid) feeders insects i.e. sucking insects.

Filter chamber:

- In fluid feeder (most homopteran) due to unusual modification of alimentary canal an organ is developed known as **filter chamber** which allow elimination of excess water and soluble carbohydrates to prevent excess dilution of haemolymph.

2. The respiratory system:

- The lungs are absent in insects therefore the respiration is carried out by a number of hollow tubes (air sacs) found in thoracic and abdominal segmentsof insect which is known ***tracheae,*** and are floating in **haemoceal** and the fine branches of tracheae are known as ***tracheoles***.

- The spiral thickening of trachea is known as ***taenidia,*** it prevents trachea to collapse during respiration but taenidia is absent in tracheoles.
- The air generally enter into trachea through paired lateral outer opening known as ***spiracles*** which are situated in the plural side of thoracic and abdominal segments.

On the basis of presence of functional spiracles respiratory system classified as following types:

1. **Holopneustic type**: Where ten pair of functional spiracles (2 thoracic + 8 abdominal) are present e.g.**Grasshopper and cockroach (ARS Exam 07).**
2. **Peripneustic type**: where at least nine (1+ 8) pairs of functional spiracles are present e.g. lepidopteron caterpillars.
3. **Hemipneustic**: at least 8 pairs of functional spiracles are present
4. **Apmhipneustic**: one mesothoracic and one abdominal spiracle is functional e.g. maggots.
5. **Propneustic**: only one thoracic spiracle is functional
6. **Metapneustic**: only one abdominal spiracle is functional
7. **Aneustic**: no single functional spiracle is present e.g. *Chironomid larvae*.

3. The circulatory system:

- In insects the **open circulatory system** is found while in mammals found closed circulatory system.
- The insect blood is known as **haemolymph** which is colorless fluid having plasma and blood cells i.e. **haemocytes**
- The body cavity that contains circulatory fluid i.e. haemolymph is known as ***haemocoel.***
- Basic functions of haemocytes are generally same as the functions of ***RBS*** and ***WBC*** in mammalian blood.
- The principle function of haemocytes is ***phagocytosis*** (ingestion of harmful foreign materials)
- The major blood sugar in insects is a disaccharide i.e. ***trehalose***.
- In insect circulatory system the main pump is pulsatile dorsal vessel which is anterior part called as ***aorta*** and posteriorly ***hearts.***

4. Excretory system:

Excretion:

- The process of elimination of nitrogenous waste (harmful) products from insect body is known as excretion. It maintains internal environment of insect body and ionic balance of haemolymph.
- In terrestrial insects main excretory product is **uric acid**
- While in aquatic insects main excretory product is **ammonia**

- And in mammals main excretory product is **urea**

The main excretory organs:

1. **Malpighian tubules (MT):** It is principle organ of excretion in insects.
2. Integument: (through molting)
3. Tracheal system: (it eliminates CO_2)
4. Wall of alimentary canal:

Other organs of excretion:

1. Nephrocytes:
2. Fat bodies:
3. Oenocytes:
4. Cephalic gland:

Malpighian tubules (MT):

- In insects it is **chief organ** of excretion
- It is present in all insects except **collembolan** and **aphids**
- The distal end of MT are closely joined to rectal glands of hind gut which is known as **cryptonephridial condition** found in larvae of Lepidoptera and larva & adult of coleopteran to enable to conserve moisture from rectum.
- The primitive number of malpighian tubules is **six**
- In Lepidoptera, MT are found only six while in isopteran 2-8, in orthoptera 2-200 in number and in collembolan and aphids these are absent.

5. Endocrine system of insects:

- It consists of ductless gland that secretes chemicals called **hormones,** it released directly into haemolymph of insects.

Main organs:

1. **Neuro secretory cells** (NSC):
    - They release **brain hormone** which is known as activation hormone
2. **Corpora cardiaca** (CC):
    - It is small paired structure known as neurohaemal organ
    - Major function is storage and release of hormones
3. **Copora allata** (CA):
    - Paired glands located behind corpora cardiaca
    - It produces **juvenile hormones** (JH) or neotenin
4. **Prothoracic glands** (PTG):
    - It is a paired organ near prothorax
    - It produces **ecdysone**, a steroid called as moulting hormone.

Insect growth regulators (IGR) in pest management:

- **IGR** refers to as chemicals (natural & synthetic) that regulate growth in insects are known as insect growth regulators. They are two types-

1. **Insect Hormones**:

(a) Brain hormone:

- It secreted by neuro-secretory cells (NSC) and liberated in to haemolymph by corpora cardiaca
- It is known as activation hormone

(b) Juvenile hormone (JH):

- It is terpens in nature and secreted by **corpora allata**
- It inhibits metamorphosis and keep larvae in to juvenile form
- Prof. William in 1967 gave the term **'third generation of pesticide'** to these chemicals.
- Idea of JH insecticides given by Williams in 1956
- When JH is present in high concentration the new cuticle are larvae.
- When JH present in low concentration the new cuticle is forms pupa while adult cuticle is formed in absence of JH
- Example of JH: **Methoprene** (Altosid) is the first IGR (first bio rational) insecticide
- Other are: **Fenoxycarb** (Logic), **Pyriproxyfen** (Admiral)

(c) Anti-juvenile hormone:

- They inhibit function of corpora allata and biosynthesis of JH e.g. **percocens**

(d) Moulting hormone (ecdysone):

- It is a steroid produced by prothoracic gland (PTG)
- Moulting of insects is influenced by presence of moulting hormone i.e. ecdysone
- In insect cholesterol acts as the precursor for biosynthesis of **ecdysone**

2. **Chitin synthesis inhibitors (CSI):**

- They inhibit chitin (cuticle) synthesis and deposition
- Common examples are as:
- **Buprofezin** (Applaud): used against homopteran insects
- **Teflubenzuron** (Nomalt): used against coleopteran and lepidopteron pests
- **Lufenuron** (Match): used against thrips.

Use of semiochemicals in pest management:

Semiochemicals:

- These are chemicals that convey message between organisms or natural chemicals released by an organism that affect the behaviors of other individuals.
- These are two types (i) pheromones and (ii) allelochemicals
- Pheromones are **Intraspecific** chemicals that secreted by an organism to environment and cause specific reaction to recipients of the same species.
- **Interspecific** signals are known as allomones and kairomones (allelochemicals).

1. Pheromones: (intraspecific)

- The term pheromone is coined by **Karlson** and **Luscher** in **1959**
- Pheromones are ***intraspecific*** signals that aid in finding of same species members for mates, food and habitat resources, warning of enemies, and avoiding competition.

(a) **Sex pheromone:**

- First sex pheromone discovered by **Butenandt in 1959**, from female silk worm moth i.e. ***Bombykol***
- Majority of sex pheromones used in pest management are released by female species because that are most powerful and have great potential for pest management.
- The sex pheromones are used in following ways 1. Monitoring, 2. Mass trapping 3. Mating disruption

Other sex pheromones:

- **Gossyplure**: for pink boll worm, ***Pectinophora gossypiella***
- **Hexalure**: it is a para pheromone (synthetic) for pink boll worm
- **Helilure**: to attract male gram pod borer, ***Helicoverpa armigera***
- **Grandlure:** cotton boll weevil, ***Antonomus grandis***

(b) **Alarm pheromone: e.g.** In aphids secreted by cornicles.

(c) **Aggregation pheromones: e.g.** found inbark beetles

(d) **Trail pheromone: e.g.** Cephalic glands in ants

2. Allelochemicals (these are *interspecific* chemicals):

- Chemicals that are secreted by an organism and cause specific reactions to receiver organism of different species.

(a) Allomones:

- Chemicals that is beneficial to releaser and not favorable to receiver.

(b) Kairomone: Chemicals that is favorable to receiver but not to releaser.

Insect taxonomy:

Brief history

- **Carl Linnaeus** gave binomial nomenclature in 10th edition of his historical book ***'Systema naturae'*** in 1758. Hence he called as ***Father of Taxonomy.***
- **Aristole** (384-322 BC) was known as ***'Father of biological classification'***
- **Lionel de Niceville** was appointed as the first entomologist to govt. of India in 1901.
- **H. M. Lefroy** was appointed as second entomologist in 1903.
- **Hem Singh Pruthi** was the founder of Entomological Society of India in 1938.
- **William Kirby** is considered as the father of Entomology.
- First legal law in India was enacted: **DIPA** (Destructive Insect-Pest Act, **1914**)
- The first quarantine institute was established as DPPQS (Directorate of plant protection, quarantine and storage in Faridabad, HR in 1946.
- **Insecticide act** enacted by govt. of India in 1968 and it came in to force from 1st Jan., 1971.
- The totally parasitic insect order is **Strepsiptera** but parasitoids that are mainly utilized as crop pest management is belong to insect order ***hymenoptera*** followed by ***dipteral.***
- The smallest insect order is **zoroptera** (it is primitive hemipteroid order)
- The highly evolved (developed) insect order is ***Hymenoptera***

A. Phylum: **Arthropoda**

- Arthropoda is a Greek word which means **segmented legs.**
- 97 % populations of arthropods are insects

B. Class:

- There are six classes i.e. Onychophora (e.g. peripetus), Crustacea (crabs & prawns), Diplopoda (millipedes), Chilopda (centipedes), Arachnida (ticks, mites & spiders) and Hexapoda (Insecta) among these later two are mainly agricultural importance.

1. Class: **Arachnida**

- Head and thorax are fused and known as ***cephalothorax***.
- In most of members **four pairs** of walking legs are present and legs are unsegmented.
- They have **no antenna** e.g. **Mites, ticks, spiders** and **scorpion**.

- Mites and ticks belong to the order acarina.

Order: Acarina:

- It has many families of phytophagous and predatory mites
- Mites have piercing and sucking or chewing type mouth parts
- Most of mites feed on lower surface of leaves
- Some important families of mites are-

Family: Tetranychidae

- It is well known phytophagous mites family.
- They have four pairs of legs.
- E.g. two spotted red spider mite: *Tetranychus urticae*
- Cucurbitaceous mite: *Tetranychus cucurbitae*

Family: Eriophyidae

- Eriophyid mites are very small, and are only visible using a microscope.
- The mites are worm like, and **have** only **two pairs of legs**.
- Some are acts as vector for viral diseases of plants e.g. sterility mosaic of arhar is spread with eriophyid mite *Aceria cajani*.
- Interaction of the mite, ***Aceria mangiferae*** with fungus, ***Fusarium mangiferae***, the causal agent of mango malformation disease.

Common **acaricide/miticides**: Poison that kills mites

- Dicofol 18.5 EC (Kelthane AP -35 @ 1-2ml/l spray)
- Ethion 50 EC (Fosmite @ 1.5-2.0 liter/ha.)
- Sulphur (Wokovit 80 % WDG @ 1.8-2.5 kg/ha.)
- Propargite 57 % EC (Omite @ 1.0-1.5 liter/ha. for vegetables)
- Abamectin 1.9 EC @ 0.5 ml/liter for vegetables)

2. **Class: Hexapoda/Insecta**

✓ Body is divided into three viz. head, thorax and abdomen.
✓ Presence of single pair of antenna
✓ Thorax carries three pair of legs and generally two pair of wings.
✓ E.g. **Insects** are **97**% of total population of the phylum arthropod.

What is insect?

- The insects are tracheate arthropods in which the body is divided in to head, thorax and abdomen.
- A single pair of antennae is present and the head bears a pair of mandibles and two pairs of maxillae, the second pair fused medially to form the labium.
- The thorax carries three pairs of legs and usually two pairs of wings.

- The abdomen is devoid of ambulatory appendages and the genital opening is situated near the posterior end of the body.
- Postembryonic developments rarely direct and metamorphosis usually occurs.
- The widely accepted classification of insects was given by the taxonomist: ***A.D Imms.***
  - ✓ According to A.D. Imms there are **29 insects order**
  - ✓ The class insecta has two subclasses viz. apterygota (4 orders) and pterygota (remaining 25 orders).
  - ✓ Pterygota further divided in to: -exopterygota and endopterygota.

SN	Order	Example	Features
Apterygota:- Primarily wingless, metamorphosis absent, pregenital appendages present			
1	Protura	Telsontails	Antennae and eyes absent, ***anomorphosis*** present (12 segmented abdomen)
2	Diplura	Japygids	Malpighian tubules absent or represented as ***papillae,*** abdomen with lateral ***styliform*** appendages.
3	Collembola	Springtails	Abdomen **six segmented**, Malpighian tubules absent, structures like ***collophore***, ***retinaculum, furcula*** present on abdominal segments for adhesing and leaping purpose.
4	Thysanura	Silverfish	A pair of many segmented cerci ending in a segmented median process.
Pterygota: - Winged or evolved from winged ancestors (e.g. fleas, head louse, bugs etc.), Metamorphosis present, pregenital abdominal appendages absent.			
A	**Exopterygota:** Wing developed external, metamorphosis incomplete, pupaabsent, life cycle with 3 stages *i.e.* eggs, nymphs, adult and immaturestage is ***nymph***.		
5	Ephemeropetra	Mayflies	Primitive because it came before odonata, presence of ***monocondylic*** type mandibles
6	Odonata	Dragon/damselfly	Wing can't be moved, ***dicondylic*** mandibles so it advanced than above, nodus wings and presence of pterostigma. They feed mosquitoes larvae on water surface.
7	Plecoptera	Stonefly	Soft bodied, ***tracheal gills*** in nymphs.
8	Grylloblatodia	Rock crawlers	Apterous, well developed ovipositor
9	Orthropetra	Grasshopper, locust	Fore wing forming ***tagmina,*** hind legs modified for jumping in grasshoppers
10	Phasmida	Stick & leaf insects	Presence of compound eyes
11	Dermaptera	Earwing	Cerci modified in to a ***sclerotized forceps***

12	Embioptera	Websprinners	Living in silken tunnels, anterior tarsi greatly ***swollen*** for webbing
13	Dictyoptera	Cockroach, mantis	Fore wing modified to ***tagmina,*** ocelli represented as ***fenestrae,*** walking type legs
14	Isoptera	Termites/white ants	Social and polymorphic, rectal pouch present, live in ***termitaria***, it is **cellulose** feeder.
15	Zorapetra	Zorapterans	Y-shaped epicranial suture present
16	Psocoptera	Book lice	Head with Y-shaped suture
17	Mallophaga	Bird lice	Biting lice, apterous, ***actoparasitic*** on birds,
18	Siphonculata	Head, body louse	Apterous insects, transmits **typhus** disease
19	Hemipetra	Bugs	Rhynchota, cerci absent, ***scutellum*** on dorsal, scent glands present.
20	Thysanoptera	Thrips	Right mandible absent, fringed wings
B	**Endopterygota:** Wing developed internal, metamorphosis complete *i.e.* life cycle having four stages- eggs, larva, pupa, adult, immature stage is known as **larva.**		
21	Hymenoptera	Bees, braconid wasp (parasitoids), ants	Presence of ***propodium,*** venation greatly reduced, acuteness of vision, ***corbicula*** or pollen basket on hind legs of worker bees.
22	Lepidoptera	Butterflies, moths	Scale wings, suctorial proboscis formed by the maxillae of ***galeae***, thorax possess ***patagia***, wings with ***trigamma***.
23	Coleoptera	Beetles, weevils	Largest insect order, fore wings as forming leathry **elytra**.
24	Diptera	Mosquitoes, house fly, gall fly, shoot flies, midges	One pair of wings, presence of ***halters***, ***puparium,*** known as true flies order
25	Strepsiptera	Stylopids	Totally endoparasitic on other insects, fore wings modified as ***pseudohalters.***
26	Neuroptera	Antlions, lace wings, Owlflies,	Green lace wing: ***Chrysoperla carnea*** is predator of cotton sucking pests
27	Mecoptera	Scorpionflies	Males carry the terminal segment of the abdomen upward curved.
28	Trichoptera	Caddisfly	***Thyridium*** present on wings, ***Gilsons glands*** on thorax.
29	Siphonoptera	Fleas	Apterous, actoparasitic on warm blooded animals.

- Out of 29 insect orders **eight orders** are mostly agricultural importance e.g. Hymenoptera, Diptera, Coleoptera, Lepidoptera, Thysanoptera, Orthoptera, Hemiptera and Isoptera.

1. Order: **Isoptera** (e.g. Termites or white ants):

- ***Iso*** = equal, ***Ptera*** = Wings
- These are polymorphic and **social insects.**
- **Frontal gland** is present in soldier
- **Rectal pounch:** It is a Flagellates (protozoan) colony found in ileum of hindgut not rectum, it helps in the **digestion of cellulose** in primitive termites.
- **Fungal garden:** habitation of termites contain fungal garden composed of a spongy, dark brown reddish coral like comb constructed by workers and fed to nymphs.
- In termites, except reproduction and defense of colony workers perform all duties i.e. care of nymph, fed queen; construct fungal garden and main function is to destroy the woods and crops.
- Family: **Termitidae**: e.g. Termite (***Odontotermes obesus***).

2. Order: **Orthopetra** (e.g. Grasshoppers, locusts, crickets)

- **Ortho** = Straight, **Ptera** =Wings
- Head is hypognathous type
- All true ring like 11 segments of abdomen are visible
- Presence of 10 pairs of functional spiracles that is known as ***Holopneustic*** type respiratory system.
- In grasshoppers hind legs are jumping type but in crickets modified for sound producing (stridulatorial) and in mole crickets fore legs are strongly fossorial (digging type).
- **Stridulation**: in members of orthoptera stridulation is the act of producing sound by rubbing together certain body parts.
- Eggs are laid in ***egg pods***.

Important families:

(a) **Acrididae** (e.g. locusts and short horned grasshoppers)

- Rice grasshopper: ***Hieroglypus banian***
- Kharif grasshopper: ***Hieroglyphus nigrorepletus***
- Desert locust: ***Schistocerca gregaria***
- Bombay locust: ***Nomadacris (Pantanga) succincta***
- Migratory locust: ***Locusta migratoria***

(b) **Tettigoniidae**: (Cricket family) their members commonly called **katydids** or **bush crickets** also known as **long-horned grasshoppers**.

(c) **Gryllotalpidae**: Mole crickets family

3. Order: **Hemiptera** (Hemi = half, Ptera = wing) e.g. bugs, whiteflies, aphids, jassids, leaf hopper, mealy bugs etc.

- Mesothorax represented dorsally by ***Scutellum*** (a triangular structure in bugs).
- Mouth parts are piercing and sucking type-
- Labium (rostrum) has four stylets (2 mandible + 2 maxillae) to forming food canal especially in bugs.
- Apical portion of fore wings are membranous forming ***hemielytra***
- Metamorphosis is incomplete (gradual) i.e. having 3 stages in their life cycle: - eggs-nymphs-adults.

It is divided into two suborders:

(A) **Heteroptera** (plant bugs): scent glands are present in some bugs

(B) **Homoptera:** (aphids, leaf hoppers, whiteflies, psyllids, jassids, mealy bugs etc. soft bodied sucking insects)

(A) Heteroptera:

Important families

(a) **Miridae** (capsidae):

- Sorghum ear head bug: ***Calocoris angustatus***
- Tea mosquito bug: ***Helopeltis theivora/antonii***
- (It cause wart like outgrowth on guava fruits)

(b) **Pyrrhocoridae**:

- Red cotton bug (cotton strainers): ***Dysdercus cingulatus***

(c) **Coreidae** (leaf footed bug, Presence of scent glands):

- Rice gundhy bug: ***Leptocorisa acuta***
- Pulse pod or bean bug: ***Riptortus species***
- Tur pod bug***: Clavigralla gibbosa***

(d) **Pentatomidae**: (body is shield like)

- Painted bug of cruciferous: ***Bagrada cruciferarum***

(B) Sub-order: **Homoptera** (Homo = uniform, Ptera = wings i.e. uniform texture of the front wings

- All are terrestrial, purely phytophagus, soft bodied insects
- In some insects wax glands are present
- **Filter chamber** is present in alimentary canal.

Important families

i. **Aphididae** (aphids/ plant lice)-

- The body is pear shaped
- In aphids reproduction chiefly by parthenogenetically and also found viviparity.
- In one year about 45 generations may be completed.

- On dorsal side of 5 or 6th abdominal segments found a pair of honey tubes (**cornicles**) which secretes **honey dew substances**.
- Aphids feed on phloem and leaf hoppers on cortical cells.
- Mustard aphid: ***Lipaphis erysimi***
- Green peach (or potato) aphid: ***Myzus persicae***
- Cotton aphid: ***Aphis gossypii***
- Maize (corn leaf) aphid: ***Rhopalosiphum maidis***
- Wheat aphid: ***Macrosiphum miscanthi***
- Apple woolly aphid: ***Eriosoma lanigerum***

ii. **Lacciferidae**:
- Lac insect: ***Laccifer lacca***, mainly female produces ***lac***

iii. **Margarodidae** (scale insects):
- Cottony cushion scale on citrus: ***Icerya purchasi***
- Mango mealy bug: ***Drosicha mangiferae***

iv. **Pseudococcidae** (mealy bugs):
- Female oval or elongated in shape and body coated with white waxy secretions.
- Citrus mealy bug: ***Planococcus citri***
- Cotton mealy bug: ***Maconellicoccus hirsutus***

v. **Psyllidae**: known as plant jumping lice
- Citrus psylla: ***Diaphorina citri***

vi. **Cicadellidae** (Jassidae e.g. leaf hoppers)
- Body is wedge shaped
- Mango leaf hopper: ***Idioscopus species*** or ***Amritodus atkinsoni***
- Paddy leaf hopper (GLH): ***Nephotettix virescence***
- Cotton jassid: ***Amrasca biguttula biguttula***

vii. **Delphacidae** (plant hoppers):
- White backed plant hopper (WBPH): ***Sogatella furcifera***
- Brown plant hopper (BPH): ***Nilaparvata lugens***
- BPH also transmit *Rice Ragged Stunt* and *Rice Grassy Stunt* viral diseases in rice

viii. Fulgoridae or **Lophopidae** (aeroplane bug)
- Sugarcane leaf hopper: ***Pyrilla perpusilla.***

4. Order: **Thysanoptera** (Thysan = fringed, ptera = wing,)

- It is link between incomplete and complete metamorphosis because pupa may or may not present in life cycle (generally metamorphosis is considered as incomplete).
- Left mandible is present as stylet but right mandible absent (hence, only 3 stylets present) so mouth parts are ***asymmetrical type.***
- Mouthy parts are **rasping** and **sucking type**
- Mostly members are act as vector for spreading viral diseases.

Family: **Thripidae:**

- Chilli thrips: ***Scirtothrips dorsalis***
- Onion thrips: ***Thrips tabaci***
- Rice thrips: ***Stenchaetothrips biformis***

5. Order: **Diptera** (Di = Two, Ptera = wing or ***true flies***) e.g. flies, mosquitoes, midges etc.

- Mouth parts are sponging type in house fly and piecing and blood sucking in female mosquitoes
- Presence of single pair of wings
- Hind wings modified in to ***halters***
- Labium forms- ***proboscis***
- Some families are parasitic e.g. Tachnidae, Bombylidae
- Some families are predatory: e.g. Syrphidae
- Metamorphosis generally complete (eggs-maggots-pupa-adult)
- Maggots (larvae) are legless (apodous).

Important families:

(i) **Culicidae** (e.g. mosquitoes family):

- Malaria mosquito: Female, ***Anapheles stephensi***
- Filiariasis mosquitoes: ***Culex fatigans/pipiens***
- Yellow fever/Dengue/Chikungunya mosquito: ***Aedes aegypti***

(ii) **Cecidomyiidae** (gall midges):

- Sorghum midge: ***Contarinia sorghicola***
- Rice gall midge: ***Orseolia oryzae***

(iii) **Agromyzidae** (leaf minors, stem pod flies):

- Red gram pod fly: ***Melanagromyza obtusa***
- Serpentine leaf minor: ***Liriomyza trifolii***

(iv) **Muscidae** (house fly & shoot fly)

- House fly: ***Musca domestica***
- Sorghum shoot fly: ***Atherigona soccata***

(v) **Tephritidae** (fruit flies family):

- Cucurbit fruit fly, ***Bactrocera cucurbitae***

- Ber fruit fly, ***Carpomyia vesuviana***
- Mango fruit fly: ***Bactrocera dorsalis***

(vi) Syrphidae (Hover flies):

- Resemble like bees and wasp e.g. **syrphid fly**
- Predacious on ***aphids***

(vii) Tachnidae: their member are parasite on other insects

6. Order: **Lepidoptera** (Lepidos = **scales**, ptera = **wing**), body covered with scales e.g. moths, butterflies, loopers, semiloopers, borers, boll worms etc.

- Mouth parts: maxillae modified to form suctorial proboscis known as ***galae***
- Abdomen bears 2 -5 pairs of **prolegs**
- Metamorphosis is complete (eggs- larva (caterpillar)-pupa-adults)

Important families:

1. Noctuidae: (semi-loopers)
   - Gram cut worm: ***Agrotis ipsilon***
   - Gram pod borer: ***Helicoverpa armigera***
   - Pink borer of rice/wheat/maize: ***Sesamia inferens***
   - Tobacco cutworm: ***Spodoptera litura***
   - Castor semilopper: ***Achaea janata***
   - Cabbage semilooper: ***Trichoplusiani***
2. Gelechiidae:
   - Pink boll worm: ***Pectinophora gossypiella***
   - Potato tuber moth: ***Phthorimaea operculella***
   - Angoumois grain moth: ***Sitotroga cerealella***
3. Plutellidae:
   - Diamond back moth (DBM): ***Plutella xylostella***
4. Bombycidae:
   - Silk worm: ***Bombyx mori***
5. Papilionidae (butterfly family)
   - Lemon butter fly: ***Papilio demoleus***
6. Lycaenidae:
   - Anar butter fly: ***Virachola isocrates***
7. Arctiidae (tiger moth)
   - Red hairy caterpillar: ***Amsacta moorei***
8. Pyralidae **(Crambidae)**:
   - Yellow stem borer of rice: *Scirpophaga* ***(Tryporyza)*** *incertulas*
   - Maize stem borer: ***Chilo partellus***
   - Top borer of sugarcane: ***Scirpophaga nivella or Trporyza novella***

9. Pyralidae (sub family: gallerinae):
    - Rice moth: ***Corcyra cephalonica***

7. Order: **Coleoptera** (Cole = sheath, ptera = wing) e.g. grubs, beetles, weevils, bruchids etc.

- Fore wings modified to form sclerotized ***elytra***
- Metamorphosis is complete (eggs, grub, pupa, adults)
- It is largest insect order having 40% population of total insects (occupies nearly 3.30,000 species).

Important families:

1. Coccinellidae:
    - Most of member of this family are predators on soft bodied insects except **Hadda beetle** which is pest of brinjal.
    - Hada bettle: ***Henosepilachna vigintioctopunctata***
    - Lady-bird-beetle: ***Coccinella septempunctata/Menochilus sexmaculatus,*** it is predator of aphids, jassids, mealy bugs etc.
2. Scarabaeidae (sub family: Melalonthiniae)
    - White grub: ***Holotrichia consanguinea***
3. Curculionidae: (weevils)
    - Presence of characteristic **snout** or **beak**
    - Rice weevil: ***Sitophilus oryzae***
    - Mango nut weevil: ***Sternochetus mangiferae***
4. Cerambycidae:
    - Mango stem borer: ***Batocera rufomaculata***
5. Chrysomelidae (Galerucidae):
    - Red pumpkin beetle: ***Raphidopalpa foveicollis***
6. Lampyridae (fire flies): common name is 'Jugnu'
    - Photogenic organ is found on 6-7th abdominal segments
7. Anobiidae:
    - Cigarette beetle: ***Lasioderma serricorne***
    - It is most serious pest of tobacco products in factories and stored place.
8. Bruchidae: (presence of ***pygidium*** i.e. half exposed abdomen)
    - Pulse beetle (gram dhora): ***Callosobruchus chinensis***
    - Moong dhora: ***Callosobruchus analis***
9. Meloidae:
    - Blister beetles produce ***cantharidin***, a highly toxic chemical that causes blistering upon contact with skin and mucus membranes.
    - Blister beetle: ***Mylabris pustulata***

10. **Apionidae**:
    - Sweet potato weevil: ***Cylas formicarius***
11. **Dermistidae**:
    - Khapra beetle: ***Trogoderma granarium***
12. **Bostrichidae:**
    - Lesser grain borer: ***Rhyzopertha dominica***
13. **Tenebrionidae**:
    - Red rust flour beetle: ***Tribolium castaneum***
14. **Dynastidae**:
    - Rhinoceros beetle of coconut: ***Oryctes rhinoceros***

8. Order: **Hymenoptera** (Hymen = membrane, ptera = wing) e.g. bees, wasp, saw fly, (it is parasitoids order)

- In larvae (grubs) mouth parts are biting and chewing type
- But in adult mouth parts are **chewing and lapping type**
- Characteristic feature in hymenoptera is the fusion of the first abdominal segment with metathorax which is known as ***Propodeum***.
- Larvae of this order are apodous and known as grub
- Most of member of this order are used as bio agent as a parasitoids for pest suppression in crops.

Important families

1. **Tethredinidae** :
    - It is single known family belong to hymenoptera that is harmful for crops.
    - In female ovipositor is saw like so name saw fly is given
    - In larvae eight pair of abdominal prolegs found and is without crochets.
    - Mustard saw fly: ***Athalia lugens proxima***
2. **Ichneumonidae**: (parasitic on many insects):
    - ***Isotima javensis***: it is pre-pupal parasitiod of sugarcane top borer (***Scirpophaga novella***)
    - ***Campoletis chlorideae***: larval parasitoid of *Helicoverpa armigera*
3. **Braconidae:**
    - ***Cotesia (apanteles) flavis,*** it is important larval parasitoid of maize stem borer (***Chilo partellus***)
    - ***Bracon hebetor***: on coconut black headed caterpillar (CBHC): ***Opisina arenosella***
4. **Trichogrammatidae**: (minute insects, their members are egg parasitoids on lepidopteron insects)
    - ***Trichogramma chilonis***: egg parasitoids on sugarcane shoot borer

- ***Trichogramma japonicum***: egg parasitoids on rice stem borer and cotton boll worms.

5. Apidae: (bees family, these are pollinators of most crops)
   - ***Apis dorsata*** (giant or rock bee): wild type and largest bee species among commercially cultivated honey bees.
   - ***Apis cerana indica*** (Indian bees): fully domesticated and resemble to *Apis melifera,* good species for honey production among Indian honey bees.
   - ***Apis mellifera*** (Italian bee): exotic, fully domesticated and **best** for honey production in India.
   - ***Apis florea*** (little bee): Smallest in size and produce very little honey (no economic to rearing for honey production)
6. Vespidae (wasp family):
7. Formicidae (ants family):

Economic decision levels for pest population or Concepts of threshold:

a. **Damage Boundary (DB):**
   - Also called the damage threshold (Pedigo et al. 1986).
   - The damage boundary is the lowest level of injury that can be measured
   - This level of injury occurs before economic loss

b. **Economic Threshold Level** (ETL):
   - The pest density at which **control measures should be applied** to prevent an increasing pest population from reaching the economic injury level. At this level insecticidal spray should be done.
   - It is also referred as '***Action Threshold Level***'.

c. **Economic Injury Level** (EIL):
   - It defined by Stern *et al.* as the lowest population density that will cause **economic** damage is known as economic injury level.
   - It is also known as '***Damage Threshold Level*** (DTL). At this level spray of pesticide should be avoided.

d. **General Equilibrium Position** (GEP):
   - GEP is the **average population density** of an insect population over a long period of time, unaffected by the temporary, interventions of pest control.

Chronology of Entomology:

1951: R.H. Painter coined the term **'Resistance'** and published his book **'Insect Resistance in Crop Plants'**. He has known as **'father of host plant resistance'**.

1962: Publication of the book ***"Silent spring"*** by Rachel Carson

1964: Publication of the book ***'Biological Control of Insect Pest and Weeds'*** by **Paul de bach.**

1973: Development of first photostable pyrethroid "***Permethrin***".

1975: ***Elcar*** (*Helicoverpa,* NPV) registered for control of bollworm & tobacco bud worm on Cotton.

1987: Development of first transgenic plant reported by M. Vaeck and co-workers of 'Belgian biotechnology company' plant genetic system by transferring ***B.t*** (endotoxin) gene to tobacco for the control of ***Manduca sexta***.

What is a pest?

- An insect or any living being which cause **economic losses** to crop or a **nuisance** and health hazard to man and his livestock is known as a pest. **(NET-2013)**
- **Pests** are unwanted plants, animals, insects, pathogens or other organisms that interfere with human activity. They may bite, destroy food crops, damage property, or make our lives difficult.

Categories of pests:

a. **Key pest:**

- These are most **severe** and **damaging** pest.
- GEP lies well **above** the DB and EIL.
- E.g. **Cotton boll worm, DBM, gram pod borer**.

b. **Major pest:**

- GEP is close to EIL but economically damage avoided by **timely interventions.**
- Sucking pests of cotton and rice.

c. **Minor pest:**

- **GEP** lies below **EIL** and **Damage boundary**.
- Under favorable environmental condition the population may cross EIL and DB for a short interval and a single application of insecticides is usually enough to prevent damage.
- E.g. **Thrips, mites and sugar cane mealy bug.**

d. **Regular pest**:

- Affect specific seasonal crops like cereals, pulses, fruits, passes through many generations during the crop period.

- E.g. **rice stem borer**

e. **Sporadic pest:**

- Population of these insect is usually **negligible** but in certain year under favorable environmental condition they appear epidemic and crossing many times over EIL and DB.
- Required suitable cultural tactics to reduce population of these pests.
- White grub, hairy caterpillar, cut worm, grass hopper.

f. **Potential pest:**

- These insects are presently not causing any economic damage.
- Therefore as such should not be labeled as pests.
- The GEP lies below the DB and doesn't cross EIL even under favorable conditions.
- But any change in ecosystem may push their GEP higher and there is a danger of economic damage from these pests if control operations against the other categories of pests are under taken in an indiscriminate manner.
- ***Spodoptera litura*** on cotton/Soyabean.
- **Army worm** on wheat are example of potential pest.

Integrated Pest Management (IPM):

- Integrated Pest Management (**IPM**) is an ecosystem approach to crop production and protection that combines different management strategies and practices to grow healthy crops and minimize the use of pesticides (FAO).

Components of IPM or common methods of pest management:

1. **Cultural methods:** (it includes certain cultural practices viz.)

a. **Crop rotation:**

✓ This practice is effective against limited host range and relatively immobile in some stages of their development.

b. **Deep ploughing:**

✓ The resting stages of soil dwelling insects, weeds and pathogen are destroyed by summer deep ploughing.

c. **Clean cultivation:**

✓ Field sanitation and destroy plant debris of previous crop

d. **Use of trap crop:**

✓ Grow trap crops around the main crop to attract the pest of main crop and then easily destroyed through spraying any effective pesticides.

- ✓ E.g. sowing okra in cotton crop as a trap crop for boll worms, castor in soyabean for spodoptera and mustard in cabbage for DBM.

e. **Pruning and thinning**:

- ✓ Infected parts should be pruned out.
- ✓ Thinning also increase the vigour of plants.

f. **Growing resistant plant varieties:(NET-2013)**

- ✓ E.g. **CoRH-1** varietyof rice are BPH and gall midge resistant
- ✓ **Suraksha:** gall midge resistance.

2. Mechanical method:

a. **Hand picking:**

- ✓ Large sized insects i.e. caterpillars, bugs, and beetles are handpicked and destroyed.

b. **Shaking and beating of branches:**

- ✓ Done by hands or mechanically to drop larvae & eggs of insects from plants.

c. **Banding:**

- ✓ Prevent climbing pests e.g. mango mealy bug

d. **Wire gauge screen**:

- ✓ Protects fruits from borer

e. **Trench digging**:

- ✓ Against nymphs of locust and larvae of red hairy caterpillar

f. **Trapping:**

- ✓ Various type traps are used e.g. light trap, bait trap, sticky trap, pheromone trap and sound trap etc.

g. **Flooding and draining:**

- ✓ To destroy soil dwelling insects and nematode by aphy.

3. Physical method:

- ✓ Involve manipulation of temperature, humidity & use of radiations.
- ✓ Flame throwers against grass hoppers
- ✓ Drying of grains at minimum moisture level to reduce storage pest's infestation.
- ✓ e.g. In **cereals** should be kept at moisture level 10-12 % for safe storage.
- ✓ In pulses moisture level should be kept 8-10 % and **oil seeds** moisture level should be **at least 6-8 %** for safe storage.

4. Legal control:

✓ Adopted various quarantine measures at various frontier area of country.

✓ Domestic quarantine also adopted for some insect-pests.

(i) **DIPA**: Destructive Insect-Pest Act-1914.

(ii) **DPPQS**: Directorate of Plant Protection Quarantine & Storage, Faridabad (HR) in 1946.

5. Biological control:

Definition:

- The utilization of parasitoids, predators and pathogens for the regulation of host population density is called as ***biological control***.
- The term biological control given by H.S. Smith, in 1919.

a. Parasitoids:

- A parasitoid is a special kind of parasite which is often about the same size as its host, kills its host and requires only one host for developing into a free living adult.
- Majority of parasitoids utilized in biological control of insect pest belong to order **Hymenoptera**.

List of common parasitoids utilizing for crop pests suppression-

SN	Parasitoids	Target pest	Crop
1	***Isotima javensis***	Top shoot borer	Sugarcane
2	***Tetrastichus Israeli***	Shoot borer & black headed caterpillar (BHC)	Sugarcane & coconut
3	***Trichospilus pupivora***	Black headed caterpillar	Coconut
4	***Apanteles flavis*** **(larval parasitoid)**	Stem borer (Chilo partellus)	Maize
5	***Platygaster oryzae***	Gall midge (*Orselia oryzae*)	Rice
6	***Telenomus remus***	*Spodoptera litura*	Tobacco
7	***Chelonus blackkburni***	*Spotted boll worm (Earias sp.)*	Cotton
8	***Bracon kirkpotricki***	Pink boll worm & top shoot borer	Cotton & sugarcane
9	***Bracon brevicornis***	Top shoot bore & BHC	Sugarcane & coconut
10	***Brcaon hebetor***	*Helicoverpa armigera, Pectinophora gossypiella*	Cotton
11	***Copidosoma koehleri***	Tuber moth *(Phthorimaea operculella)*	Potato
12	***Cotesia plutella***	DBM (*Plutella xylostella*)	Cabbage
13	***Campoletis chloridae***	Pod borer (*Helicoverpa*	Gram, cotton

		armigera)	
14	*Trichogramma chilonis*	Stem borer & top borer	Sugarcane
15	*Trichogramma chilonis/brasilensis*	Boll worms (*Helicoverpa armigera, Earias insulena, pectinophora gossipiella)*	Cotton
16	*Trichogramma japonicum*	Stem borer	Rice
17	*Encarsia Formosa*	White flies	Green house crop
18	*Epiricania melanoleuca*	*Pyrilla perpusella*	Sugarcane
19	*Encarsia perniciosi*	Sanjose scale (*quadraspidiotus perniciosus*)	Apple
20	*Aphelinus mali*	Wooly aphid (*Eriosoma lanigerum*)	Apple
22	*Leptomastix dactylopii*	Mealy bug (*Planococcus citri*)	Citrus

b. Predators:

- Predator is a free living organism throughout life and usually larger in size, kill their prey and require more than one pray for developing free living adult. Common examples are follows-

1. **Lady Bird Beetle** (***Coccinella septumpuctata***), coccinellidae, coleopterafeed on aphids of various crops
2. **Lady bird beetle** (*Menochilus sexmaculata*), coccinellidae, coleopteran predator of mealy bugs and scale insects
3. **Lace wing** (***Chrysoperla carnea***), Chrysopidae, Neuroptera feed on sucking pests (white flies aphids, thrips) and bollworms of cotton
4. ***Cryptolaemus moutrouzieri*** (beetle) for control of citrus mealy bug
5. ***Rodalia cardanalis*** (beetle) for cottony cushion scale *(Icerya purchasi)* of citrus
6. **Mirid bug** *(Cyrtorhinus levipedinis)* effectively control brown plant hopper (BPH) of rice
7. **Reduvid bug** (*Platymeris laevicollis*), Reduviidae, Hemiptera is a predator of coconut rhinoceros beetle (*Oryctes rhinoceros*)
8. **Predatory mite** (*Phytoseiulus persinilis)* for control of phytophagus mite, *Tetranychus spp.*

- Other:- **Hover fly** (Syrphidae), **Robber fly** (Asilidae), diptera

Classical control:

- The searching, importing, colonizing and dispersal of a natural enemy in to an area where they don't already exist for the suppression of pest population.

C. Microbial control: term given by stern hauns, 1949.

- The control of pests by the use of microorganisms such as viruses, bacteria, protozoa, fungi, rickettsia and nematode or their by-products are known as ***microbial control***.
- Father of insect pathology: ***Agastino bassi.***

1. Bacteria:

- Entomopathogenic bacteria belong to family Bacillaceae and genera *Bacillus* and *Clostridium* in which *Bacillus* is most important (important species is *Bacillus thuriengensis*) these are gram positive mostly produces endospores.
- **Bt (*Bacillus thuriengensis*) bacteria**: it is spore producing, crystalliferous, facultative bacteria isolated by Japanese scientist Ishiwata from diseased larvae of *Bombyx mori* (silk worm) in 1902, *Bt.* strains produces different crystal (Cry) proteins with insecticidal activity toward specific insect pests (mostly larvae of Lepidoptera)

Common Bt. strains are follows:-

1. *Bt. kurstaki*: presently Bt. kurstaki strain HD-1 based products are registered for lepidopteran larvae control
2. *Bt. israelensis*: used mosquitoes and black flies control
3. *Bt. tenebriosis:* used against beetles of coleopteran
4. *Bt. israelensis*: effective against black flies (diptera)

Toxins:

- Mainly produces **delta endotoxin** (also produces a number of exotoxins in which **β-exotoxin** or **thuringensin** is a thermostable toxin and has effect on insects as broadspectrum)
- First commercial product called **Dipel** (isolated from *Bt.* strain kurstaki), other commercial formulations are Biobit, Halt, Thuricide, M-Trak, Spicturin, Delfin and **Javelin** (new)

Bacillus papillae: spore forming obligate bacteria

- It cause **milky disease** on grubs of Japanese beetle (*Papillia japonica*) reported in 1933.
- *Bacilus papillae* is also naturally occurs on white gurb in India
- **Doom** is the commercial formulation of *Bacillus papillae.* It promising against white gurb *(Holotrichia consaingunea)* in groundnut.

Bacillus sphaerius:

- It produces spherical spores and used for the very good contol of mosquito (*Culex sp.)* but not for *Aedes sp.* and other species
- Non spore forming bacteria: *Pseudomonas fluroscence* it is useful for management of pathogenic fungus and nematodes in plants.

2. Viruses in pest management:

- Well known family of viruses for insect pest's management is: ***Baculoviridae.***
- Both genus **NPV** and **GV** belong to family baculoviridae and ideal for integrated pest management being safe to humans, the environment and non target beneficial organisms. They have good potential for development of microbial insecticides especially for lepidopteran larvae.
- Virus pesticides are compatible with other methods of pest control including chemical insecticides.

1. **NPV** (*Nuclear polyhedrosis viruses*):
   - They contains many virus particles in each occlusion body and composed of a polyhedron protein
   - These are generally genus or species specific e.g. Ha NPV (for *Helicoverpa armigera*), Hz NPV (for *Helicoverpa zea*), Sl NPV (for *Spodoptera litura*) etc.
2. **GV** (*Granulosis viruses*):
   - These viruses contain only one virus particle in each occlusion boody and composed of granule protein
   - GVs are more specific than NPV as they have been reported only from Lepidoptera e.g. CpGV: for codling moth (*Cydia pomonella*), PiGV: for indian meal moth (*Plodia interpuctella*), PoGV: for potato tuber moth (*Phthorimaea operculella*), GmGV: for *Galleria melanauca,* CiGV *(Chilo infuscatellus),* AjGV *(Achaea janata)*

Mode of **action**

- Both NPV and Bt. act as stomach poison (at alkaline gut pH)
- NPV is mostly used in pest management specially to kill lepidopteran larvae
- Because these larvae has high gut pH (> 10) where protein coat of viruses breakdown and occlusion bodies dissolve in mid gut epithelium cells where pinkish polyhedral particles of viruses are accumulated on ventral side of insect body, larvae become flaccid, finally rupture the skin.

Dose of NPV:

- **250-500 LE/ha** @ 0.5 litre/ha (LE = Larval equivalent)
- **1LE = 6 x 10^{-9} POB** (1 LE obtained from 3 dead larvae through virus infection)
- First commercial formulation of NPV is: ***Elcar*** *(Hz NPV)*
- *Other- Gemstar (HzNPV)*

Demerit of NPV and Bt.:

- Slow speed of action
- They are easily degraded in sunlight (UV rays), so NPV and Bt. should be sprayed in the evening hours and their self life can be increased by use of **tinapol** or charcoal, ink, molasses or **egg albumin** with mixing in the spray solution.

Symptoms of virus and bacteria infested larvae:

SN	Virus infested	Bacteria infested
1	Infested larvae stop feeding	Larvae reduce feeding
2	the larvae turn into pinkish white in ventral side because of accumulation of polyhedral bodies	Body become dark or black in color and reduce its activity
3	Larvae become flaccid and skin become fragile & finally ruptures.	Fluid discharge from anus & mouth and ***septicemia*** (blood poisoning) occurs.
4	Larval crawls top of plant & hang upside down by prologs, this symptoms known as ***tree top disease***	Nervous system dis-coordinated and finally death of larvae.

3. Use of fungi in insect pests management:

- **Agostino bassi** (1835) firstly demonstrated that the fungus *Beauveria basssiana* caused white muscardine disease in silk worm (*Bombyx mori*)

List of important insect pathogenic fungi used in agriculture pest management-

SN	Name of fungi	Trade name	Target pets
1	***Beauveria bassiana*** (it cause white muscardine disease)	Boverin, Mycotrol, Naturalis	Lepidopteron and coleopteran insects
2	***Beauveria brongniartii***	-	All species of **white gurb**
3	***Metarhizium anisopliae*** (it produces green muscardine disease)	Biomax, Biopath, Bioblast	Beetles of **white grub** & Bioblast for termite
4	***Metarhizium flavoviridae***	-	Locust & grasshopper
5	***Verticillium lecanii***	Vertilec	Aphids, jassids, thrips, white flies & scale insects
6	***Hirsutella thopmsonii***	Mycar	Phytophagous mites
7	***Nomuraea rileyi***	-	*Helicoverpa armigera*

D. Use of nematodes in insect pests management:

- Nematodes that kill insects are knwon as entomopathogenic nematodes (**EPN**).
- **DD-136** –a mixture of nematode (*Steinernema spp.*) and bacteria which sold in the market under trade name of **Geen commando** for control of boll worms, caterpillars, white gurbs and shoot borers.

Advantages of biological control:

- It's environmentally safe and non-hazardous to men and domestic animals.
- Supplement the natural balance.
- Long lasting control, may establish slowly but once establishment gives permanent control
- Economical: initiate cost may be high.
- Self-perpetuate and permanent.

Disadvantages:

- Slow control and not guaranteed result.
- Research intensive & require large budget.
- Less efficient compare to chemical and mechanical control.
- Since method is depends on environmental factors, failure **can't be unavoided.**

6. Genetic control:

1. Sterile insect control
2. Sterile insect release methods **(SIRM)**

7. Chemical control:

- Conventional insecticides with their trade names are classified as:

a. Organochlorines:

- **DDT**: Dicloro-dipheyl trichloroethane. It was the first commercial organoclorines discovered its insecticidal property by **Paul Muller** in 1939 and first used in agriculture in 1941. But due to their long residual (poisonous) effect it was first banned in US in 1972 and in India, 1989.
- **BHC**: Its pure name is Hexa-Cyclo-Hexane.
- Endosulfan (35 EC)
- Dicofol (18.5 EC) and Kelthane
- Presently all organochlorine pesticides banned in India..

b. Organophosphate:

Examples of common organophosphates:

SN	Name	Trade Name	Formulation	Target pests
	Contact organophosphate			
1	**Diclorvos/ DDVP**	Vapon	76 EC	It is effective against fruitlies, aphids, spider mites, caterpillars, thrips, and whiteflies in greenhouse vegetable crops
2	**Malathion**	Cythion	50EC, 5% Dust	Mosquito control, safest insecticides of vegetables among organophosphospahtes
3	**Chlorpyriphos**	Dharusban, Dermate	20EC, 50EC	It is used for seed treatment and in standing crops for termite control
4	**Quinolphos**	Ekalux	25 EC	It is used for soil pests i.e. white gurb control
5	**Methyl parathion**	Metacid	2% Dust	It is used for locust and red hairy caterpillar control
	Systemic Organophosphates			
1	**Monocrotophos**	Nuvacron	36 SL	In India it is restricted used only in agriculture field crops but ban in vegetables
2	**Dimethoate**	Rogor	30 EC	It is used against sucking pests e.g. beans aphids, thrips, mites and whiteflies control
3	**Acephate**	Acetaf	75% SP	It is used primarily for control of aphids and onion and chilli thrips control
4	**Phorate**	Thimet	10 G	It is used against whitegrub and stem borers, shootfly of field crops
5	**Methyl demeton**	Metasystox	25/30EC	It is effective against mustard aphid

Most of organophosphate pesticides presently banned in India.

c. Carbamate insecticides:

- Carbaryl 75 WP Sevin (contact & broadspectrum)
- Carbofuron 3 G Furadon (systemic, also as nematicide)
- Aldicarb 10 G Temik (systemic used as seed treatment)
- Presently all carbamates banned.

d. Synthetic pyrethroids:

- These are synthetic derivatives of natural pyrethrum
- They have quick knock down effcet but easily break down (degraded) in sunheat.
- Allethrin is used in mosquito coils & liquid mosquitoes killers.

Examples:

- **Allethrin**: it is first synthetic pyrethroid (1949)
- **Permethrin**: it is first photostable synthetic pyrethoid (1970)
- Cypermethrin:
- Deltamethrin: it is used in stored grain pests management
- **Fenvalerate**: 0.4% dust particles (DP).
- Beta- Cyfluthrin (Bulldok)
- Etofenprox (non ester synthetic pyrethroids)

E. Fumigant insecticides:

(i) **Hydrogen cyanide (HCN):**

- Tarde name Calcid, Cynogas, Cynag, highly inflamble, non toxic to plants, but highly toxic to mammals.

(ii) **Carbon disulphite (CS_2):**

- Highly penetrating, so used in stored grains and other materials, highly inflamble so mixed with CCl_4, first discovered by Kuhn (1871) for sugar beet cyst nematode control.

(iii) **Methyl bromide (CH_3Br):-**

- It is colourless, odour less gas, highly penetrating (volatile), used as packing material for export purpose (quarantine station), non inflammable but toxic to mammals, it cause ozone depletion.

(iv) **Ethylene dichloride ($C_2H_4Cl_2$):**

- Low mammalian toxicity, inflamable so miixed with carbon tetrachloride (CCl_4), used for stored grain pest control

(v) **Ethylene dichloride-carbon tetrachloride (ED: CT) mixer:**

- Used in 3:1 ratio (both gase and liquid form)
- It safely used in stored grains because it is safest among all fumigants.

(vi) **CarbonTetra Chloride (CCl_4):**

- It is non inflammable, non explosive so mainly used as a fumigant additive to reduce inflammability
- It is a chloroform like gas/liquid has low insect toxicity

(vii) **DD mixer (1, 3 Dichloropropene & 1, 2 dicloroprapane):**

- Discovered by **Carter (1943)**
- Control of plant nematodes and soil insects

(viii) **EDB (Ethylene dibromide):**

- Discovered by **Chritie (1945).** It controls of plant nematodes and soil insects.

(ix) **Chloropicrin:**

- Colourless gas, it is limited used in stored grains
- Also called as **tear gas** because it is highly lachrymatry and poisonous to mammals

(x) **Hydrogen phosphide (PH_3):**

- Trade name-**Celphos**
- Highly toxic to mammals, it has strong garlic smell
- Pellets containing aluminium phosphide 50% + ammonium carbamate (47%) and release phosphine gas with contact of moisture.
- It is used as grain fumigants in bags or bins @ 3tab. /ton of grains or 140 tab./100m^2 area .

F. Insecticides of plant origin/botanicals:

a. **Rotenone**: Derived from roots of ***Lonchocarpus/Derris elyptica.***

b. **Pyrethrum**: Derived from flowers of ***Chrysantheum cinerarifolium.***

c. **Azadirachtins**: From seeds & leaves of Neem, ***Azadiracta indica.***

d. **Nicotine**: Derieved from roots of tobacco: ***Nicotiana spp.***

G. New group of insecticides:

(i) Neonicotinoids: Imidacloprid is the most widely used insecticide in the world.

- **Important member of neonicotinoid insecticides-**

Name	Products	Target pests
Imidacloprid	Confidor, Admire, Gaucho, Advocate	Sucking insect-pests like aphid, whitelies, jassids, thrips
Thiamethoxam	Actara, Platinum, Cruiser	Cotton white fly
Clothianidin	Poncho, Dantosu, Dantop, Belay	Groundnut white gurb
Acetamiprid	Pride, Assail, Chipco, Tristar	Sucking insect-pests of cotton, rice, vegetables

(ii) **Ketoenols:**

- **Spirotetramat (Momento)**: - its chemical class of ketoenols, subclass tetramic acid derivatives. (for sucking pests control).

(iii) **Oxadiazine group:**

- Indoxacarb: Avaunt (14.5 SC) (for caterpillars)

(iv) **Phenylpyrazole:**

- Fipronil: Regent @ 5% (effective against termite)

H. New insecticides from microorganism:

- Abamectin: **Vertimac** (for mite & leaf minors)
- Spinosad: **Tracer** (for lepidopteran larvae)

I. Insecticides from animal origin:

- Neries toxin is poison originated from a marine annelid, ***Lumbrineris heteropoda*** isolated by Nita in 1934.
- Synthetic derivatives of neries toxin are Cartap hydrochloride (Padan).

Animal systemic insecticides:

- **Co-ral**
- **Ronnel**
- **Rodene**

Management control:

- Method which aid in the control of livestock pests are preferably referred to as management control.
- Cattle grub, screw worm, helminthes (internal pest) and external pest like flies, mites, lice, ticks that are killed by these above insecticides without producing any adverse effect on host.

Toxicity (in order) of insecticides:

- Phorate >Dimeton> Parathion >Aldicarb

Safer (selective) order of insecticides:

- Methoprene >Permethrin>Methoxychlor>Malathion

MRL: **Maximum residues limit: (ARS/NET-10)**

- It is the **optimum safe concentration** of the left over pesticides or its degradation products.
- Expressed in **mg/kg body weight.**

Propesticides:

- It is a pesticidally active material or compound which in its original form is inactive and is transformed into an active state by a plant, animal or microorganism.
- ***E.g. Acephate, Cartap hydrochloride.***

Plant protection equipments:

a. **Dusters:**

- It consists essentially of a **fan** or **blower** and a type of container named **'hopper'**.
- To produce **air blasts** is the **basic principle of duster**.
- The **efficiency of dusting is ten times more than that of spraying** but the average job of spray is equivalent to very good dusting.

b. **Sprayers:**

- Sprayers are the most commonly used in the pesticides application.

Spray	Particle size (in micron/μ)
Coarse spray	400 or more
Fine spray	100-400
Mist	50-150 **(NET-2013)**
Fog	1-50
Smoke	0.001-0.1

IPM Modules for Rice:

Pest	ETL
Hispa	1-2 adult/hill
BPH	5-10 hopper/hill

- **Avoiding** use of **excess nitrogen** which increases population of Brown plant hopper (BPH) and **leaf folder**.
- Alternate wetting & drying for **BPH**.
- Avoided close planting in case of **BPH** epidemic area.
- Passing the rope on the crop and draining of water for **case worm**.
- Using the light traps to monitor BPH, Green plant hopper and stem borer.
- Clipping of seedling tips for **yellow stem borer and thrips**
- Use of **NSKE** (Neem seed kernel extract) 50% or Neem oil for **ear head bug**.

IPM Modules for cotton:

- ETL values of different pest of cotton.

Pest	ETL
Leaf hopper	1-2/leaf
White flies	5-10/leaf for adult or 20 nymphs/leaf
Boll worms	10% shoots/bolls damaged
Stem weevil	10% infested plants
Spodoptera	8 egg masses/100 m row

- Use light trap for hoppers, Spodoptera & bollworms
- Use pheromone trap for monitoring and destruction
- Collection and destruction of plant parts & infected bolls.
- Use of Natural enemies of cotton pest i.e. ***Chrysoperla carnea***, ***Trichogramma spp.***

Insect- pests of major field crops:
Rice crop pests:

1. Yellow stem borer: ***Tryporyza/Scirpophaga incertulus.***

- **Family: Pyralidae** (Lepidoptera)
- **Absolute** or monophagus pest of rice.
- Full grown larvae hibernate in **rice stubbles**.
- It pupates inside the plant.
- Adultsare **dirty white** or greenish yellow front wings.

Symptoms of damage:

- **Caterpillar** alone is destructive, it bore & feed inside the stem and **construct an emergence hole** which always located **above** the water level.
- Before flowering caterpillar produces ***'dead heart'*** on affected plants by cutting growing parts of the plant from the base.
- The drying of growing part of plant is known as ***'dead heart'***
- Plants that is attacked in early stages produce ears **devoid of grains** and are known as ***white ear or white head*** due to drying of entire panicle.
- **Basmati** varieties suffer **heavy damage** then coarse varieties.

Management:

- Removal & destruction of crop stubbles.
- Since eggs of stem borer are laid **near tip of leaf** hence, **clipping** and **burning** of tips of the seedling as a routine before transplanting. (JRF-2011)
- Application of ***Cartap hydrochloride*** 4G @ 2.5kg a.i. /ha. chlorantraniliprole (0.4% GR, 18.5 SC)

2. **Rice hispa: *Diclodispa armigera***

- **Family- Chrysomelidae**, Coleoptera
- **Serious** pest of rice.
- Small **bluish black** beetle, recognized by numerous **short spines** on the body which gives a **characteristic** appearance.

Symptoms:

- Damage is done by larvae as **leaf minors,** produce **transparent blotches** on the upper surface of leaf**.**
- The adult also feed on green matter and produce **whitish streaks parallel to midrib** on leaves
- Severely infected leaves dry up and present a **white driedup** appearance in the field.

Management:

- Nursery beds are **flooded**; the floated beetles are collected by nets and killed.
- **Clip off the leaf tips** before transplanting seedlings to kill the eggs.
- **Chlorpyriphos** @ 625 ml spray with 250 liters of water.
- Larval parasitoid of hispa is: ***Eulophus femoralis.***

3. Rice gundhi bug: ***Leptocorisa acuta*** **(IARI, Ph.D. Exam-1997)**

- Family- Coreidae: Hemiptera
- Periodic pest of rice
- Adult bugs have **long legs** and are slender about **20 mm** long.

Symptoms of damage:

- The pest when disturbed **emit arepugnant or unpleasant smell** which gives to this pest the name "***Gandhi bug***".
- In early morning the nymphs and adults **suck milky juice** from developing grains at **milky stage**, causing incompletely filled panicles or panicles with ***empty of grains***.
- Black and brown spots are appearing around the holes made by the bugs on which a ***sooty mold*** may develop.
- This pest attacks the rice crop ***periodically***.

Management:

- **Clean cultivation**: Paddy field and surrounding bunds should be kept clean from weeds.
- Adult bugs captured through light traps and destroyed.
- Hand collection of bugs
- Resistant variety: BR-10, Bhadshah Bhog etc.
- Spray Malathion 50 EC @ **675 ml/ha**

4. Brown plant hopper**: *Nilaparvata lugens***

- **Family - Delphacidae**: Hemiptera
- Serious pest of high yielding varieties of paddy.
- It appears during later part of crop season.

Symptoms:

- Both nymph & adult cause damage by sucking cell sap from leaves which **turn yellow**.
- Under the favorable condition of high humidity, high nitrogen application and no wind, the population increase and due to severe damage '***hopper burn***' is obtained in various localities.
- It transmitted the "***grassy stunt***" virus disease of rice characterized by narrow, erect, yellowish green leaves.

Management:

- Closer spacing should be avoided (follow a spacing of 20 x 15 cm).

- Alternate drying & wetting of field.
- **Alley 30 cm** wide after **every 3 m** of rice planting provide proper aeration to crop
- Spray of **Imidacloprid** (200 SL) @ 0.3%
- Grow multi gene resistant varieties e.g. IR-36, IR-60.

5. Rice case worm: ***Nymphula depunctalis***
   - Family- **Pyralidae**, order- Lepidoptera
   - The larvae are **light green** with a light brownish **orange head**.
   - The presence of filamentous gills on the sides of body of the larvae is helpful for its **semi aquatic** life as it derives its **O_2 requirement** from the water found inside the leaf case.

Symptoms:

- **Tubular case** is constructed by the larvae.
- Larvae feed by **scrapping** the under surface of the **leaf blade** leaving the upper epidermis intact.

Management:

- **A rope** may be passed to over the young crop for **dislodgin**g the larval cases.
- Put **filter barrier** between irrigation channels of different field.

Other pests of rice:

a. **Green leaf hopper**: ***Nephotettex nigropictus /virescens***.
   - Order- Hemiptera: Family- **Cicadelidae**
   - This is responsible for vector of **Tungro** virus diseases, which is most serious.

b. White backed plant hopper: ***Sogatella furcifera***
   - F. **Delphacidae**: Hemiptera
   - This pest appears in early crop season (July-August)
   - It cause rust red symptoms

c. Rice gall midge: ***Orseolia oryzae***
   - F. **Cecidomyiidae**: Diptera
   - Cause **'silver shoot & onion leaves'** like symptoms because of the appearance of rice tillers as leaf of onion due to tubular galls formed on leaves.

d. Rice thrips: ***Stenochaetothrips biformis***
   - F. **Thripidae**: Thysanoptera
   - Important pest of **rice nursery** ***(JRF-09)***

e. Rice grass hopper: ***Hieroglyphus banian***
   - F. **Acrididae**: Orthoptera
   - Damage by **nibbling of rice ear head**

Insect-Pests of cotton:

1. **Cotton jassid**: ***Amrasca biguttula***, Cicadelidae: Hemiptera.
   - Destructive pest of **American cotton**
   - Nymphs are **wedge shaped** and adults are **winged** form.

Symptoms:

- Adult and nymph cause injury to crop is due to **loss of cell sap** and probably also due to the **injection of toxins.**
- **Yellowing** and **downward rolling/cupping** of leaves due to toxins.

2. **Cotton whitefly**: ***Bemisia tabaci***, **Aleyrodidae**: Hemiptera
   - The pale yellow body is slightly dusted with a **white waxy** powder.
   - They have two pairs of **pure white** wings and have prominent long hind wings.

Symptoms: damage done by two types.

a. The **vitality** of the plants is lowered through the **loss of cell sap**.

b. The normal photosynthesis is **interfered** with due to the growth of a **sooty mould** on the **honey dew** excreted by the insects.

✓ It transmit a number of viral disease including the **leaf curl disease** of tobacco, **vein curling** disease of okra and **leaf curl of sesame**.

Management:

✓ Both jassids and whiteflies can be effectively managed by spraying of **Imidacloprid 200 SL** @ 75 g a.i. /ha or Thiomethoxam @ 45 g a.i. /ha.

3. **Pink boll worm**: ***Pectinophora gossypiella***, Gelechidae: Lepidoptera
   - Most destructive pest of cotton.
   - The caterpillars are **pink** in colour and found inside the **flower buds**, **panicles** and the **bolls of cotton**.

Symptoms:

- Damage done by caterpillar in various ways there is **excessive shedding** of the fruiting bodies.
- **'Rosette flower'** is characteristic symptoms. **(IARI, Ph. D-09)**
- The total shedding is caused by all the boll worms collectively **one half** may be due to the attack of pink boll worm.
- **Double seed** formation: The two adjoining seed are joining together within damaged boll by pink boll worm.

Management:

- Growing of **Bt. cotton varieties** is the effective method for boll worm complex.
- Destruction of cotton sprouts or burning of the **plant debris.**
- Deep ploughing (with furrow turning plough) done by the end of February.

- Larval parasitoid, ***Bracon greeni, Chelonus pectinophorae.***

4. **Spotted boll worm**: ***Earias insulena/vitella*, Noctuidae:** Lepidoptera.
   - It is serious pest of **cotton** and **okra**.
   - A series of longitudinally **black spots** on the body with **dull green** colour and having tiny **stout bristles** are the main identification mark of spotted boll worm.

Symptoms:

- Cause **heavy shedding** of fruiting bodies.
- **'Flare square'** (Flare-up) is characteristic symptoms. ***(JRF-07)***
- In the attacked bolls the lint is **spoiled** by larval feeding.
- The infected bolls open **prematurely** and produce **poor lint** resulting in lower market values.

Other pests of cotton:

a. Dusky cotton bug: ***Oxycarenus laetus***, Lycaenidae: Hemiptera.

b. **Red cotton bug**: ***Dysdercus singulatus***, **Pyrrhocoridae**: Hemiptera.Red cotton bug **stain** the lint in ginning factories when it crushed with cotton.

c. **American boll worm**: ***Helicoverpa armigera***, Noctuidae: Lepidoptera.

d. **Cotton leaf roller**: ***Sylepta derogata***, Pyralidae: Lepidoptera.

Insect-pests of sugarcane:

1. **Top borer**: ***Tryporhyza/Scirpophaga novella***, Pyralidae: Lepidoptera.
   - Caterpillar are **creamy white** or **sluggish** and the moths are **pure white** in color.

Symptoms:

- Young plants attacked by caterpillars show characteristic **reddish streaks** on the **mid rib**.
- They show a number of shoot holes in leaves which ultimately cause **dead hearts**.
- After cane formation it attacks the **terminal portion** of canes causing **'bunchy tops'**.

Management:

- Collect and destroy moths and egg clusters.
- Cut the attacked shoots at ground level from April to June.
- Application of **Cartap hydrochloride** 4G @ 25 kg/ha in July.

2. **Sugar cane leaf hopper**: ***Pyrilla perpusilla*, Lophopidae:** Hemiptera.
   - **Jet** like body pasture.
   - The leaf hoppers are **very agile** and jumps around in large numbers making a **faint noise** when a person walks through **heavily infested** field.

Symptoms:

- Owing to the loss of cell sap the leaves turn **pale yellow** and **shrivel up** later.
- Leaf hoppers secrete a thick transparent liquid known as **honey dew** on which black mould is developed.
- Leaves acquire a **sickly black** appearance and the attacked crop can be **spoiled** from a distance.

Management:

- Spray Malathion 50 EC @ 1 liter/ha.
- Egg parasitoid: ***Epiricania melanoleuca*** (this is single parasitoid belonging to order **Lepidoptera**).

Other pests of sugarcane:

a. Sugar cane white fly: ***Aleurobolus barodensis***, Aleyrodidae: Hemiptera, (Only **nymphs** cause damage by sucking cell sap, yellowing streaks appears on the attacked leaves and a crop acquires *a* ***polish green*** appearance).

b. **Sugar cane mealy bug**: ***Saccharicoccus sacchari,*** Pseudococcidae, Hemiptera

c. **Sugar cane shoot borer**: ***Chilo infuscatellus***, Pyralidae: Lepidoptera.

d. **Gurdaspur borer**: ***Acigona steinella,*** Pyralidae: Lepidoptera.

e. **Sugar cane internodes borer**: ***Chilo sacchariphagus indicus***, Pyralidae, Lepidoptera.

f. **Sugar cane scale insect**: ***Melonapsis glomerats.***

Insect-pest of wheat:

1. **Army worm of wheat**: ***Mythimna separata***, Noctuidae: Lepidoptera.
   - The freshly emerged larvae **spin threads** from which they suspend themselves in the air and then with the help of air current reach from one plant to another
   - Early stage feeds on tender leaves in to **central whorl** of the plants but later caterpillars are able to feed on older leaves and **skelerotized** them totally
   - In case of severe attack by the army worm whole leaves including the mid ribs are consumed and the field looks as **'grazed by the cattle'**.
   - Spray **Quinolphos** 25 EC @ 500 ml/ha.
2. **Ghujhia weevil**: ***Tanymecus indicus,*** **Curculionidae**: Coleoptera.
   - It is specific pest of wheat **nursery.***(ARS-07)*
   - Damage caused by **adult weevil** only.
   - They cut the **germinating seedlings** at the ground level.
   - Mainly damage is done in the months of **October- November**.

Other pests of wheat:

a. **Termite**: ***Odontotermes obesus***, Termitidae: Isoptera.

b. **Wheat aphid**: ***Macrosiphum miscanthi***, Aphididae: Hemiptera.

c. **Wheat mite**: ***Petrobia latens***, Tetranychidae, Acarina. ***(ARS-05)***

Insect-Pest of Pearl millet (Bajra):-

1. **White grub:** *Holotrichia consainguinea, F. Melonthenidae*
2. **Termite:** *Odontotermes obesus, F.Termitidae*
3. **Red hairy caterpillar:** *Amsacta moori,F. Arctidae*
4. **Shoot fly:** *Atherigona soccata, F. Muscidae*

Insect-pests of maize:

1. Maize stem borer: ***Chilo partellus***, Pyralidae: Lepidoptera.

- Most **destructive** pest of maize and sorghum also.
- Caterpillars infect the crop usually **one month** after sowing.
- It cause **pin holes** on leaves.
- Growing points of central shoots are dries up, cause **'dead hearts'.**
- The infestation may be noticed **till harvesting**.
- Larvae remains hibernate in **maize stubbles**.

Management:

- Destroy **stubbles,** weeds & alternate hosts.
- Application of Emamectin benzoate 5 SG or fipronil 0.3 % **G**.
- Release of ***Cotesia (Apanteles)* spp. (larval parasitoid)**

Other pests of maize:

a. Maize aphid: ***Rhapalosiphum maydis***, Aphididae: Hemiptera

b. Pink borer: ***Sesamia inferens***, Noctuidae: Lepidoptera.

- It mainly feeds on **maize**, **rice** and **wheat**
- It causes***'dead heart'*** on central shoot in wheat. **(IARI, Ph.D-04)**

Insect-pests of sorghum:

1. Sorghum shoot fly: ***Atherigona soccata***, Muscidae: Diptera.

- The insect attacks the young crop when it is in the **six leaves** stage.
- **Six weeks** after planting the crop is **seldom** attacked.
- **Maggots** feed on the **main shoot** and **growing point** is destroyed which is **easily plucked** out.
- The young plants show typical **'dead hearts'** symptoms.**(NET-2013)**
- **Cloudy weather** favorable the multiplication of this insect.
- It is believed that infestation is also higher in **irrigated** fields.

Management:

- **Early sowing**, just after the onset of monsoon. ***(ARS-06)***
- Use of **higher seed rate. (NET-2012/13)**
- Use of fish meal trap. **(NET-2012)**
- **Removed** infested plants.

- Seed treatment by thiamethoxam 30FS or Imidaclopride 600 FS at the time of sowing is effective. **(NET-2013)**
- Foliar application of **Cartap hydrochloride** a week after emergence of crop.

Other pests of sorghum:

a. Sorghum midge: ***Contarinia sorghicola***, Cecidomyiidae: Diptera.

b. Sorghum ear head bug: ***Calocoris angustatus***, Miridae: Hemiptera. (It laid cigar shaped eggs on sorghum leaves). ***JRF-06***

c. Red spider mite: ***Oligonichu sindicus***, Tetranchydae, Acarina. ***(IARI, Ph. D Ento-05).***

Insect-Pests of pulse crops:

1. **Gram pod borer**: ***Helicoverpa armigera***, F.Noctuidae: Lepidoptera
   - It is **polyphagous** & **cosmopolitan** in nature.
   - Caterpillar feed foliage, when young and on seeds in the later stages.
   - Larvae pupate in soil.

Management:

- Timely sowing (up to mid-**October**).
- Grow **early maturity** varieties.
- Mixed or intercropping with barley, wheat, mustard.
- Application of NPV @ **250-500 LE/ha. (RPSC, AO-09)**
- Egg parasitoid, ***Trichogramma chilonis.***
- Larval parasitoid of Helicoverpa is ***Cosmopletis chloridae.***
- Use of light traps and Pheromone traps.
- Application of Indoxacarb 14.5 SC @ 75g a.i./ha or Spinosad 45% SC @ 0.5 ml/litre at evening time.

2. **Gram cut worm**: ***Agrotis ipsilon***, Noctuidae: Lepidoptera
   - **Polyphagous** pest but cause much damage to gram crop.
   - **Cannibalism** habit present in larvae.
   - The young larvae feed on the **epidermis** of leaves.
   - During day time larvae live in **cracks** and **holes** in ground and come out **at night** and fall the plants by cutting their stem either below the surface or above the ground.

Management:

- Hand picking & destruction of larvae.
- Application of Flubendiamide (Fame) @ 0.5 % sprays at **evening time**.
- Application of Malathion 5 % dust @ 25 kg/ha at evening time

3. **Red gram pod fly**: ***Melanagromyza obtusa***, Agromyzidae: Diptera.
   - The maggots **eat away** only a part of the **seed**.

- The partially damaged seeds become subjected to **bacterial** and **fungal infection**.
- Grains are **rendered,** unfit for human consumption.
- Maggots pupate in **pods**.

Management:

- Application of **dimethoate**@ 1.0 liter in 250 liter water/ha.

Other pests of pulses:

a. **Tur pod bug**: ***Clavigralla gibbosa***, Coreidae: Hemiptera
(Both nymph and adult suck the cell sap on leaves and pods of arhar, cow pea and gram).

b. **Plume moth**: ***Exelastis atomosa***, Pterophoridae: Lepidoptera.

- It is **specific** pest of arhar.
- Caterpillar **scrapes** of pods and make **hole** on pods.

c. **Girdle beetle**: ***Oberia bravis***, Cerambycidae: Coleoptera.

- It is **specific** pest of **soybean.**
- The ovipositing female beetle **girdles** the stem twice.
- It makes three **punctures** just above the lower ring before inserting a single egg through the largest hole into pith.
- **Management:** spray Malathion 50 EC 1.25 liter/ha.

d. **Bean fly**: ***Ophiomyia phaseoli***, Agromyzidae: Diptera.

- Feed on phaseolus, cow pea, soybean, arhar.

e. **Lentil pod borer**: ***Etiella zinckenella***, Phycitidae: Lepidoptera.

Insect-pest of mustard:

1. **Mustard aphid**: ***Lipaphis erysimi***, Aphididae: Hemiptera.

- Both nymph and adult **suck cell sap** from leaves, stem, inflorescence or **developing pods**.
- The leaves acquire a **curly appearance.**
- It breed parthenogenetically may be produced **45 generations** in a year.

Management:

- Spray **Imidacloprid 200 SL** @ 100 ml/ha.
- Application of dimethoate @ 625 ml/ha.
- Use of predator lady bird beetle: ***Coccinella septempunctata***

2. **Painted bug**: ***Bagrada cruciferarum***, Pantatomidae: Hemiptera.

- Both nymph and adult **suck cell sap** from leaves and pods.
- The nymph and adult bugs also excrete a short of resinous material which **spoils the pods**.
- **Over wintering** of adult stage under **heaps** of dried oil seed plants lying in field.

Management:

- Egg parasitoid: ***Gryon spp***.
- Spray **Malathion** 50 EC @ 1litre in 200 liters water/ha.

3. **Mustard saw fly**: ***Athalia lugens proxima***, Tenthredinidae: Hymenoptera.

- Damage done by larvae which are dark green and have **8 pairs of abdominal prolegs. (ARS-07)**
- Prolegs of mustard saw fly is without **crochets.**

Insect pest of Sesame:

1. **Til hawk moth**: ***Acherontiastyx***, Sphingidae: Lepidopetra.

- This insect is variously known as **hawk moth**, **sphinx moths** or **death head** moth based on its structure and behavioral characteristics.
- Thorax has a prominent **death head** mark on the prothorax.
- The moths are **swift fliers** and often make hawk like darts to a source of light at **dusk**.
- Full grown caterpillar which measure about 5 cm in length one cm width often retracts and looks like a **sphinx**.
- The **horn** like projection on the hind end of the abdomen is conspicuous.

Insect-pest of sweet potato:

1. **Sweet potato horned caterpillar**: ***Agarius convolvulus***, Sphingidae: Lepidoptera.

- Larvae with presence of characteristic **anal horne.**

2. **Sweet potato weevil**: ***Cylas formicarius***, Apioniodae: Coleoptera. ***(JRF-07)***

- It's a pest of both in **field and storage**.

Polyphagous pests:

1. **Red hairy caterpillar**: ***Amsacta moorei***, Arctiidae: Lepidoptera.

- Body of caterpillars is covered with the numerous **long hairs** arising from the fleshly tubers.
- The entire abdomen is **scarlet red** there are **black bands** and **dots** on the abdomen.
- Feed on gregariously and as they grow older **moult 6 times.**
- Feed on all kind vegetation of **kharif season**.
- Full grown caterpillar prefers to eat the **growing points** of plants.
- The caterpillars are **voracious defoliating** the plants
- During a severe attack, the caterpillars in bands destroying fields after field.

- Field after field is **devastated** by the moving army of caterpillars in the year of severe infestation there may be a **complete failure** of the kharif crop.

Management:

- Hairy caterpillars emerge with rainfall of mansoon so in initial stage of larvae can be managed by digging trenches around the field and dusted Malathion 5% dust.
- Use of light traps
- Vegetation or field where infestation can be seen dusts Quinolphos 1.5 % @ 25 kg/ha.
- Larvae can be killed by application of bio-pesticide like NPV or *Bt*

2. White grub: ***Holotrichia consanguinea***. Melolonthidae: Coleoptera.

- Polyphagous feed on all kharif crops, as Groundnut, Bajra, Jowar, Sesame, Cotton, Sugarcane etc.
- Damage done by **grub** in field crops and by **adults in trees.**
- The grub (larvae) is 'C' shaped and feed on roots of plants more severely in sandy soils.
- The damage caused by grub is so much so that entire stand crop is destroyed thereby necessitating the **resowing in field**.
- During the last few years white grub has posed such a serious and alarming situation in the country that it has been designated as '**National pest**'.
- The plant damaged by the grub gives a **wilted appearance** and finally **dries out**.
- In case of severe grub infestation crop is destroyed and infested plants **easily uprooted.**
- The full grown grubs move down deeper in the soil in search of moisture & for pupation.
- The grub constructed an **earthen cell** in which it passes a quiescent or pre-pupal stage.
- After pre monsoon rain/ first shower of monsoon the adult comes out from soil and cause damage to trees during night.
- Eggs are laid in singly in **loose moist sandy** or sandy loam soil on the **onset of monsoon.**
- Grubs feed on the roots and rootlets of almost all types of kharif crops.

Management:

- Collection and destruction of beetles during night with the help of light trap
- Sumer ploughing to expose the hibernating stages.

- In standing crop soil application of 4 liter Chlorpyriphos 20 EC @ 4 liter/ha
- Application of Quinolphos 25 EC (Ekalux) @ 3 liter/ha has been found effective.
- Seed treatment with thiamethoxam 30 FS or clothianidin (Dantotsu) @ 4-6g/kg is very effective.
- The adult can be brought under biological controlled by a fungus ***Matarrhizium anisopliae.***

3. **Locust:** Family: **Acrididae,** Orthoptera

- Locusts belonging to the Indian subcontinent are being described below:

a. Bombay locust (*Pantanga succinct L.*):

- Mostly confined to **Gujarat** and **Tamil Nadu**
- Its breed during mansoon in the Western Ghats and has **only one brood**in a year.

b. Migratory locust ***(Locusta migratoria):***

- Mainly occurs in **Rajasthan** and **Gujarat**
- Its breed twice in a year: winter-spring breeding occurs in Pakistan and summer-monsoon breeding in Rajasthan and Gujarat.
- It may, have many broods a year.

c. Desert locust (***Schistocerca gregarea***):

- This is **most destructive** of all locust

The adult of this species occurs in two phases:

a. Solitary phase (Solitaria): wherein the insects remain scattered
b. Gregarious phase (Gregarea): wherein the insects congregates as nymphs and adults.

Life-history:

- Mating occurs between sexually mature yellow adults.
- Egg-laying starts after 8-24 hours of mating, a single female can lay up to 500 eggs in about 5 egg-pods or capsule.
- Female inserts her long ovipositor in sandy soil to bore a hole 2-4 inches deep.
- The summer breeding occurs during June –September (monsoon) in this quiescent eggs are hatched out and nymphal period is about 12-15 days, within this period emerges in five nymphal instars.
- Within the period of **one month** nymphal stage emerges into adult.

Damage:

- Both nymphs and adults cause damage

Host plants:

- Mainly **pearl millet**, **sorghum,** maize but being Polyphagous nature it's eat on all kind of vegetation except a few namely, **aak, neem, dhatura, Jamun** and **sheesham**.

Management:

- Eggs can be destroyed by deep summer ploughing or flooding
- The **nymphs are the most vulnerable** stage for management point of view.
- Because at second and third nymphal stage is key stage where pest population can be checked.
- A trench can be dug around the breeding field of locust so that when nymphs emerge, they can be buried under soil, or methyl parathion 2% can be dusted to kill the emerging nymphs.
- When infestation starts on field crops or vegetation around the field dusts, Malathion 5% @ 25 kg/ha on field crops including field border and vegetation of non-cropped area.

4. Termite: *Odontotermes obesus,* Termitidae: Isoptera.

- Also known as **white ants**.
- Termite lives in **social colony** in underground nets make earthen moulds (**termatoria**).

A. Reproductive castes:

(i) **Colonizing individuals**:

- Winged individuals of both sexes and are produced in large number during rainy season.
- After mating they cast off the wings and start a new colony.

(ii) **Queen**:

- It is only perfectly developed female in the colony.
- It measures about 5.0-7.5 cm in length and she is known as a phenomenon, ***'Egg laying machine'***.
- It laying one egg/second or **70,000-80,000** eggs in 24 hours.
- Queen fed by workers, choicest food live in **"royal chamber"** which situated in centre of nest (0.5 m depth).

(iii) **King:**

- It developed from **unfertilized** eggs.
- It lives with queen in **royal chamber** & smaller than queen and **mates** with queen from time to time.

(iv) **Complementary caste:**

- Short winged or wingless of both sexes lead a subterranean life

- They are replaced in place of queen & king after ultimately death of them.

(v) **Sterile caste:**

a. **Workers:**

- They develop from fertilized eggs but remain stunted.
- Actual **members that cause damage to crops.**
- Smaller than soldier & highest number in population
- Except reproduction and defense of community, practically all other duties are performed by the workers they take care of eggs, fed queen and cultivate fungus food **ambrosia** in underground gardens.

b. **Soldier:**

- They develop from unfertilized egg and remaining underdeveloped.
- They most specialized member of the community and can be recognized by the large head & strong chitinized sickle shaped mandibles adopted for protection of colony.

Management:

- When a colony is established, it is not so easy to eradicate; only **sure method** is to reach the center of nest and kill the **queen** and the **complementary forms**.
- Before sowing granular application of **Regent** (Fipronil 5 % SC) @ 20 kg /ha.
- Shouldn't be use of **undecomposed** green manure and **FYM** at termite susceptible field.
- Seed treat with **Chlorpyriphos 20 EC** @ 450ml/100 kg of seeds in wheat.
- In standing crop application of **Chlorpyriphos 20 EC** @ 4 liter/ha
- Or Fifronil (Regent) 5 SC @ 18 kg/ha granular form effective against termite in standing crops.
- Seed treat with Imidacloprid (**Provodo**) is effective to protect from termite infestation in nursery stage.

Insect-pests of stored grains and their management:

List of storage insects:

SN	Common name	Scientific name	Family	Order
1	Rice weevil	*Sitophilus oryzae*	Curculionidae	Coleoptera
2	Grain weevil	*Sitophilus granarium*	Curculionidae	Coleoptera
3	Lesser grain borer	*Rhizopertha dominica*	Bostrichidae	Coleoptera
4	Red rust four beetle	*Tribolium castaneum*	Tenebrionidae	Coleoptera
5	Khapra beetle	*Trogoderma granarium*	Dermestidae	Coleoptera
6	Saw tooth beetle	*Oryzaephilus surinamensis*	Silvanidae	Coleoptera
7	Tobacco (cigarette) bettle	*Lasioderma serricorne*	Anobidae	Coleoptera
8	Drug store beetle	*Stagobium penisecum*	Anobidae	Coleoptera
9	Pulse beetle (gram dhora)	*Callosobruchus chinensis*	Bruchidae	Coleoptera
10	Groudnut beetle	*Caryodon cerratus*	Bruchidae	Coleoptera
11	Large head four beetle	*Latheticus oryzae*	Tenebrionidae	Coleoptera
12	Angoumois grain moth	*Sitotroga cerealella*	Gelechidae	Lepidoptera
13	Almond moth	*Ephestia cautella*	Phycitidae	Lepidoptera
14	Indian meal moth	*Plodia interpunctella*	Phycitidae	Lepidoptera
15	Rice moth	*Corcyra cephalonica*	Pyralidae	Lepidoptera
16	Potato tuber moth	*Phthorimaea operculella*	Gelechidae	Lepidoptera

Storage insects carried infestation by two types:-

1. **Internal Feeder**: Larvae/pupae remains inside the grains and insects lay eggs inside or on the seeds. E.g. Rice weevil, Pulse beetle, Sitotroga, Rhizopertha.
2. **External feeder**: Passes all life stages outside the grain and all stages are visible. E.g. Khapra beetle, Corcyra, Red flour beetle, *Oryzae philus*etc.

Other category:-

1. **Primary pests**: Which infests the whole grains and are capable of damaging all kind of storage grains e.g. Sitophilus, Rhizopertha, Bruchids etc.
2. **Secondary pests**: They infest which is already infected by other insects or they are capable of attack the broken or milled grains only e.g. *Tribolium castaneum, Oryzaephilus* etc.

Common storage pests:

A. Coleopteran pests:

1. Rice weevil: ***Sitophilus oryzae***

Host: Serious pests of wheat, rice and barley

Mark of identification:

- First reported in rice so named rice weevil
- It prefers humid climate
- Most identified structure is presence of ***rostrum***
- Head is protruded into ***snout*** *(rostrum)*
- Grub is most injurious stage of damage
- Very difficult for this weevil to survive at low moisture

2. **Lesser grain borer**: *Rhizoperthadominica*

Host:

- Prefer the dark and moist place
- Wheat, rye, jowar

Mark of identification:

- Beetle is more harmful than grubs
- Pronotum is rounded at the front whereas the transfer row of teeth centrally and posterior flattened
- Lay eggs pear shaped and white later change into pinkish color
- Pupation takes place inside the grains
- It produces profuse powdery substances in stored grains is characteristic feature of its damage.

3. **Khapra beetle**: *Trogoderma granarium, F. Demestidae, Coleoptera*

- External feeder, secondary pest

Damage:

- At high temperature & less humidity damage is done by grub (only larvae).

Mark of identification:

- Its eggs stage is highly sensitive to oxygen

- Can be live at less moisture availability
- Adult are harmless only larvae cause damage on whole grain
- It can be live without food up to 4-5 year even 7 years
- At serious infestation in wheat grain eaten completely and white portion left looked like rice.

4. **Red four beetle**: *Tribolium castaneum, F. Tenebrionidae, Coleoptera*
   - It is germ (external) feeder

   Damage:
   - Adult and larvae both are damaging stage
   - Damage the flour, *suji, meda*
   - In case of huge damage the flour turned grayish yellowish color & gives peculiar pungent smell.

Mark of identification:

- Female lay eggs singly in flour up to 400 eggs
- Eggs are white cylindrical and hatch about 4-10 days
- In adult head, thorax and abdomen is very distinct
- **Cannibalism** (one individual consuming another of the same species) is also very common in this insect.

5. **Saw tooth beetle**: *Oryzaephilus surinamensis*
   - Secondary pest

Mark of identification:

- Larvae are straw colored & dark patches on both side of the body
- In adult each side of thorax found six teeth like projection
- Generally larvae feed endosperm of the grain

6. **Pulse beetle**: *Callosobruchus chinensis*
   - First discovered in China so named ***chinensis***
   - Mostly damage the pulses in storage but it's also pest of pulse field

   Mark of identification:
   - Only eggs and adult are visible
   - Antennae are serrate in female but pectinate in male
   - Adults are small chocolate dark brown colored
   - **Only grubs** are destructive.

B. Lepidopteran pests:

1. **Angoumois grain moth**: ***Sitotroga cerealella***
   - It also pest of field
   - In storage the infestation is restricted by surface
   - Damage is done by larvae

2. Rice moth: ***Corcyra cephalonica***: Pyralidae, Lepidoptera
    - Present in humid area
    - External feeder (larvae)
    - Heavy infestation recognized by **webbing**
3. Almond moth: ***Ephestia cautella***
    - External feeder, Only larvae are germ feeder

Management of storage pests:

1. Seed treatment: **Malathion 5% dust @ 250g/qt**. of seeds is recommended or 2ml of fenvalerate in 500 ml of water for one qt. of seeds.
2. Surface treatment**:** Bags dipped into **0.0125% fenvalerate** 20 EC or cypermethrin 25 EC for 10 minutes and drying them in shade or **Spray 0.05% malathion** 50 EC emulsion on floor, wall and ceilings.
3. Fumigation: Fumigats the stored grains with **Ethylene Dichlorine + Carbon Tetrachloride (EDCT 3:1) @ 1 lit/20 qt.** of grains or **Methyl bromide @ 3.5 kg/100 m^3** for 10-12 hours or **Hydrogen phosphide** or **aluminium phosphide 1 tab. per metric tons** of grains.

Entomology terminology:-

Aute ecology:

- The study of individual organisms or species in relation to its environment.

Synecology:

- The study of groups of organism comprising different species in relation to their environment.

Population:

- A group of organisms of the same species occupying a particular area at a specific time is called a **population.**

Biome:

- Biome is an ecosystem of a large area in which the nature of stable communities is similar in appearance.

Ecological niche:

- Every organisms occupies a particular physical location at a given time and does a given thing (**eats, copulates** and **rests** etc) at that place and time at the location is known as **niche** of a particular organism.

Homeostasis:

- The term generally applied to the tendency for biological system to a resist changes and to remain in a state of equilibrium.
- Also known as ***stability in ecosystem or balance of nature.***

Bioluminescence:

- The **production of light by living organism** is called bioluminescence. It has one or more functions viz. illumination, attracting potential prey and matting behavior e.g. spring tails, cicadas and certain families of coleoptera.

Sibling species:

- Group of similar closely related species which are reproductively isolated but morphologically identical or nearly so.

Life table:

- The tabular representation of **death rate**, **causes of death**, **chances of survival** and **life expectancy** is called life table.

Law of tolerance:

- Given by **V. E. Shelford in 1913.**
- Law of tolerance deals with the limiting effect of maximum as well as minimizes temperature on the presence and success of organisms.

6. PLANT PATHOLOGY

Important points:

- ✓ Father of plant pathology: ***Anton de Bary***
- ✓ Bordeaux mixture was developed by: ***P.M.A. Millardet*** *(1885)*
- ✓ Wart disease of potato is caused by: ***Synchytrium endobioticum***
- ✓ Bunchy top of banana is caused by: ***Virus (BBTV)***
- ✓ Late blight of potato is caused by: ***Phytophthora infestans***
- ✓ Club root of cauliflower is caused by: ***Plasmodiophora brassicae.***
- ✓ Father of microbiology: ***Louis Pasteur***
- ✓ Ooze test is done for detecting: ***Bacteria***
- ✓ White rust of cruciferous is caused by: ***Albugo candida***
- ✓ Papaya mosaic disease is: ***Viral***
- ✓ Brown spot of rice caused by: ***Helminthosporium oryzae*** **(ARS-10)**
- ✓ Systemic fungicide is discovered by: ***Van Schmeling*** *and* ***Kulka*** **(1966)**
- ✓ Caulflower mosaic virus contain: ds ***DNA***
- ✓ Bacteria used in biological control: ***Pseudomonas fluorescens***
- ✓ Most of plant viruses are: ***ss-RNA(+)*** 78% of all plant viruses
- ✓ The largest single group of plant viruses family is: **Potyviridae**
- ✓ Carboxin (Vitavax)is very effective against: ***Smut diseases***
- ✓ Downy mildew of bajra is caused by: ***Sclerospora graminicola***
- ✓ Generally viruses are composed of: ***Nucleic acid and protein***
- ✓ The causal organism of black rust of wheat is: ***Puccinia graminis tritici***
- ✓ The '**prions**' are considred to be ***infectious protienaceous particles*** which have no nucleic acids.
- ✓ Agar agar is commercially obtained by: ***Gelidium amansii*** *(Red sea algae)*
- ✓ Wart disease of potato in the country is restricted (i.e. endemic disease) in: ***Darjeeling hills (W. Bengal)***
- ✓ Mycoplasma is sensitive to: ***Tetracycline***
- ✓ First transgenic plant was developed in the world is: ***Tobacco***
- ✓ The downy mildew of pearl millet is suppressed by: ***Chlorine*** *(RAS Pre-09)*
- ✓ Loose smut pathogen of wheat is: ***Internally seed borne***
- ✓ One of factors causing malformation of mango is: ***Fungus (Fusarium)***
- ✓ The most effective fungicide for the control of powdry mildew: ***Sulphur fungicide*** or Dinocap (**Karathane**)
- ✓ The vector responsible for spread of potato leaf roll virus disease is: ***Aphids***
- ✓ **Alexander Fleming (1929)** first time discovered antibiotic: ***Penicillin***

- ✓ The MLO's when, appeared in spiral (helical) shape and are motile known as: ***Spiroplasma.***
- ✓ The sum total of all symptoms and signs is: ***Syndrome***
- ✓ The first bacterial disease in plant was discovered: ***Fire blight of pear***.
- ✓ The fungus causing late blight of potato and white rust of crucifers belong to order ***Peronosporales***
- ✓ The ***cybrids*** contain: nucleus of one and the cytoplasm of both the parents.
- ✓ In rust fungi repeating spores are: ***Uredospores (n +n = dikaryotic)***
- ✓ Sexual spores in downy mildew fungus is: ***Oospores (2n = diploid)***
- ✓ The incidence of blast disease can be reduced by application of: ***Silicon in rice field***
- ✓ Smut fungi belongs to the order: ***Ustilaginales***
- ✓ Vertical resistance is also known as: ***Major gene/qualitative/race specific resistance***
- ✓ Examples of prokaryotes are: ***BGA, bacteria and mycoplasma/phytoplasma***
- ✓ Moko disease of banana is caused by: ***Pseudomonas solanacearum***
- ✓ 'Pine-apple disease' is associated with sugarcane caused by: ***Ceratocystis paradoxa (fungi).***
- ✓ Agrobacterium mediated gene transfer is only effective for: ***Dicots***
- ✓ Antibiotic used to kill MLOs (no cell wall in MLOs) is: ***Tetracycline*** (because it acts on ribosomes)
- ✓ Coffee rust occurred in Sri Lanka 1867 by: ***Hemileia vastatrix***
- ✓ Crystallization of virus particles (TMV) first obtained by: ***Stanley 1935***
- ✓ Downey mildew fungi belongs to order: ***Peronosporales***
- ✓ ELISA (Enzyme-Linked Immuno-Sorbent Assay) test is mainly done for: ***Viral diseases.***
- ✓ Ergot is a mycelia form called: ***Sclerotia***
- ✓ The most common vectors of plant viruses are: ***Aphids***
- ✓ In downy mildew of Bajra, Oospores are resting spores, which are responsible for primary infection
- ✓ Gram staining method in bacteria was given by: ***Christian Gram 1884***
- ✓ The effective and new fungicide for the control of Oomycetous fungi is: ***Metalaxyl*** (marketed under the trade names of ***Ridomil*** & ***Apron SD***)
- ✓ Most dangerous and poisonous mushroom is: ***Amanita phalloides***
- ✓ Purple blotch of onion is caused by: ***Alternaria porri***.
- ✓ ***Pathogenesis:*** The chain of metabolic events that bring about the disease is pathogenesis.
- ✓ ***Pathogenicity***: The relative capability of a pathogen to cause disease
- ✓ ***Virulence:*** The degree of infectivity of a given pathogen.

- ✓ ***Virion:*** it consist of nucleic acid (may be DNA or RNA) and a surrounding protein coat (known as a **capsid**).
- ✓ ***Infection:*** establishment of pathogen after penetration, has host parasitic relationship
- ✓ ***Inoculum:*** Part of pathogen, which on contact can cause infection in the host, is called inoculum e.g. conidia, spores, sclerotia etc. serves as inoculum.
- ✓ ***Invasion:*** The penetration and spread of a pathogen in the host
- ✓ ***Incubation period***: The period of time between penetration of a pathogen to the host and the first appearance of symptoms on the plant.
- ✓ ***Disease potential***: It is defined as the ability of the host to contract disease.
- ✓ ***Turkey X Disease***: discovered in the early 1960s, was an epidemic causing mass deaths in turkeys, ducklings, and chicks, due to toxic aflatoxins produced by *Aspergillus flavus* mold, often found in contaminated groundnut (peanut) meal, leading to severe liver damage, immunosuppression, and cancer in animals and humans.

Unit -1

Plant pathology

- Name derived from two Greek words.

  Pathos = Suffering

  Logus = Study
- Plant pathology is the study of causes of suffering plants and their management.

Historical background:

a. **Iris famine: 1845:**
- Occurred due to **late blight of potato** (***Phytophthora infestans***). This disease destroyed the entire potato crop.
- In Ireland **20 lacks** people were died due to **starvation**.

b. **Coffee rust 1867**:
- Occurred in Sri Lanka due to ***Hemilia vestetricola***.

c. **Downy mildew of grapes** in France:
- It was caused by ***Plasmapora viticola***.
- **Millardet** discovered **Bordeaux mixture** in **1885** for the control of **downy mildew** of grapes.
- Composition: **Copper sulphate** (5 pound): **lime** (5 pound): **water** (50 gallon).

d. **Bengal famine**, **1943:**
- Due to **brown spot of rice** (*Helminthosporium oryzae*).

e. **Theophrastus:** (380-287 B.C.)
- Wrote two books: **'The nature of plants'** and **'Reasons of vegetable growth'.**
- For this works Theophrastus called as ***'Father of Botany'***.

f. **Anton van Leeuwenhoek** (1675):
- Developed the first **microscope.**

g. **P.A Micheli** (1729):
- He first time studied fungi and saw fungal spores on watermelon
- He known as **Father of Mycology**.
- He wrote a book ***'Nova Plantarum Genera'***

h. **C. Linnaeus** (1753):
- ✓ Gave **'Binomial system of nomenclature'** in 10th Edition of his famous book *Systema Naturae* (1958)
- ✓ He wrote a famous book **'Species Plantarium'.**
- ✓ He known as ***'Father of Zoological Classification'***.

i. **Tillet** (1755):
- Published a paper on **bunt** or stinking **smut of wheat**

j. **Prevost** (1807):
- He proved that bunt of wheat caused by a fungus.
- He used copper sulphate to controlled bunt of wheat.
- So credit for **discovering** the life cycle of the **bunt fungus** goes to **Prevost**.

k. **Anton de Bary** (1861):
- Known as father of **Father of Plant Pathology**
- **He** conducted detail study of **potato late blight** and **proved** that fungus *phytophthora infestans* was the cause of late blight.
- The discovery of **heteroecious** nature of black rust fungi was reported by him in 1885.

l. **Brefeld** (1875):
- Developed methods of **artificial culture** of microbes.
- Father of **pure culture technique**.

m. **Robert Koch** (1876):
- He known as **'Father of Microbial Technique of Bacteriology'**
- He discovered first bacterial etiology of **anthrax disease**
- He gave the **Koch's postulates** (theory) for proving that a particular organism caused a particular disease.
- Koch's postulate's fourth step was given by **Smith (1905)**

n. **T. J. Burrill** (1882): (USA)

- First discovered bacterial causes for plant disease i.e. **fire blight of pear** caused by ***Erwinia amylovora***.

o. **H.H. Flor** (1946):

- He proposed **'Gene for Gene Concept' of disease resistance** and susceptibility in linseed rust (*Melanspora lini*).

p. **Vanderplank (1963):**

- He gave the concept of two kinds of resistance i.e. **vertical** resistance and **horizontal** resistance in plants.
- He is known as **Father of Epidemiology**.

q. **Kassanis** (1962):

- **Satellite viruses** were discovered by him. He observed that some strains of Tobacco Necrosis Virus (TNV) contained smaller particles of virus (satellite virus) that was dependent for its replication on TNV.

r. **Shepherd** *et al*. (1968):

- First reported **double-stranded DNA** in cauliflower mosaic virus, followed by the discovery of single stranded DNA Gemini viruses by **Goodman** (1971).

s. **T.O. Diener** (1971):

- He gave the term **'Viroid'** and first time discovered viroids causal agent of **potato spindle tuber disease**.
- Viroids don't have protein coat, they are consisted of circular naked RNA.

History of plant pathology in India:

a. **K.R. Kirtikar**:

- First Indian scientists who collected and identified the fungi in India.

b. **E.J. Butler**:

- He known as **Father of Plant Pathology** (in India)
- He is also known as **Father of Mycology** (in India)
- He wrote a book ***'Fungi and Disease in Plants'*** in 1918

c. **Dr. K.C. Mehta**:

- He is known as **Father of Wheat Rust diseases in India**.
- He studied epidemiology of cereal rusts at Agra, and discovered the disease cycle of **wheat rust** in India.

d. **B.B. Mundkur**: (1948)

- He started work on control of **cotton wilt** through **varietal resistance.**

- He established **'Indian Phytopathological Society' at IARI, New Delhi** with its journal ***'Indian phytopathology'.***
- He authored a text book **'Fungi and Plant Disease'.**

Classification of plant disease: based on occurrence.

a. **Endemic disease**:
- When a disease is more or less constantly present from year to year in moderate to severe form in a particular part of country or state it is classified as endemic disease to that area.
- The word "***endemic***" means prevalent in, and confined to, a particular country, district, or location.
- E.g. Wart disease of potato.

b. **Epidemic disease**:
- A disease which occurs widely in severe form but periodically
- It assumes severity only on occasions.
- Examples: **Red rot of sugar cane**, **Late blight of potato**, **Wheat rusts.**

c. **Sporadic disease**:
- Occur at very irregular intervals and locations e.g. **Green ear disease of Bajra**.

d. **Pandemic disease**:
- When a disease prevalent throughout the country, continent or the world is known as ***pandemic disease e.g. Late blight of potato***, ***Covid-19*** in human beings (due to corona virus)

General symptoms of diseases:

SN	Symptoms	Character
1	**Downy mildew**	The downy growth of the pathogen on lower surface of leaves.
2	**Powdery mildew**	Dusty or powdery appearance (growth) of fungus on upper surface of plants.
3	**Rust** (i) (ii)	Rusty appearance and appear as relatively small pustules of spores. ***Autoecious rust***: Complete its life cycle **in same** host e.g. **Linseed rust (*Melampsora lini*)**. ***Heteroecious rust***: The rust which **requires alternate host** or completes their life cycle on more than one host e.g. **Wheat rusts (*Puccinia spp.*).**
4	**Anthracnose**	Dark ulcer-like lesions on stem and pods.
5	**Smut**	**Sooty** or **charcoal-**like powder. Plant shows a black or purplish **black dusty mass** composed of the

		fungus spores particularly in **ovulary part of floral** organs.
6	White blisters	Generally on leaves of **cruciferous** plants there may be found numerous white, blister-like pustules which break open and expose white powdery mass of spores is known as ***white rust***.
7	Scab	Refers to a roughened or **crust-like lesion** or to a freckled appearance of the diseased organ.
8	Sclerotia	A compact, often hard, mass of dormant fungus mycelium.
9	Phyllody	It is a serious problem of **sesame.** Transformation of floral part in green leafy structure.
10	Etiolation	**Loss of green colour** due to prolonged exposure to darkness, the condition is called ***etiolation.***
11	Exudation	Masses of bacterial cells ooze out to the surface of affected plant parts and form some drops or smear, it is called ***exudation.***
12	Chlorosis	**Yellowing** of leaves due to **lack of chlorophyll**.
13	Necrosis	Refers to **death of tissues or cells.**
14	Hyperplasia	Overgrowth due to ***increase in number of cells*** of organ (due to cell division).
15	Hypertrophy	Overgrowth due to ***increase in size of cells*** (owing to nucleus division)
16	Curl	Leaves are arched, twisted and curled
17	Wilt	Drying or drooping of leaves due to loss of turgor
18	Damping off	Stem affected the near of soil at seedling stage by ***Pythium spp***.
19	Blight	**Burnt** like appearance on plant parts.
20	Canker	Dead area in the bark or cortex of stem of woody plants.
21	Die-back	Drying of twigs and branches of plants from the top backwards.

UNIT -2

Fungi

Fungi are filamentous, multi-cellular or some may be uni-cellular (yeast) organisms, having organized nucleus with nuclear membrane, always lacking of chlorophyll, but presence of cell wall in true fungi and multiplication by production of asexual or sexual spores of vegetative cells.

- The fungal branches are known as ***hypae*** and the entire vegetative body is known as ***mycelium.***

Vegetative structure of fungi:

- In the true fungi cell wall is present whether the thallus is a unicellular or a multicellular branching body.

Reproduction of fungi:

a. **Asexual reproduction**:

- It is repeated several times in the life cycle of a fungus.
- In asexual reproduction the hyphae cut off in minute spores.
- The structure and origin of these spores vary greatly and each types has been given different names i.e. ***chlamydospores, conidiospores*** (conidia), in mastigomycotina *sporangium*, ***(pl., sporangia)*** are formed, in aquatic forms, sporangia liberate naked protoplasmic bodied, the *swarmspore* or if the sporangiospores are motile they called **zoospores.**
- In deuteromycotina, the conidia may be produced in special structures known as ***pycnidia, sporodochia, acervuli.***

b. **Sexual reproduction**:

- It occurs only once in the life cycle of a fungus.
- Involves two separate gametes.
- The sexual reproduction by union of two morphologically dissimilar gametes is known as **oogamy** (leads to development of simple structure **oospores**
- In most of higher form of fungi the sexual process involves the union of two similar gametangia. This is known as **isogamy** (lead to formation of **zygospore or zygote**).
- The sex organs of fungi are called ***gametangia.***

Common orders of plant diseases fungi are:

1.	**Ustilaginales** :	Smut fungi
2.	**Uredinales** :	Rust fungi
3.	**Erysiphales** :	Powdery mildews
4.	**Peronosporales** :	Downy mildews

Fungicides: these are chemicals used to kill fungi.

1. **Sulphur fungicides:**
   - Two main groups:

   a. **Elemental sulphur**: it is used as **dust** @ 25kg/ha.

   b. **Wettable powder:** it is more popular and used as spray @ 2.5% e.g. Sulfex,
   - Sulphur fungicides are effective against **powdery mildews.**
   - Use of sulphur in summer (above 30 °C) may cause injury to melons because of soft and alkaline nature of cucurbits.

2. **Copper compound fungicides:**
   - The copper fungicides are inorganic nature
   - These are effective against both fungi and bacterial diseases

   a. **Copper sulphate:**
   - It was first discovered by **Prevost 1807** and used against bunt of wheat.

   b. **Bordeaux mixture:**
   - PMA Millardet, Prof. of botany at Bordeaux University in France (1885) discovered Bordeaux mixture from the mixture of copper sulphate and lime.
   - It was used against **downy mildew of grapes**
   - Composition = Copper sulphate + lime + water (5: 5: 50: ratio).

   c. **Burgundy mixture: Mason, 1887**
   - In this mixture lime is replaced by sodium carbonate or washing soda. ***(ARS-10)***
   - Burgundy mixture: Copper sulphate + sodium carbonate + water (5:6:50)

   d. **Chaubattia paste: V.B. Singh, 1942**
   - This was used as a wound dressing fungicides against disease of apple and pear as black stem, brown stem etc.
   - **Composition**: Copper carbonate + red lead + raw linseed oil or linolin (8:8:10)

3. **Dithiocarbamates fungicides:** (Organic sulphur fungicides)

   a. **Thiram:** (first dithiocarbate)
   - It is commonly used as a **seed protectant**

   b. **Zineb:**
   - It is safe & widely used in soil and as foliar spray mostly for fruit diseases.

   c. **Mancozeb:**
   - It contains both zinc and Mn.

- **It is best** and **widely** used foliar spray fungicide.

d. **Metham-sodium (Vapam):**
- It is soil fungicides.
- Used for *soil* sterilization, acting as fungicides, nematicides and herbicides. (RPSC AO-2010)

4. **Systemic fungicides:**

a. **Oxathiin**:
- The first systemic fungicide 'Carboxin' was developed by **Von Schmeling and Kulka** in **1966.**

(i) **Carboxin/DMO:**
- Trade name is ***Vitavax***.
- Used against **smuts** and rusts
- Specially used against loose smut of wheat and *Rhizoctonia* disease of cotton.

(ii) **Oxycarboxin**:
- Trade name is ***Plantvax***
- used against **rusts** of wheat
- It can be used in controlling disease related to ***fungi imperfecti.***

5. **Benzimidazoles:**
- It comes under the trade names of Benomyl and **Bavistin**.
- It is a **superior systemic fungicide** which acts as a good eradicant and protectant.

6. **Phenylamides:**
- **Metalaxyl M** was the superior fungicides introduced in 1977.
- Active on oomycetous fungi and inhibit rRNA synthesis.

7. **Organo-mercurial fungicides:**
- It is effective as fungicide and bactericide.
- Organic mercurial are basically used as eradicant.
- Due to high toxicity now it banned in India.
- E.g. Agrosan GN, Aretan, Ceresan, Agallol.

8. **Other fungicides:**

a. **Captan:**
- Seed treatment and soil drenching @ 0.25% for protection against damping off disease.

b. **Karathane (Dinocap)**:
- Used against **powdery mildew** diseases in plants and also used for controlling mites.

c. **Antibiotics:**

- Streptomycin at 100 ppm or more has been found effective against bacterial seed borne pathogen.
- To reduce the possibility of development of resistant strain in bacteria, streptomycin is mixed with Oxy-tetracycline (**Terramycin**). This mixture is called **Agrimycin**.

d. **Triazoles:**

- This is most active on powdery mildews, rusts and leaf spot fungi. E.g. **Tebuconazole, Hexaconazole, Propaconazole**.
- This is important member of **sterol biosynthesis inhibitors** (SBIs).

Fungicides and their trade name:

S.N.	Class/Common Name	Trade name/composition
1.	**Copper fungicides**	
	Bordeaux mixture	Cu_2SO_4: Lime: Water
	Burgundy mixture	$Cu_2SO_4 + Na_2CO_3$ + Water
	Copper oxychloride	**Blitox-50**, Fytolan
	Copper sulphate	**Blue copper**
2.	**Organo mercurial**	
	Ethyl mercury chloride	**Ceresan**
	Methoxy ethyl mercuric chloride	Aretan, Agallol
	Phenyl mercury acetate + ethyl mercuric chloride	**Agrosan G N**
3.	**Benzimidazoles**	
	Benomyl	Benlate
	Carbendazim	**Bavistin, Derosal**
4.	**Dithiocarbamate**	
	Maneb	Dithane M-22
	Mancozeb	**Dithane M-45, Indofil M-45**
	Zineb	**Dithane Z-78**
5.	**Triazoles**	
	Difenoconazole	**Score**
	Tebuconazole	**Folicur**
	Hexaconazole	**Contaf, Taqat**
	Propaconazole	**Tilt**
6.	**Other fungicides**	
	Thiram	Arasan, Thiride
	Phthalimides	Captan
	Dinocap	Karathane, Arathane

	Oxycarboxin	Plantvax
	Carboxin	Vitavax
	Metalaxyl	Ridomil, Apron
	Metham N sodium	Vapam
	Azoxystrobin	Amistar, Quadris
	Fosetyl Al	Alietle

* In burgundy mixture only lime is replaced by Sodium carbonate other component are same

General Principal of plant disease control

1. **Preventive measures**:
   (i) Avoidance of the pathogen
   (ii) Exclusion
   (iii) Eradication
   (iv) Protection
   (v) Resistance
2. **Therapy (curative)**
   (i) Chemotherapy
   (ii) Thermotherapy
   (iii) Surgery
   (iv) Pruning

Exclusion: Method which prevent a disease from entering into a new area. E.g. Quarantine inspection and embargo certification etc.

Eradication:

- ✓ This measure is taken for the elimination of the pathogen after its establishment on the host.
- ✓ It includes crop rotation, sanitation, elimination of alternate and over wintering hosts, elimination of pathogen from the seeds, tubers etc.

Seed treatment (by physical methods):

1. **Hot water treatment method**:
   - ✓ It was developed by **Jensen** in 1887 (RPSC AAO-2013)
   - ✓ Provide hot water to seeds about **54°** C for **10** minutes.
   - ✓ E.g. **Loose smut** of wheat controlled by hot water treatment of seeds.
2. **Solar energy treatment**
   - ✓ To control loose smut of wheat, it was first developed by **Luthra & Sattar**, 1934 (Pre PG-2011)
   - ✓ In this method seeds are soaked **in water for 4-5 hours**, and dried in sun in the month of May –June by the keeping on cemented floor or metal- sheet.
   - ✓ This method is used to control **loose smut** of wheat and barley.

UNIT -3

Classification of plant diseases

1. Bacterial diseases:

- Phylum **Proteobacteria** is the biggest phylum containing the gram negative bacteria e.g. *Agrobacterium, Rhizobium, Xanthomonas, Pseudomonas, Erwinia, Serratia etc.*

SN	Common name	Causal bacteria	Vector
1	Citrus canker	*Xanthomonas axonopodis pv. Citri*	-
2	Black rot of crucifers	*Xanthomonascompestris pv. Compestris*	-
3	Bacterial blight of rice	*Xanthomonasoryzae pv. Oryzae*	-
4	Angular leaf spot/black arm of cotton	*Xanthomonas compastris pv. Malvacearum*	-
5	Greening disease of Citrus	***Canditatus Liberobacter*** (former MLOs)	Transmitted by Psyllids (***Diaphorina citri***)
6	Fire blight of apple/pear	*Erwinia amylovora*	-
7	Scab disease of potato	*Streptomyces scabies*	-
8	Red stripe of sugarcane	*Pseudomonas rubrineans*	-
9	Ratoon stunting disease of sugarcane (RSD)	*Clavibacter xyli (now Leifsonia xyli sub sp. Xyli)*	*Mechanical transmission*
10	Yellow vine of cucurbits	*Serrata marcescens*	*Transmitted by Squash bug (**Anasa tritis**)*
11	Corn stunt	*Spiroplasma kunkelii**	*Transmitted by Dalbulus maidis*

2. Mycoplasma /Phytoplasma diseases

SN	Name of disease	Host plant	Vector
1	Yellow dwarf disease	Rice	Green leaf hopper (GLH): *Nephotettix virescens*
2	Phyllody	Sesame	Leaf hopper: *Orosius albicinctus* (ARS Ento-2013)

3	**Little leaf (LLB)**	Brinjal	Leaf hopper: *Cestius (Hishimonas) phycitis*
5	**Grassy shoot disease (GSD)**	Sugarcane	Aphids
6	**Purple top roll**	Potato	Leaf hopper
7	**Sandal spike disease**	Sandal	Leaf hopper: *Jassus indicus*

3. Viral diseases:

Definition of virus: Viruses are elementary particles that possess some properties of living system, such as having a genome and being able to adapt to changing environment **(by ICTV).**

SN	Host plant	Viral disease	Vector
1	**Wheat**	Streak mosaic	Eriophyid mite: *Aceria tulipae*
2	**Maize**	Mosaic	Sorghum bug: *Peregrinus maidis*
3	**Rice**	Tungro disease	GLH: *Nephotettix virscence*
4	**Rice**	Grassy stunt	BPH: *Nilaparvata lugens*
4	**Groundnut**	Rosette virus	Aphid: *Aphis craccivora*
5	**Groundnut**	Bud necrosis	Thrips: *Frankliniella scultzei*
6	**Pigeon pea**	Sterility mosaic	Eriophyid mite: *Aceria cajani*
7	**Bhindi**	Yellow vein mosaic	Whitefly: *Bemisia tabaci*
8	**Tomato**	Leaf curl	Whitefly: *Bemisia tabaci*
9	**Chili**	Leaf curl	White fly: *Bemisia tabaci* Thrips: *Scirtothrips dorsalis* (ARS Ento 2013)
	Chilli	Mosaic	Whitefly: *Bemisia tabaci*
10	**Potato**	Potato leaf roll	Aphid: *Myzus persicae*
11	**Banana**	Bunchy top (BBTV)	Aphid: *Pentalonia nigronervosa*
12	**Citrus**	Tristeza	Aphid: *Toxoptera aurantii*
13	**Papaya**	Mosaic	Aphid: *Aphis craccivora*
14	**Cardamum**	Kattee/Foorkey	Aphid: *Pentalonia nigronervosa*
15	**Cotton**	Leaf curl	Whitefly: *Bemisia tabaci*
16	**Cow pea**	Mosaic	Aphid: *Aphis craccivora*
17	**Mango**	Malformation	Bud mite: *Aceria mangiferae*

UNIT -4

Crop Diseases

1. RICE

1. **Blast disease**: ***Pyricularia oryzae*** *or (Magnaporthe grisea)*
   - Most severe disease of rice also called as **rich men's disease** or rotten neck.
   - **Air borne** disease, conidia spread by wind cause secondary infection.

Symptoms:

- Brown spindle **eye or boat** shaped lesion on leaf, and leaf sheath.
- At the end of season **rotten neck** or collar rot symptoms appear.
- The infected pedicels result in non-production of seeds is called ***blanking.***

Management:

- Seed treatment with Thiram @ 2.5 g/kg of seeds
- Spray **Mancozeb** @ 0.3 % on standing crops.

2. **Bacterial leaf blight of rice** (BLB): ***Xanthomonas oryzae* pv. *oryzae***
   - Initially symtoms as **'Kresek'** (seedling wilt) occurs in nursery, plant withers and dries up.
   - Later at 4-6 week old plants in field **'blighting'** starts from tip of leaves to the base.
   - **Yellowing bacterial ooze** appears on the surface of legions.

Management:

- Seed treatment with **Streptocycline** 0.015 %
- 3- 4 spray of Agrimycin 100g + 500g of copper oxychloride in 500 liter water/ha.

3. **Brown leaf spot**: ***Helminthosporium oryzae*** (Syn.= *Drechslera oryzae*)
   - This disease is responsible for the great Bengal famine in 1943.
   - Perfcet stage- ***Cochliobolus miyabeanus***, it is known as perfect stage or teleomorph.

Symptoms:

- Pathogen infects the coleoptile of seedlings and cause blighting
- Many dark brown circular to oval spots on upper surface of leaves
- This disease is severe in **potash deficient soils.**

4. **Khaira disease:** **(IARI, Ph. D-1997, JET-2009, BHU-2013)**
   - Due to **zinc** deficiency
   - Usually in nursery chloratic or yellow patches at leaf base on both sides of midrib.

Management:

- Spray **$ZnSO_4$** @ 5 kg/ha or application of zinc sulphate @ 25kg/ha at the time of sowing in zinc deficient soils.

5. **Bakanae disease:**
   - Foot rot or Bakanae disease caused due to ***Gibberella fujikorai*** isolated by a Japanese scientist, **Kurasova.**
   - This disease is also known as **'foolish seedling disease'**.
6. Pan Sukh: **Physiological disease**.
   - Management: drain excess water, apply **ammonium sulphate** @ 15-20 kg/ha.
7. **Tungro disease**: Rice **Tungro virus (NET Ento-2013)**
   - Vector is: **Green Leaf Hopper** (*Nephottetix virscence)*
   - The rice tungro is unique disease which is caused by two viruses, as RNA virus (RTSV), and a DNA virus (RTBV).
   - Stunting of plants and yellowing of leaves from tip and margins are two main symptoms of tungro.

Other minor diseases of rice:

- Sheath blight: Due to fungus*: **Rhizoctonia solani***
- Grassy stunt: **Virus disease transmitted** by 'Brown plant Hopper **(BPH):** *Nilaparvata lugens*
- Yellow dwarf: Due to ***Phytoplasma***
- Udbatta disease: Due to fungus '***Ephelis oryzae'***
- False smut: *Claviceps oryzae-sativae* ***(ICS Exam-1995)***
- Symptoms of false smut appears after flowering when infected spikelet is transformed in to a large velvety, yellow to orange mass which later changed to greenish in color.

2. WHEAT

1. **Rusts:** these are three types, caused by different species of *Puccinia.*

a. **Brown or leaf rusts**: ***Puccinia recondite tritici***
   - It's most widespread and **most damaging** in our country.
   - It is **earliest in appearance** on wheat in India.

Symptoms:

- Bright orange, uredinial pustules, scattered mainly on leaves so called leaf rust.

b. **Yellow or stripe rust**: ***Puccinia striiformis***
   - The yellow rust is major disease in N.W. part of country but in temperate areas it is restricted.

Symptoms:

- Damage is done in early stage of crop.
- Lemon yellow pustules develop in rows on leaf, leaf sheath, and glumes is characteristic feature of yellow rust.

c. **Black/stem rust**: ***Puccinia graminis tritici***

- Known as **'Killer'** disease of wheat because it kills wheat plant.
- Because of late appearance the damage is less than other rusts
- Severe in Southern, Central and Eastern part of India

Symptoms:

- The stem is most severely affected so it was named as **stem rust**
- Elongated brown pustules of uredinia appear on stem, leaf sheath and leaves.

Management of rusts:

- Avoid late sowing
- Use of resistance varieties
- Use high dose of potash fertilizer
- After appearance of symptoms spray of Mancozeb @ 2 kg/ha
- Dusting of Sulphur @ 25 kg/ha is effective against all three rusts infestation.

2. **Karnal bunt or partial bunt**: ***Neovossia indica***

- 1st time reported in Karnal by **Mitra (1931)**
- Called as **'Cancer'** of wheat and it is new emerging disease at International level.
- Many countries have imposed restrictions to import wheat from diseased area (quarantine importance disease).

Symptoms:

- Infection occurs at flowering stage by air-borne sporidia.
- Few grains are partially converted into **black sooty powder**.
- Rotten black powder gives a **foul smell** due to the presence of ***tri-methyl-amine***.
- The flour prepared from infected grains of this disease becomes dark colored and gives a ***fishy smell*** which is unfit for human consumption.
- Disease is seed, soil and air borne.

Management:

- It is soil, seed and air borne disease hence use of resistance varieties is only option
- Avoid excessive irrigation particularly at flowering.

3. **Loose smut**: ***Ustilago nuda tritici*** (Syn= *U. segetum var. tritici*)

- Internally seed borne disease (JRF-09)

Symptoms:

- Infected plants produces early emergence of head than healthy.
- Entire 'ear' except the rachis and awns, are smutted
- Initially the spore masses are covered by a silvery membrane and later on membrane is ruptured
- Finally production of **black powder** in place of grains in the ear

Management:

- Seed treatment with **vitavax** @ 2.5 g/kg of seeds before sowing
- Solar heat treatment of seeds: given by **Luthara and Suttar**(1934)
- Hot water treatment: 54 °c for 30 minutes given by Swingle
- Rouging of diseased plants.

4. **Powdery mildew**: ***Erysiphe graminis tritici*** (now *Blumeria tritici*)
   - White powdery growth of fungus on the upper surface of leaves.
   - It can be managed by sulphur dusting @ 25kg/ha.
5. **Tundu disease/sehu** or yellow ear rot disease: ***Clavibacter tritici*** + ***Anguina tritici*** (Bacteria + Nematode associated disease)

Symptoms:

- Curling of affected plant leaves
- Bright **yellow slimy ooze** on leaves and inflorescence
- Agglutinated inflorescence

Management:

- Floatation of seeds in **20 % salt solution** (i.e. Brine solution)

6. **Other diseases of wheat :**
   a. Flag smut: ***Urocystis tritici***
   b. Hill bunt: ***Tilletia foetida*/caries.**
   c. Ear–cockle of wheat: due to ***Anguina tritici.*** It is managed by seed floatation in **20% salt solution** before sowing.

3. BARLEY

1. **Covered smut**: ***Ustilago hordei***

Symptoms:

- Smutted ears and grains replaced by black agglutinated spore masses.

Management:

- Externally seed borne hence easy to control,
- Seed dressing with vitavax or thiram @ 2.5 g/kg seeds.

2. **Loose smut**: *Ustilago nuda*, same as wheat
3. **Powdery mildew**: ***Erysiphe graminis*** var. ***hordei***
   - **Cottony growth** on both the leaf surface is major symptom
   - Spray of **sulphur** fungicides or **Karathane**

4. PEARL MILLET

1. **Downy mildew/green ear disease**: ***Sclerospora graminicola*** *(order- Peronosporales)*

    - Primary infection is carried by round shape golden colored oospores present in seeds and secondary infection by air borne sporangia which producing zoopores.

Symptoms- recognized by two stages-

1. **Downy stage**- white fungus growth can be seen on the lower surface of leaves in humid weather hence the name given **downy mildew**.
2. **Green ear stage**- Later ear/floral portion transformed into **green leaf like** or leafy whorl type structure so named green ear disease.

Management:

- Roughing of green ear heads
- The disease is soil borne, so adopt crop rotation about 4-5 year
- Seed treatment with **Apron SD 35** @ 6 g/kg seed before sowing.
- Spray **Mancozeb** @ 2kg/ha 21 days after sowing.

2. **Ergot: *Claviceps fusiformis*** *(order- hypocreales)*

    - Fungus is seed and soil borne, symptoms has two stage
        1. **Honey dew stage**- Symptoms first appear on the ears in the form of honey like pinkish liquid.
        2. **Sclerotial stage**- on later stage pinkish liquid appear as **brown to black *Sclerotia.*** Infected grains have toxic alkaloids **'ergots'** which is injurious to human and cattle health. The disease caused by eating peral millet grains contaminated with the parasitic fungus, *Claviceps fusiformis* is known as ***ergotism.***

Management:

- Avoid late planting.
- Seed treatment with Captan or Thiram @ 3g/kg seeds
- Dip the seeds in **20 % salt solution** and floating seeds should be collected and destroyed.
- At the time of ear emergence two spray of **Mancozeb @ 2 kg/ha,** at 3-3 days interval.

3. **Smut: *Tolyposporium penicillariae***

    - In the beginning affected grains become green and larger in size than healthy but later turn to ***black sori***.
    - Seed treatment with Thiram @ 3g/kg seeds.
    - Spray Vitavax @ 0.25 %.

5. MAIZE

1. **Downy mildew of maize**: *Peronosclerospora sorghi*
   - It is considered as major disease of maize.
   - At lower leaf surface cottony growth of fungus is seen.
   - Spray mancozeb or ridomil MZ.
2. **White bud of maize**: ***Zn** deficiency*
   - Apical portion of leaf becomes white.
   - Apply $ZnSO_4$ @ 20-25 kg/ha at sowing time.
3. **Charcoal rot or stalk rot**: ***Macrophomina phaseolina.***
   - Avoid water stress after the flowering of the crop.
4. **Seed Rot**/Seedling Blight: ***Pythium aphanidermatum.***
5. **Smut of maize**: *Ustilago maydis.*

6. GRAM

1. **Fusarium wilt**: ***Fusarium oxysporium*** f. sp. ***ciceri***
   - Soil borne
   - Most serious in central India.

Symptoms:
- Drooping of leaves and
- Browning of xylem vessels.

Management:
- Avoid early sowing (early sowing of gram protect the crop from *H. armigera* but may be severe infestation of *Fusarium* wilt)
- Grow resistance variety i.e. ***Avrodhi***
- Seed treatment with carbendazim 2g/kg or *Trichoderma viridae* @ 6-8 g/kg of seed.
- Crop rotation and deep sowing

2. **Ascochyta blight**: ***Ascochyta rabiei***
   - Seed and soil borne
   - Most potential disease of North West plains
   - All plant parts except roots are affected

Moong, urd, moth beans-

1. **Web blight**- *Rhizoctonia solani*
   - Symptoms- yellowing aand drooping of leaves and produce small necrotic lesions.
2. **Yellow mosaic**- yellow mosaic virus.
   - It belongs to Gemini virus group (ssDNA) and transmitted by white fly.

7. SORGHUM

1. **Downy mildew**: ***Sclerospora sorghi***

Symptoms:

- Downy white growth on lower surface with yellowing on upper surface on young leaves i.e. chlorosis at downy spot

Management:

- Seed treatment with metalaxyl (Apron 35 SD) @ 0.35g a.i./kg
- Spray Mancozeb @ 0.2 %

2. Other:
   - **Grain smut**: ***Sphacelotheca sorghi*** (RPSC-AO Exam-2012)
   - **Leaf rust**: ***Puccinia purpurea***
   - **Head smut**: ***Sphacelotheca reiliana,*** portion of entire floral structure transformed into **smutted galls**
   - **Long smut**: ***Tolyposporium ehrenbergii***
   - **Loose smut**: ***Sphacelotheca cruenta***
   - **Sugary disease** of sorghum: ***Sphacelia sorghi***.

8. ARHAR

1. **Fusarium wilt: *Fusarium oxysporium*** f. sp. **udum**
   - Most destructive soil borne fungal disease

Management:

- Mixed cropping and crop rotation, no chemical control found effective

2. **Sterility mosaic**: due to sterility mosaic virus
   - Vector of sterility mosaic is, Eriophyid mite: ***Aceria cajani***
   - It renders the plants sterile by inhibiting flower formation.
   - Due to viral infection the plants grow luxuriantly, and the disease is called ***green plague***.

9. SOYBEAN

1. **Yellow mosaic**: due to yellow mosaic virus
   - Vector is ***Bemisia tabaci*** (whitefly)
2. **Charcoal rot**: ***Rhizoctonia bataticola***
   - Reddish black to brown necrotic lesion at crown portion
   - Seed treatment with Carbendazim, summer ploughing and crop rotation.
3. **Rhizoctonia blight**: ***Rhizoctonia solani***.

10. GROUNDNUT

1. **Tikka or leaf spot disease**: two type spot-
    1. ***Early leaf spot-*** *Cercospora arachidichola*
    2. ***Late leaf spot -*** *Cercospora personata (more harmful)*

Symptoms:

- Small dark brown spots and pre mature leaf shedding.
- In case of *C. personata*, brown spots are regular and not more than 0.6 in diameter but in *C. arachidichola*, there are irregular spots of larger diameter.
- Late leaf spot more dangerous because they produce much more numerous spots and their rate of conidia formation is much higher.

Management:

- Spray Mancozeb @ 1.5 kg/ha 40 days after sowing

2. **Collar rot or crown rot disease**: ***Aspergillus niger***
    - **Symptoms**: attack at seedling stage at the base, black conidiophores (spores) is seen at root, it's a common saprophyte fungus (live on dead organic matter).
    - ***Aspergillus flavus*** produces ***Aflatoxin*** in stored kernels.

Management:

- Seed treatment with Mancozeb 3g or carbendazim 2g/kg of kernels
- Soil drenching with carbendazim 0.3 %
- Application of ***Trichoderma viride*** @ 4 kg/ha mixed with 15-20 kg well-rotted FYM and subsequently irrigate the field (it has antagonistic effect on pathogenic fungi).

3. **Pea nut clump virus disease**- Indian pea nut clump virus (IPCV, ssRNA), transmitted by a fungus- *Polymyxa graminis*).
    - Infected plants remain small, stunted and clumped.

11. MUSTARD

1. **White rust/blister**: ***Albugo candida***
    - The fungus belongs to order: **Peronosporales**
    - It is obligate parasite i.e. can live only in living plants.

Symptoms:

- White or yellow pustules with variable sizes & shapes on lower surface of leaf.
- Infection covers all parts except roots.
- Resting spores are ***oospores.***

Management:

- Spray Blitox -50 @ 1.5 kg/ha at 15 days interval.
- Seed treatment with Apron 35 SD @ 6 g/kg seeds.
- 2 -3 spray Mancozeb @ 2 kg/ha at 20 days intervals.

2. **Alternaria blight**: ***Alternaria brassicae***
   - **Symptoms**: Concentrate black spots on leaves, stem and pods
   - **Management**: Spray Mancozeb @ 2kg/1000 liter water at 15 days intervals starting from 40-45 days after sowing.
3. **Downy mildew**: ***Pernospora brassicae***
   - Yellow & irregular spots on upper surface & white growth on under surface of leaves malformed inflorescence.

12. SUGARCANE

1. **Red rot**: ***Colletotrichum falcatum***
   - Most destructive disease in both tropical and subtropical India.

Symptoms:

- Infestation starts from July.
- Typical symptoms of red rot are seen inside the stalk i.e. reddened pith which is responsible for quick decline of many of the sugarcane varieties.
- Infested stalks gives alcoholic smell from field.

Management:

- Setts are dipped in 0.1% solution of carbendazim or thiophanate methyl @ 0.1 % for five minute.
- Dipping setts in hot water at 52^{0}C for 18 minutes can be effective.

2. **Grassy shoot of sugarcane**: Phytoplasma (MLO)
   - Symptoms: Excessive tillering, sprouting of lateral buds.
3. **Smut**: ***Ustilago scitaminea*** (IARI, Ph.D-1997)
   - It is also known as '***culmicolous***' smut of sugarcane because it affects the stalk of cane.
   - Severe from April to July, 10% loss in juice

Symptoms:

- The appearance of black whip like structure at the apex of the stalk, it is also known as '***Whip smut***'.

4. **Red stripe**: ***Xanthomonas rubrilineans***
   - Long narrow red stripe appear on leaf, only leaves and top shoots are affected.
5. **RSD** (Ratoon Stunt Disease of sugarcane)

- Caused by ***Clavibacter xyli*** sub sp. *Xyli* (now *Leifsonia xyli*)
- It is called **'ratoon stunting'** because the disease is more severe on ratoon crops than on 'planted as sets of cane.
- Symptoms: Pink discoloration from the vascular bundles at growing points (nodes) is seen.

Other diseases:

1. **Gummosis**: ***Xanthomonas axonopodis*** *pv.vasculorum*: the fibro-vascular becomes red.
2. **Grassy shoot disease** is caused by **Phytoplasma,** vector is aphid (*Aphis sacchari*)
3. **Sett rot** or **pineapple disease** is caused by: ***Ceratosystis paradoxa.*** Initially it causes rotting of sugarcane seed pieces or setts.

13. POTATO

1. **Late blight**: ***Phytophthora infestans***
    - It is most destructive disease of potato and tomato also.
    - Symptoms appear late in season, at the flowering stage.
    - Bright brown spots on leaves and irregular patches starts from leaf tip later turned in brownish black patches.
    - First, leaves near ground show symptoms and in favorable condition entire vegetative parts are killed within a day hence called ***blight***.

Management:

- Foliar application of copper oxychloride @ 1.5 kg/ha through spaying
- Spray Metalaxyl or Mancozeb 2.0 kg/ha.2-3 sprays at 15 days interval.

2. **Early blight**: ***Alternaria solani***
    - It occurs earlier than the late blight, so named ***early blight***
    - Concentric ring or ***'target board'*** appeared on leaf lamina is most characteristic symptoms, spots are irregularly distributed.
    - Spray Mancozeb @ 0.2%
3. **Black scurf**: ***Rhizoctonia solani***
    - Symptoms: surface of tubers covered with black incrustations (Sclerotia of pathogen), growing tips of tuber sprouts are killed.
4. **Wart: *Synchytrium endobioticum***
    - Appearance of tumors or warts on tubers, stems and stolon.
5. **Black heart:**
    - Due to **oxygen starvation** (O_2 deficiency in storage) and high temp.
6. **Scab: *Streptomyces scabies.***

14. SESAME

1. **Phyllody disease** is caused by **Phytoplasma (MLO)**
   - Transmitted by leaf hoppers, *Orosius albicinctus*

Symptoms: Green leafy floral parts and profuse branching

Management: Spray dimethoate 1 ml/lit.ofwater.

15. COTTON

1. **Bacterial blight**/angular leaf spot/black arm: *Xantomonas campestris* pv. ***malvacearum***
   - Most serious disease of cotton in India
   - Angular and water soaked lesion on leaf & stem.
   - On the stem develops peripheral cracks called '***black arm***'.

Management:

- Spray a mixer of Streptocyclin + Copper oxychloride @ 50 + 250 g/ha
- Seed treatment with **Agrimycine**

2. **Tirak/bad opening of bolls**: **Physiological disorder**
   - Premature defective opening of bolls & shedding of leaves

Management:

- Follow late sowing
- Apply extra water at flowering & fruiting in sandy soil.

3. **Fusarium wilt:** ***F. monilliform*****e** f. sp. **vasinfectum.**
4. **Stenosis/small leaf**: **Viral.**

16. TOMATO

1. **Late blight and early** blight: **Fungal diseases**
   - Burnt like appearance on leaves

Management:

- Spray of Copper oxychloride @ 2 g/liter or Mancozeb @ 2g/liter
- Spray Bacteriomycin @ 0.2 g/liter of water

2. **Leaf curl**: Viral
   - Transmitted by white flies, *Bemisia tabaci*

Management:

- Acephate 1.5 ml/liter
- Acetamiprid @ 60 g a.i. /ha

3. **Blossom end rot**: **Calcium deficiency**
4. **Fruit cracking**: **Boron deficiency.**

17. BRINJAL

1. **Phomopsis blight or fruit rot**: ***Phomopsis vexans***
   - Soft & water soaked on fruit, blight on foliage, fruit rot
   - Seed treat with Carbendazim @ 1 g/kg of seed
   - Spray Bavistin 0.05 %
2. **Little leaf of brinjal**: Phytoplasma (MLO)
   - Vectors is leaf hoppers (*Hishimonus phycitis*)

18. CAPSICUM

1. Damping off: ***Pythium aphanidermatum* or *Phytophthora sp.***
2. Anthracnose/ripe rot or die back: ***Colletotrichum capsici***
3. Leaf curl: **Viral**
   - Downward cupping/curling of leaves
   - Transmitted by **thrips** and **whiteflies**
   - Managed by seed treatment with **Acetamiprid** @ 45 g a.i. /ha.
   - Spray of Acephate @ 1.5 ml/liter + Streptocycline @ 0.15 g/liter.

19. CRUCIFERS

1. **White blister** same as **mustard**
2. **Browning of cauliflower**: **Boron deficiency**
   - Brown curd, hollow stem
   - Apply Borax @ 10-15 kg/ha in soil or spray 0.2 %
3. **Whiptail of cauliflower**: **Mo deficiency**
   - Thick, green & leathery leaf and no curd formation.
   - Apply 1-2 kg sodium or aluminum molybdate
4. **Club-root** (finger & toe disease): ***Plasmodiophora brassicae***
   - It is plasmodium disease, formed club-shaped roots.
5. **Black rot of cabbage**: *Xanthomonas campestris*
   - Leaf margin as chlorotic or yellow area. The yellow area extends to veins & mid rib turning characteristic 'V' shaped chlorotic spots. And finally blackening of vascular bundles of roots occurs.

20. BANANA

1. **Panama wilt**: ***Fusarium oxysporium*** var. **Cubense**
2. **Bunchy top**: **Virus (BBTV)**
   - Vector is aphid: ***Pentalonia nigronervosa.***

21. CITRUS

1. **Citrus canker**: ***Xanthomonas compestris*p.v.*citri***
   - Small brown corky outgrowths on leaves, twinges and fruits.
2. **Die-back/wither tip**: ***Colletotrichum gloeosporioides***
3. **Gummosis**: ***Phytophthora palmivora***
   - Ruptured lengthwise bark, exudes gum
4. **Greening disease: *Candidatus Liberobacter***
   - A gram negative, walled phloem-colonizing bacterium is responsible for the destructive citrus-greening disease (earlier known as **mycoplasma disease**).
   - In China called as **citrus Huanglongbing** (HLB) or **Yellow dragon disease.**
   - Transmitted by vector Asian citrus psylla: ***Diaphorina citri***
   - Drench with Tetracycline @ 500 ppm
5. **Tristeza**: **viral**
   - Vector is aphids: ***Toxoptera citricida***

22. CUMIN

1. **Alternaria blight**: *Alternaria burnsii*
   - It occurs after flowering stage in cumin
2. **Wilt:** *Fusarium oxysporium* pv. *cumini*
   - Wilt is most severe disease of cumin

23. CARDAMUM

1. **Damping-off (Rhizome rot):** *Pythium vexans or Rhizoctonia solani*
2. **Katte disease (mosaic or marble disease)**: caused by Cardamom mosaic virus (CdMV), Transmitted by banana aphid.

   Management:- check the plantation for symptoms of disease at least twice a year; remove and destroy affected plants; replant with healthy material.
3. **Azhukal (Capsule rot) disease**: *Phytophthora spp.*

   Symptoms: -Water-soaked lesions on rachis; plant dries up beyond point of infection; rotting capsules cause foul smell.

7. NEMATOLOGY

Important points:

- ✓ Father of nematology: **H.C. Bastian**.
- ✓ Father of American or modern nematology: **N.A. Cobb**
- ✓ Abnormal increase in the number of cells is known as: **Hyperplasia**
- ✓ Abnormal increase in the size of cells (due to nucleus dividion) of a tissue is known as: **Hypertrophy**
- ✓ Hypertrophy and Hyperplasia are formed at feeding site of: ***Meloidogyne***
- ✓ **Mentake** disease of rice is caused by: **Rice root nematode.**
- ✓ The entire names of nematode corrected by **Filipjev,** 1936.
- ✓ Root lesion nematode, ***Pratylenchus coffee*** is a serious pest of coffee in S. India.
- ✓ Borrowing nematode '***Rodopholus similis'*** cause severe damage to fruit crops, spices and it's responsible for ***'spreading decline'*** of citrus in Florida.
- ✓ Citrus nematode cause **'slow decline & die back'** of citrus.
- ✓ Reniform nematode is as **second** important nematode of vegetables after root-knot.
- ✓ The nematode body is basically ***bilaterally symmetrical***
- ✓ **Curvature** of nematode body is always towards **ventral side** except in case of **Dorsella** which curved in a **'C'** shape towards dorsal side.
- ✓ All the natural openings of nematode are situated in **ventral side**
- ✓ **Amphids:** Paired chemosensory organ, located near lip region and play important role to finding host cell.
- ✓ **Phasmids:** these are paired, chemoreceptor in function and found on lateral side in posterior region of nematode.
- ✓ **Deirid**: are paired structures apposite to excretory pore and are mechanoreceptors in function.
- ✓ **Circulatory** and **respiratory** system are absent in nematodes. **(JRF-2011)**
- ✓ Number of moulting in nematode generally four (it means nematode have four larval stages)**: JRF- 2011**
- ✓ In plant parasitic nematodes the first molt (**J_1**) occurs within the eggs and **stylet** bearing 2nd stage larvae (**J_2**) are released.
- ✓ ***Anguina tritici*** undergoes **quiescent** in J_2 for up to **28-32 Years**.
- ✓ In ***Ditylenchus dipsaii*** the **J_4** (fourth) stage undergoes quiescence.
- ✓ Majority of the plant pathogenic nematodes belong to class-***Secernentea,*** and order ***Tylenchida.***

- ✓ In Tylenchids cuticle is marked with transverse striation known as ***striae*** these are deep in criconematidae and called ***annuli*** or spine on posterior margin, and are useful for taxonomic importance.
- ✓ Only hypodermis is **metabolic active** & unicellular layer.
- ✓ In **tylenchids:** Present **stoma** or **baccal cavity** or **stylet,** hence tylenchids named **stomatostylet**, it used for piercing into plant cells.
- ✓ Stomatostylet developed from fusion of walls of **baccal cavity**.
- ✓ In some nematodes rectal glands **three in female** and **six in males** are present; they are responsible for production of **gelatinous matrix** in which eggs are laid. It is normally feature of ***Dorylaimida*** but it has also been reported in ***Meloidogyne.***
- ✓ ***Aphelenchoides sp***. associated with strawberry disease known as **'spring crimp'** and **'summer crimp'**.
- ✓ Root rots, causes on potato tuber by: ***Ditylenchus destructor***.
- ✓ Lesion & necrosis caused by ***Pratylenchus*** & ***Rodopholus similis***.
- ✓ First journal of nematology in the world was: ***Nematologia***
- ✓ Nematode possess six stages in the life cycle i.e. ***eggs, 4 larval or juvenile stages and the adult***
- ✓ Most of Plat Parasitic Nematode (PPN) generally attacked: ***on roots***
- ✓ Esophagus of nematode is consists of 3 parts known as: ***triradial symmetry***
- ✓ The oral aperature is basically surrounded by six lips known as: ***hexaradial symmetry***.
- ✓ Marrygold (Tagetes) and garlic act as antagonist to: ***root knot nematodes***
- ✓ All genera of virus vector nematodes belong to order: ***Dorylaimida***
- ✓ All genera of virus vector nematodes have: ***autoparasitically mode of feeding***
- ✓ Nematodes transmitted only: ***RNA viruses.***
- ✓ By growing of a non-host crop in between two susceptible ones the population level of the nematode is brought down below damaging threshold is example of: ***Cultural control (crop rotation)***

Unit-1. Nematology

Study of free living and plant parasitic nematodes is known as ***nematology.***

Nematode: It derived from two **Greek** words:

- *Nema* = Thread
- *Oides* = Like

Means thread like organisms also known as Eelworm or **round worms** because round in cross section.

Definition:-

- Nematodes are ***triploblastic***, ***bilaterally symmetrical***, unsegmented, ***pseudocoelomatic*** invertibrates with a ***triradial oesophagus*** and lacking respiratory and circulatory system.
- Nematodes are **pseudo coelomate:** only ectoderm is linked with the mesoderm while it is not lined to endoderm or alimentary canal of nematode.

Morphology:-

- The nematode body is consists of a tubular structure having two tubes i.e. the outer tube is known as body wall and the inner tubes is known as alimentary canal, the space between the two tubes is called as body cavity which is ***pseudocoelom*** because it is devoid of mesodermal linings.

The body wall (outer tube) of nematode is consists of three layers i.e. cuticle, hypodermis or epidermis and the muscle layer.

(a) Cuticle:-

- It is non-cellular, non-living layer of outer body wall
- It is made up from structural protein i.e. chitin is not present in nematode cuticle while in insect, cuticle is made of 25-60% chitin and 25-40% protein (on dry wt. basis).

(b) Epidermis:-

- The epidermis of nematode is unicellular layer and its major function is to secretion of cuticle.

(c) Muscle layer:-Internal layeron which muscles are attached

The alimentary canal (inner tube) of nematode body is divided into three parts i.e. stomodeum, mesentron, proctodeum.

1. Stomodaeum:-

- It is the anterior parts of alimentary canal.
- In Plant Parasitic Nematodes (PPN), the stoma walls form hypodermic needle like structure known as stylet or ***stomatostylets.***
- In Tylenchid (PPN) stomatostylet is comprises of three parts i.e. ***conus, shaft and knobs.***

Unit-2. Helminthology

- Study of nematodes which are **parasitic** exclusively on **man** and **domesticated animals** is known as ***Helminthology.***

Animal parasitic nematodes (APN):

SN	Nematode (APN)	Disease
1	*Placenta gigentissima*	It is largest APN (about 8m long) found in **whale fish**
2	*Dracunculus medineusis*	**Guinea worm**, cause **'Naru bala'** disease in human
3	*Wucheraria bancrofti*	Cause **'Elephantiasis'** or **Filariasis** disease in man. **(JRF Ento -11)** The vector for **'Filariasis'** disease is *Aedes agypti* (mosquitoes*)*. **(RPSC-AAO-09)**
4	*Ascaris lumbricoides*	Known as **round worms,** intestinal parasite of human
5	*Enterobius vermicularis*	It causes severe itching in anal area particularly while childerens are sleeping.
6	*Onchocera volvulus*	It causes **'Onchocerciasis'** or river blindness, commonly in Africa, transmitted by black fly, *Simulium sp.*

Unit-3. History of nematology

1743: **J.T. Needham**: discovered first plant parasitic nematode (PPN) ***Anguina tritici*** and it was named ɛs ***'Vibrio tritici.***

1855: **Berkeley,** reported root knot nematode (***Meloidogyne spp***.) from roots of greenhouse cucumber in England.

1857: **Kuhn** reported stem & bulb nematode, ***Ditylenchus dipsaii***.

1859: **Schacht**, reported sugar beet cyst nematode (***Heterodera schachtii***).

1865: **H.C. Bastian**, he known as **Father of Nematology** and wrote monograph of the, ***Anguillulidae***

1871: **Kuhn** first used **CS_2** gas for control of Sugar beet cyst nematode.

1884: **J.G. Deman**, gave **'Biometry'** and taxonomic monograph (**Morphometry**).

1907: **N. A. Cobb** known as **Father of American Nematology**.

1933: **Tom Godey**: He published two books;

1. **Plant parasitic nematodes and disease they cause** in 1933.
2. **Soil and fresh water nematodes** in 1951.

1937: **Chitwood & Chitwood** wrote a book: ***'An Introduction to***

Nematology'.

1943: **Carter** discovered nematicidal properties of **DD**. **(JRF- 09)**

1945: **Christie**, discovered nematicidal properties of **EDB**.

1956: Nematologica-first journal of European Society of Nematologist.

1958: **Hewitt**, **Raski** and **Goheen**, first reported the **transmission** of viruses by nematode ***(Xiphinema index)*** transmitting 'grape vine fan leaf' virus disease in USA.

1961: **G. Thorne** wrote book ***'Principle of Nematology'*** (known as Bible of nematology).

History of Nematology in India:

1901: **Barber** reported **'root knot nematode'** (Meloidogyne spp.) infesting tea in Keralaas a first PPN in India.

1913: **Butler** reported **'Ufra'** disease of rice caused by ***Ditylenchus angustus*** from Bengal.

1919: **Milne**, reported ***Anguina tritici*** (seed gall) of wheat in Punjab.

1936: **Dastur**, reported **white tip disease** of rice by ***'Aphelenchoides besseyi'***.

1958: **Vasudeva** reported **Molya disease** of wheat (***H. anenae***) from **Neem ka Thana**, Sikar, Rajasthan, later by **Sesadhari (1959)**.

1961: **Siddiqi**, reported **citrus nematode** from **UP** (Aligarh).

1961: **FGW Jones** reported **potato cyst** (*Globodera rostochiensis)*. Nematode from Nilgiri (**TN**).

1966: **Division of Nematology** established at **IARI**, New Delhi.

1969: **Nematological Society of India** was founded at **IARI, New Delhi**.

1971: First time **Indian Journal of Nematology** published.

1977: All India Co-ordinate Research Project (AICRP) on Nematode pests of crops and their control started.

Unit-4. Classification of nematodes

The word taxonomy coined by: **De Candole in 1875**

- Phylum = **Nematoda**

Class: Two important classes: **Secernentia** and **Adenophorea.**

A. Secernentia: (known as phasmida)

- ✓ Phasmids are present
- ✓ Caudal gland are absent

Class secernentia has two orders, namely tylenchida, aphelenchida.

1. **Order: Tylenchida.**
   - **Tylenchida** is highly evolved & sedentary parasite in nature.
   - It includes sub-order **tylenchina**

- Tylenchina is having super families of: Tylenchoidea, Haplaimoidea and Heterodoidea

a. **Tylenchoidea**:

Families: Tylenchidae (eg. *Tylenchus,*)

: Belonolaimidae (eg. *Tylenchorhynchus, Belonomus*)

: Anguinidae (eg. *Anguina, Ditylenchus*)

b. **Haplolaimoidea**:

Families: Pratylenchidae (eg. *Pratylenchus, Radopholus, Nacobbus*)

: Haploimidae (eg. *Haplolaimus, Rotylenchulus*)

c. **Heterodoidea:**

Families: Heteroderidae (eg. *Heterodera, Globodera*)

: Meloidogynidae (eg. *Meloidogyne*)

B. Class: Adenophorea (aphasmida).

✓ Phasmids absent

✓ Caudal glands are present

1. Order: Dorylaimida.

Super family: Longidoroidea

Family: Longidoridae (e.g. *Longidorus, Paralongidorus*)

: Xiphinematidae (e.g. *Xiphinema*)

Super family: Trichodoroidea

Family: Trichodoridae (e.g. Trichodorus, Paratrichodorus)

Difference between tylenchida and dorylaimida:-

Tylenchida	**Dorylaimida**
Presence of three parts of esophagus	Two part esophagus is present
Bacal cavity is known as ***Stomatostylet***	Bacal cavity is known as ***Odontostylet***
Cuticle possess distinct annulations or transverse striations	Cuticle annulations is not clearly observed

Unit-5. Parasitism of nematodes

1. **Ectoparasite:** Living and feeding on a host from outside.
   - **Migratory** parasite eg. ***Xiphinema*** **(JRF-2010)**, ***Longidorus***
   - **Sedentary** e.g.: *Hemicycliophoro*.
2. **Semi-endoparasite**:
   - Migratory e.g. *Haplolaimus*, ***Tylenchorhynchus***.
   - Sedentary e.g. ***Rotylenchulus***, ***Tylenchulus***.
3. **Endoparasite**: Enters the host tissue and feeds from within.
   - Migratory e.g. ***Pratylenchus***, ***Rodopholus***.
   - Sedentary e.g. ***Meloidogyne***, ***Heterodera***.

4. **Both Ecto and Endo parasite:**
- Ditylenchus
- Aphelenchoides
- Anguina

Unit-6. Important plant parasitic nematodes:

1. Seed gall/ ear cockle nematode of wheat: ***Anguina tritici***
2. Tundu/ Sehu or yellow ear rot of wheat: ***Anguina tritici + Clavibacter tritici.*** (**Yellow slime** or gum on leaves is major symptoms of Tundu)
3. Cereal cyst nematode: ***Heterodera avenae***.
4. Root knot nematode: ***Meloidogyne spp.***
5. Citrus nematode: ***Tylenchulus semipenetrans***.
   : (Formation of 6-10 cortical cells, known as **'nurse cells'** is specialized symptoms, cause slow decline or die back of citrus).
6. Reniform (pulse) nematode: ***Rotylenchulus reniformis***.
7. Stem & bulb nematode: ***Ditylenchus dipaci***.
   : (**'Onion bloat'** in onion is major symptoms of *D. dipsaii*)
8. Stem nematode of rice: ***Ditylenchus angustus.***
   : (Cause **'Ufra disease'** of rice)
9. Potato root nematode: ***Ditylenchus destructor***.
10. Mushroom nematode: ***Ditylenchus mycelophagus***.
    : (it reported as **First mushroom** nematode)
    : Second mushroom nematode is- ***Aphelenchoides composticola***
11. Bud and leaf nematode (foliar): ***Aphelenchoides sp***.
    : Rice leaf or white tip nematode: ***Aphelenchoides besseyi.***
12. Coconut red ring nematode: ***Rhadenaphalenchus cocophilus***.
    : (Vector for this nematode is **palm weevil: *Rhyncophorus feruginonsis***).
13. Potato cyst nematode: ***Globodera rostochiensis***.
    : (a characteristic symptom ***Syncytium*** is developed)
14. Root lesion nematode: ***Pratylenchus sp***. **(JRF-2011)**
    : (also referred as **'Meadow'** nematode)
15. Rice root nematode: ***Hirschnynniella sp***.
    : Cause **'Mentek disease'** of rice. **(JRF-2011)**
16. Burrowing nematode: ***Rodopholus sp***.
    : Infective stage: **all**
17. Blackhead disease, rot root or rhizome root of banana: ***Rodopholus similis*** (also cause **'Yellow pepper'** disease in black pepper),
18. Spreading decline of citrus: ***Rodopholus citrophilus***.
19. False root knot nematode: ***Naccobus sp.*** (because symptoms

resemble to as root knot nematode by enlargement of cells)

20. Stunt or stylet nematode: ***Tylenchorhynchus sp***.
21. Ring nematode: ***Criconema sp***.
22. Pin nematode: ***Paratylenchus sp.*** **(JRF-2011)**
23. Needle nematode: ***Longidorus sp***. **(JRF-2011)**
24. Dagger nematode: ***Xiphinema index*** **(JRF 2011)**
25. Stubby root: ***Trichodorus sp.***
27. Lance nematode: *Haplolaimus sp*. **(JRF- 2011)**

Unit-7. Nematodes of common crops

1. Seed gall/ear cockle nematode: *Anguina tritici*

- ✓ First reported by J.T. Needham in 1743.
- ✓ He called **Vibrios**; name *Anguina* given by Filipjev in 1936.
- ✓ In India first reported by **Milne from Punjabin 1919**.
- ✓ This nematode cause **epidemic in Bihar, MP** where the farmers burnt the standing crop of wheat.
- ✓ This **nematode association** with a bacterium ***Clavibacter*** (Corynebacterium) ***tritici*** is caused **yellow ear rot** or **Tundu** disease of wheat.
- ✓ The seed galls have **3000-12000** second stage juveniles **(J_2)** in a **quiescent stage**.
- ✓ It first feeds as **ectoparasitically** & after three moults it becomes **endoparasitic.**
- ✓ Some species of *Anguina* produces galls in leaf surface for e.g. ***Anguina tumafaciens*** on *Cynodon transvalense*. **(JRF-11).**

Symptoms:

- Seedling leaves are **twisted** and **crinkled.**
- Seedling may **stunted** and **die**.
- Plant show **profuse tellering** and infested plant produce ears almost 30-40 days before healthy plants.
- Finally seeds are replaced by ***galls***.
- The most important character of this nematode is the conversion of grains in to ***cockles or galls.***

Management:

- **Salt floatation: Use of 20 % salt** (brine) solution for washing of seeds is very effective control method.
- Hot water treatment: **54 ºC for 10 minutes**.
- **Crop rotation** also useful method to escape infestation of nematode.

2. Cereal cyst nematode: *Heterodera avenae*

- ✓ Cause "**Molya disease'** in wheat and barley.
- ✓ ***Heteros*** = different and *deras* = skin means different types of skin & **lemon shaped** female.
- ✓ Females have ***zig-zag pattern*** or lace like pattern on body known as ***cyst wall pattern.***
- ✓ **Cyst nematodes**: Includes genus, ***Heterodera*** and ***Globodera*** both have **unique character** of retention eggs & larvae within mature female body which turn into brown or black hard body after its death; that why known as ***cysts***. (JET Exam -2014)
- ✓ **Syncytium** is formed at feeding site in cyst nematodes.
- ✓ Cyst nematodes are generally host specific.
- ✓ Pigeonpea cyst nematode (***H.*** *cajani)* also infests other pulses.
- ✓ Sugarbeet cyst nematode is: ***Heterodera scchatii***
- ✓ Soybean cyst nematode: ***H. glycine*** (it cause **yellow dwarf disease** in soybean)
- ✓ ***Heterodera oryzicola*** (paddy cyst nematode) serious pests of rice

Damage:

- J_2 is infective stage
- ETL = **3 eggs/g of soil.**

Management of cereal cyst nematode:

- Crop rotation at least 3 years with mustard, chick pea is most effective control because it is very host specific.
- Early sowing escape to damage
- Deep ploughing is destroyed quiescent eggs of nematodes.
- Grow resistant varieties e.g. for wheat: **Raj MR-1**
- For barley: **RD 2052,** RD-2035 and **Rajkiran** ***(JRF-11)***

3. Potato cyst nematode: *Globodera rostochiensis.*

- ✓ Golden nematode of potato first reported by **FGW Jones** in 1961 from Nilgiri Hills (TN).
- ✓ It feed pericycle cortex with results formation of nurse cells i.e. ***Syncytium*** develop which provides nutrients to female throughout development.

Management:

- **Legislative methods: Domestic quarantine** was introduced against **Globodera**, not carry infested potato tubers without phytosanitory certificate issued from **Nilgiri to other part** of country. ***(JRF-06)***
- **Resistant variety**: **Kufri swarna** is most resistant to *Globodera sp.*
- **Fumigants**: **Methyl bromide** exposer for 5-6 hrs.

4. Root-knot nematode: *Meloidogyne spp.*

- ✓ Means **melon or apple** shaped female.
- ✓ First reported by **Barkley 1855** from **cucumber.**
- ✓ In India it reported by **Barber in 1901** on tea (Kerala).
- ✓ In meloidogyne the ***Perineal pattern*** present on **vulva-anus area** which is helpful in **species identification.** **(JRF-07)**
- ✓ Eggs lay in gelatinous matrix (200-500 eggs).
- ✓ **Sex differentiates** occurs in late second juvenile stage (**J_2**) where the genital primordial is become 'V' shaped in female and 'I' shaped in male.
- ✓ **Second juvenile (J_2) is only infective** stage in Meloidogyne. **(JRF -11)**
- ✓ ***'Spike tail'*** in Meloidogyne: it is succeeding molting of second, third and fourth stage juveniles of *Meloidogyne spp.*
- ✓ ***Sexual reversal:*** In ***Meloidogyne*** some time under adverse conditions male changed in tofemale.
- ✓ Most of species of Meloidogyne completed life cycle within 25 days at ***25-*** 30°C and 7-8 overlapping generation completed in a year.

Classification:

- *M. incognita*: Present throughout India on **vegetables.**
- *M. exigua*: Found on coffee in **Karnataka.**
- *M. hapla*: Restricted to **temperate parts** (on hilly area).
- *M. arenaria*: It is a serious pest of **ground nut** in Gujarat. **(JRF-11)**
- *M. brevicauda*:Found on tea in Assam
- *M. graminicola*: Found on rice in eastern part of country.**(JRF-2010)**

Symptoms:

- The reaction to the entrance of the larvae is the development of ***hypertrophy*** (enlargement of cell size due to nucleus division) of cortical cells and formed ***6-10 Giant cells*** from which it derives food.
- Nematode secrete protease enzyme results break down protein into amino acid (**tryptophan**) which is a **'precursor of IAA'** is important and responsible for creating hormonal imbalance at site of reaction. Due to excessIAA formation at feeding site cause ***gall/knot*** on **roots.**

Difference between **root knot** and **rhizobium nodules**:

Root knot	Rhizobium nodules
In case of meloidogyne infestation root knots are hard in consistency and deep seated	Rhizobium nodules are spongy consistency, found along side of roots as side appendages
They don't easily rubbed off	Soft and can be easily detached from roots

Management:

- Deep ploughing in May-June is very effective to kill quiescent J_2 stage.
- Interculture of marigold (*tagetes sp.*) as a trap crop.

Grow resistant varieties:

- For tomato **Hissar lalit, SL- 120**
- For brinjal: **Black beauty, Neelkantha**.
- For: chilli: **Pusa jwala**.

Biological control of *Meloidogyne*:

- Fungus: ***Paecilomyces lilacinus, Pochonia chlamydosporium***
- Bacteria: ***Pasteuria penetrans, Pseudomonas fluroscence***
- ***Bt. Kurstaki***: also effective against *Meloidogyne spp.*

5. Citrus nematode: *Tylenchulus semipenetrans*

- ✓ Reported by **Cobb** 1912 (USA).
- ✓ In India by Siddiqui, 1961 in Aligarh (UP).
- ✓ Juvenile second (J2) is infective stage.
- ✓ At feeding site formed ***6-10 nurse cells***. **(JRF-11)**
- ✓ It is cortical feeder as sedentary semi-endoparastic.

Symptoms:

- Slow decline/citrus decline and die back. **(JRF -2011)**

6. Reniform nematode: *Rotylenchulus reniformis*

- ✓ Mature female is **'Kidney shaped'**.
- ✓ Reported by **Linford** and **Oliveria** from USA in 1940.
- ✓ In India by **Seshadhari** & **Shivkumar** 1963 on cotton.
- ✓ Pre adult (immature female) is infective stage.

Two races:

- **Race 'A':** Multiply on cow pea, castor & cotton.
- **Race 'B':** Reproduces only on cowpea. **(JRF-11)**

7. Stunt or stylet nematode: *Tylenchorhynchus spp.*

- It is migratory semi-endoparasite
- *Tylenchorhynchus brassicae*: Feeds on **Cruciferae.**
- *T. brevilineatus:* Cause **'Kalahasti disease'** in ground nut in AP.
- *T. vulgaris:* Feeds on **cereals.**
- *T. claytoni:* Known as **'tobacco stunt'** nematode.

8. Stem & bulb nematode: *Ditylenchus sp.*

- ***Nemawool***: **A** frothy substances secreted through saliva of pre adult (J_4) nematode responsible for quiescent in ***Ditylenchus***.

a. Rice stem nematode: ***Ditylenchus angustus***

- ✓ In India only ***D. angustus*** is reported.
- ✓ It is reported by **Butler, 1913**

- ✓ It causes **'Ufra'** disease
- ✓ Infestation mainly in deep water paddy.

Symptoms:

- **Twisting of leaf** & leaf sheath.
- The symptoms of **Ufra** are well noticed in **two monthsold crops**.

b. Stem and bulb nematode: ***Ditylenchus dipsaii***

- ✓ By infestation of this nematode onion suffers heavy losses as it cause **'onion bloat'** disease.

c. Potato rot nematode: ***Ditylenchus destructor***

- ✓ It found in temperate region of Europe & N. America
- ✓ Cause **'rots'** on potato roots.

d. Mushroom nematode: ***Ditylenchus myceliophagus***

- ✓ This nematode was first reported that to cause damage on **mushroom.** It is also found in India.

9. Leaf and bud nematode: *Aphelenchoides sp.*

- ✓ Also known as **foliar nematode**

1. **White tip or rice leaf nematode**: ***Aphelenchoides besseyi***
   - ✓ This nematode associated with **serious epidemic of rice in MP.**
   - ✓ It feed **actoparasitically**
   - ✓ Humidity favors infestation.
   - ✓ This is **seed borne** nematode
2. **Chrysanthemum/foliar nematode**: ***Aphelenchoides ritzemabosi***
   - ✓ Known as **Spring crimp** or straw berry nematode
3. ***Aphelenchoides composticola***:
   - ✓ Important nematode of mushroom and known as mycophagus nematode

10. Red ring nematode:*Rhadinaphelenchus cocophilus*

- ✓ Cause serious disease of coconut palm is called **'red-ring'**.
- ✓ This nematode is transmitted mechanically through palm weevil, ***Rhynocophorus palmarum***.

11. Root lesion nematode: *Pratylenchus sp.*

- ✓ Also referred as **'Meadow nematode'**.

12. Rice root nematode: *Hirschrmunniella sp.*

- ✓ This nematode responsible for causing **'Mentek'** disease of rice in 1949.
- ✓ **Infective stage**: all Juveniles (J) and adult also. **(JRF-2011)**
- ✓ It can be managed by soil treatment of Carbofuran @ 4kg a.i/ha before sowing.

13. Burrowing nematode: *Rodopholus sp.*

a. **Root and rhizome rot of banana** : ***Rodopholus similis***

- ✓ It is commonly found in India.
- ✓ Other names are **'banana decline' and 'black head disease** 'of banana.
- ✓ This nematode also helpful in causing **panama wilt** of banana in association with fungus ***Fusarium oxysporium*** var. ***cubease***
- ✓ It is caused **'Yellow disease'** of pepper.

b. ***Rodopholus citrophilus:***

- ✓ Responsible for causing **'spreading decline'** of citrus in Florida, USA

14. False root knot nematode: *Naccobus sp.* (JRF-11)

- ✓ Develops **hypertrophy**

15. Ring nematode: *Criconema sp.*

- ✓ It has large annuli.

16. Pin nematode: *Paratylenchus sp.*

- ✓ This is **smallest** plant parasitic nematode with size of **< 0.5 mm.**

17. Dagger nematode: *Xiphinema sp.*

- ✓ Presence of characteristic **sword** like stylet
- ✓ It transmit **fan leaf virus** of grape virus.

Nematode and microbes interaction:-

- The incidence of cotton wilt (*Fusarium oxysporium*) was more severe in presence of root knot nematode
- In strawberry by the infestation of *Aphelenchoides frageriae* (nematode) and *clavibacter fusciens* (bacteria) cause ***'cauliflower disease'***.

Unit-8: Management of nematodes

1. **Chemical control:**

A. **Fumigants:** e.g. **DD-** a mixture of 1, 3-dichloropropene, 1, 2-dichloropropane (**Pre PG Exam-2015**), **EDB-** ethylene dibromide, MBr-methyl bromide, Nemagone (DBCP dibromochloropropane)

B. **Non-fumigants:** are grouped in to:-

(i) **Carbamates:** - Aldicarb (Temik), Carbofuran (Furadan), Oxymyl (Vydate), Metham sodium (Vapam).

(ii) **Organophosphate:**-Thionazin (**Nemaphos**),Phenamiphos (**Nemacur**)

2. **Physical control:** Use of Heat, hot water treatment (54^0C for 10-15 min.) Irradiation, Osmatic pressure and Ultrasonic.
3. **Regulatory methods:** such as adoption of legislative quarantine regulations e.g. potato tubers used as seed materials restricted use from Nilgiri hills (TN) to other part of country.

4. **Biological control:**

   (i) **Fungi**- Trchoderma ***spp., Pochonia chlamydosporia*** *(egg parasitic),* ***Paecilomyces lilacinus*** *(Oviparasite),* ***Arthrobotrys dactyloides,*** *Metarhizium anisopliae and Beauveria bassiana* are also found effective especially against root knot nematode.

   (ii) **Bacteria**: *Pasteuria penetrans (larval parasite), Pseudomonas fluroscence, B. subtilis and Bacillus cereus, B. velezensis (VB7), Bacillus thuriengensis (Bt)* are also used for controlling root knot nematode.

5. **Cultural methods:**

   a. **Crop rotation:** It is most effective for ***Molya disease*** control by growing of mustard and gram.

   b. **Trap/Antagonistic crop:**

      (i) **Mustard:** contains Allyl isothiocynate toxic for **cyst** nematode.

      (ii) **Marigold**: contains **α- terthienyl,** reduce population of *Meloidogyne spp., Paratylenchus* and *Rotylenchulus.*

   c. **Flooding**: it kills nematode by asphyxiation.

   d. Propagation through healthy planting material.

   e. **Soil solarization**: in summer it is effective by 2-3 cross deep ploughing or use of transparent polythene sheet for solarization.

Unit-9 Entomopathogenic nematode (EPN):

- Nematode families of Steirnematidae, heterorhabditidae have potential as biocontrol agents. Because these are ***Entomopathogenic*** *nematode* and are facultative parasites they associated with symbiotic bacteria e.g. The ***Xenorhabdus*** *bacteria* associated with ***Steinernema*** and the ***Photorhabdus bacteria*** *is associated with nematode* ***Heterorhabditid***.

EPN Products:

- In India a commercially formulation ***"Ecomax"*** is available, it's prepared from nematode ***S. carpocapsae*** and ***H. bacteriophora.***
- The product of ***Neoplectana carpocapsae*** with Bactria, ***Achromobacter nematophilus*** is commonly named as **DD-136.** It is also available in several labs under the Trade name of ***'Black Commando'***.

Unit-10. Virus vector nematodes

1. **NEPO viruses**: Nematode-transmitted polyhedral particles of viruses.

   - Tomato black ring (TBR): ***Longidorus sp.***
   - Raspberry ring spot: ***Longidorus elogatus***
   - Grapevine fan leaf (GFLV): ***Xiphinema index.*** **(JRF-07)**

- Tobacco/tomato ring spot (TRSV): *Xiphinema americana*

2. **NETU or TOBRA viruses**: Nematode transmitted tubular particles.
   - Tobacco rattle virus (TRV): *Trichodorus similis*. **(JRF-04)**
   - Pea early browning (PEBV): *Trichodorus sp*.
   - Pepper ring spot virus (PRV): *Paratrichodorus minor*

Some terms-

Mesentron of nematodes:

- **Initima** (cuticular lining) is absent and fine finger like projection known as **microvilli** present around lumen.
- Microvilli **increases the surface area** of intestine and are both secretary & absorptive in function.

Female ovaries:

- **Monodelphic**: Only one ovary (gonad) is functionally.
- **Didelphic**: Two ovaries are functionally in female
- **Mono prodelphic**: -The female gonoduct is one in number and projected anteriorly.
- **Amphidelphic**: - In female one gonoduct is projected anteriorly and other posteriorly.

Male testis:

- **Monorchic:** Only one testis is present
- **Diorchic:** Two testes are present.
- The spicules, gubernaculums and bursa are secondary sexual organ in: male nematode
- Most of nematode generally possess a single genital tract is known as: ***monorchic***

Voltinisms: refers to number of generations in a year

1. **Multivoltine nematodes**: many generations are completed in a year. Examples are: ***Meloidogyne incognita, Pratylenchus, Rotylenchulus and Tylenchulus sp.***
2. **Univoltine:** only one generation is completed in a year. E.g. ***Anguina* tritici** and ***Heterodera avenae*** is univoltine.

8. PLANT BREEDING & GENETICS

Important points:

- ✓ Father of plant tissue culture**:** ***G. Haberlandt (1902)***
- ✓ Father of Botany: ***Theophrastus***
- ✓ Father of Zoology: ***Aristotle***
- ✓ Father of Genetics: ***G.J. Mendel***
- ✓ The longest mitotic phase is: ***Prophase***
- ✓ Segregation occurs during: ***Meiosis only***
- ✓ The most abundant form of RNA is known as: *r-* ***RNA (80%)***
- ✓ Bacteriophages are: ***Single strain DNA (virus)***
- ✓ In mitochondria found **only 70 s** types of: ***Ribosomes***
- ✓ Alternative form of a gene is known as: ***Allele***
- ✓ Tift 60 is an important source of male sterility in: ***Sorghum***
- ✓ Often cross pollinated crops are: ***Cotton, sorghum, Arhar***
- ✓ The theory of inheritance was proposed by: ***G. Mendel***
- ✓ The shortest phase of all the mitosis phase is: ***Anaphase***
- ✓ The site of protein synthesis is: ***Ribosome***
- ✓ Physical basis of life is: ***Protoplasm***
- ✓ Onion is a modified form of: ***Stem (tunicate bulb)***
- ✓ Onion & garlic are example of: ***Bulb***
- ✓ Edible part of apple: ***Thalamus***
- ✓ Edible part of mango: ***Mesocarp***
- ✓ Edible part of coconut: ***Endosperm***
- ✓ An agent that causes gene mutation mutation is called as: ***Mutagen***
- ✓ Most mutant tolerant are: **Polyploids**
- ✓ Characters which express in one sex only is: ***Sex limited traits***
- ✓ RNA synthesis is in which organ: ***Nucleolus***
- ✓ The inflorescence of banana is known as: ***Spadix***
- ✓ The theory of evolution through natural selection was given by **Charles Darwin**, in his book on the **origin of species,**1859.
- ✓ The theory of acquired characteristics was proposed by: ***Lamarck***
- ✓ Single gene affecting more than one character is known as: ***Pleiotropy***
- ✓ Garlic & turmeric are examples of modified stem of: ***Rhizome***
- ✓ The structure of chromosome can be best observed during: ***Metaphase***
- ✓ **Ribosomes** are small cellular particles which are the sites of: ***Protein synthesis.***
- ✓ Closest form of inbreeding possible in self- incompatible sp. is: ***Sib- mating***

- ✓ Physical chromosomal exchange (crossing over) takes place in: ***Pachytene***
- ✓ Re-discovery of Mendel laws in 1900 by: ***de Vries, Carl Correns & tschemark***
- ✓ **Only about 0.**1% of the induced mutations are useful in crop improvement.
- ✓ The enzymes associated with Krebs cycle are also present in the: ***Matrix***
- ✓ Life was first originated in water the theory of genes in the chromosomes was introduced by: ***Morgan***
- ✓ **In 1944**, Avery, Mac Leod and Mc Carty demonstrated that DNA was the genetic material.
- ✓ In CAM plants transpiration is negligible since stomata are closed during the day.
- ✓ Examples of prokaryotes: *cells of* ***Mycoplama, Bacteria and Blue green algae.***
- ✓ Cytoplasm: includes all the structures of outside the nucleus.
- ✓ Protoplasm*:* ***Cytoplasm + Nucleus.***
- ✓ Centrosome is present only in animal cell.
- ✓ Openings are left at plants between the adjacent cells known as ***plasmodesma*** and thread like lining between two cells is known as ***Plasmodesmata***.
- ✓ A common crop of natural auto-tetraploids grown in India is: ***Potato***
- ✓ Cell wall of fungi is chiefly made up of: ***Chitin***
- ✓ Plant showing less than 5% cross pollination is considered as: ***Self pollinated***
- ✓ A variety resistant to more than one race of pathogen is due to: ***Horizontal resistance***
- ✓ Type of male sterility used in hybrid seed production in pearl millet: ***Cytoplasmic geneticmale sterility***
- ✓ If gamete is fertile but seed is not produce due to: ***Self incompability***
- ✓ An individual having both male and female reproductive organs is known as: ***Hermaphrodite***
- ✓ Aseptic conditions means: ***Pathogen free environment***
- ✓ ***Pedigree:*** record of ancestry of an individual selected plant for its various generations
- ✓ The study of interactions between antigens and antibodies is called: ***Serology.***

- ✓ Transfer of genetic material from one cell to another cell by means of viral vector: ***Transduction*** (discovered by Lederberg & Zinder, 1952)
- ✓ PCR techniques provides: **Several copies of specific DNA sequence**
- ✓ Science of identifying the sequence of DNA in species: ***Genomics.***
- ✓ The lines are homozygous and homogenous in nature called as: ***Pure line***
- ✓ Autotriploidy found in: ***Banana***
- ✓ If the development of fruit without fertilization the process called: ***Parthenocapry***
- ✓ Repeated crossing of hybrid progeny back to one of its parents called: ***Back cross***
- ✓ The ***Tift 23-A*** and ***Kafir-60*** are the important sources of male sterility in **pearl millet** and **sorghum** respectively.
- ✓ *Ethidium bromide* has been found effective in inducing cytoplasmic male sterility in **pearl millet** and **barley**.
- ✓ *Over-dominance*: is the phenomenon where the intensity of a character's expression is greater in the heterozygote than in either of the two concerned homozygotes.
- ✓ *Endoreduplication*: in plants is a cell cycle variation where cells replicate their DNA multiple times without dividing (mitosis), leading to polyploidy (more chromosome sets) and larger cells.
- ✓ *Holandric character* refers to a genetic feature determined by genes located only on the Y chromosome, leading to its inheritance exclusively from father to son, never to daughters, as only males possess the Y chromosome; examples include male fertility factors or hairy ears (hypertrichosis pinnae).
- ✓ *The Sodium hydroxide (NaOH) test:* it is seed health test, used to detect **Karnal bunt disease** in wheat and **kernel smut** in paddy.

SN	Organ	Discovered/Named by
1	**Living cell**	Antony von Leeuwenhoek ,1723
2	**Cell wall**	Robert Hook, 1665
3	**Nucleus**	Robert Brown, 1851
4	**Nucleolus**	Fontana, 1781, seen nucleolus But termed by Bowman, 1840
5	**Nucleic acid first discovered by**	E. Meischer , 1869 (he gave the term **nuclein**)
6	**Term 'Nucleic acid' given by**	Richard Altmann (1889)
7	**Cell theory**	Schleiden & Schwann 1839

8	**Copy –choice theory** (also known as classical theory)	John Belling(1928-1931)
9	**Genic balance theory of sex determination**	Calvin Bridge (1926)
10	**Term 'Mitosis'**	Flemming 1882
11	**Term 'Meosis'**	Former and Moore
12	**Plasmodesmata**	E. Strasburger, 1901
13	**Protoplasm**	J. E. Purkinje 1840
14	**Protoplast**	Hanstein, 1880
15	**Plasmalemma**	J.Q. Plowe 1931
16	**Unit membrane concept of cell membrane**	David Robertson 1962
17	**Fluid mosaic model**	Singer & Nicholson, 1972
18	**Endoplasmic reticulum**	K. R. Porter, 1945
19	**Plastid**	E. Haeckel, 1865
20	**Chloroplast**	A.F.W. Schimper
21	**Mitochondria**	C. Benda, 1867
22	**Golgi body**	Camilo Golgi
23	**RNA discovered by**	Palade 1955
24	**Ribosome**	First reported by **Palade,** 1953 But named by **Haguenau**, 1958
25	**Lysosomes**	De Duve
26	**Centrosome**	T. Boveri
27	**Chromosome** first observed by	E. Strasburger, 1875
28	**Term chromosomes** given by	Waldeyer, 1888
29	Term **'polytene'** coined by	P.C. Koller (1935)
30	**Giant chromosomes** (polytene) first observed by	E.G. Balbiani (1881)
31	**Double helical structure** of DNA was given by	Watson & Crick 1953
32	**Recombinant DNA** technology /DNA sequencing	Berg, Gilbert & Anger
33	**Genetic linkage** was first observed by	W. Bateson & R.C. Punnet (1906), studied on sweet pea
34	**Term Linkage & Linkage theory** given by	T.H. Morgan (1910), he worked on Drosophila)
35	Term **Phenotype, Gene & Genotype**	W. Johannsen, 1903

36	**Overdominance hypothesis** of heterosis in maize proposed by	E.M. East & G. Shull (1908)
37	**X-ray act as physical mutant** discovered and demonstrated by	H.J. Muller (1927)
38	**Bacterial transformation**	Avery, Mac leod & McCarty
39	First **inter-specific** cross was made by	Thomas Fairchild
40	**Centers of origin** of crops were first given by	Vavilon
41	Term **heterosis** was coined by	Shull,1914
42	RNA was **first synthesized** by	S. Ochoa in 1969
43	The term **genetics** coined by	W. Bateson, 1905
44	**One Gene One Enzyme** Hypothesis	Beadle and Tatum in 1914
45	**Operon Hypothesis**	Jacob and Monod in 1961
46	**Mutation** first defined by	Hugo de vries, 1901
47	**Deletion** first observed by	Bridges (1917)
48	**Translocation** first observed by	Bridges (1923)
49	Concept of **progeny testing** in plants (sugar beet) given by	L.D. Vilmorin (1857)
50	Early ideas of **progeny testing** for livestock given by	Robert Bakewell (18th Century)
51	**Raphanobrassica** (man made intergeneric hybrid of radish and carrot) developed by	G.D. Karpechenko (1928)
52	**Reciprocal recurrent selection** (RRS) concept developed by	Comstock, Robinson & Harvey (1949)
53	**Recurrent selection** was conceptually introduced by	Hayes & Garber (1919) and East & Jonees (1920)
54	The Term **recurrent selection** was coined by	F.H. Hull (1945)
55	The **Single Seed Descent** (SSD) method in plant breeding was proposed by	Goulden (1939)
56	**Single seed descent** method modified and popularized by	C.A. Brim (1966)
57	Who first demonstrated cytoplasmic inheritance in four o'	Carl Carens (1908)

	clock, *Mirabilis jalapa*	
58	Who first described cytoplasmic male sterility (**CMS**) in maize	Rhodes (1933)
59	**Chromosomal theory of inheritance** was proposed by	W. Sutton & T. Boveri (1902)
60	**Backcross method** in plant breeding was proposed by	H.V. Harlan & M.N. Pope (1922)
61	The **ear to row method in maize breeding** proposed by	C.G. Hopkins (1908)
62	The concept of **double fertilization** discovered by	S.G. Nawaschin (1898) & L. Guigngard (1899)
63	The concept of **pure line theory** was given by	W. Johannsen (1903)
64	The **Dominance hypothesis of heterosis** first proposed by	Charles Davenport (1908)
65	Concept of **multiline varieties** was first proposed by	J. Jensen (1952)
66	The **multiple factor hypothesis** (polygene theory) and **bulk method** first developed by	H. Nilsson Ehle (1908)
67	The stratified or **grid method of mass selection** suggested by	Gardner (1961)

Heredity theories:

- ✓ **Preformation theory** of heredity (**preformationism**) proposed by: **Swammerdam & Bonnet** (1720-1793).
- ✓ Theory of **Epigenesis proposed by**: C.F. ***Wolff,*** *1734-94*
- ✓ Theory of acquired character (Lamarckism): ***Lamarck,****1809*
- ✓ Theory of pangenesis (gemmules): ***Charles Darwin,****1868*
- ✓ Theory of heredity or fundamental laws of inheritance: ***G. Mendel (1866)***
- ✓ Germ plasm theory: ***Weisman,****1889*
- ✓ Chromosomal theory of inheritance: ***Sutton & Boveri,****1903*

Organisms used in genetics:

- Bacteria: *E. coli*
- Fungi: *Neurospora crassa*
- Insect: *Drosophila melongester*
- Plant: *Maize, Pea*

Unit-1. The Cell

Cell: is the structural and functional unit of all living organisms, except ***viruses***.

The size of cells:

- The smallest cell size ranges from 0.2 to 5.0μ (microns) in bacteria to up to largest 75 mm in the case of ostrich eggs.

Cell Structure:

1. **Cell wall:**

- Plant cells are surrounded by a ***non living*** and rigid coat called cell wall. It secreted & maintained by protoplasm.
- The main functions of cell walls are to provide plant cells a definite shape, mechanical support and strength to tissues and organs.

Cell wall is differentiated into three parts:-

a. ***Middle Lamella***: The walls of immediate neighbor cells are joined by middle lamella. This is composed mainly of ***pectin*** (mainly in the form of Ca^{2+}& Mg^{2+} salts).

b. ***Primary cell wall***: It is deposited after the formation of middle lamella. Its main constituents are ***hemi celluloses*** (53%) and cellulose (30%). Cells in meristematic tissues contain only primary cell wall.

c. ***Secondary cell wall***: It is the inner most layer of wall. It composed mainly of ***cellulose***.

2. **Plasma lemma or cell membrane:**

- The membrane enclosing cytoplasm of a cell is known as *plasma lemma* or ***plasma membrane***.
- It composed of lipid (60 %) and proteins (40 %).
- It acts as a permeability barrier, and acts as vehicle for the transport of substances from one organelle to another.

3. **Cytoplasm:**

- The substances, except nucleus, surrounded by the plasma lemma are known as ***cytoplasm.***

It may contain the following structures:-

a. **Endoplasmic Reticulum:** An extensive network of membrane-enclosed spaces is known as *endoplasmic reticulum* (ER).

- ER is not found in RBCs, eggs or embryonic cell and prokaryotic cells.
- ER is grouped in to two categories:

(i) ***Smooth ER***: Both outer and inner are regular and smooth. No protein synthesis is takes place in smooth ER. It provides the structural base for synthesis of certain type of steroids i.e. ***Cholesterol, progesterone*** *and* ***testosterone***.

(ii) *Rough ER*: The outer surfaces of membranes have a rough appearance due to the attachment of ribosomes. The rough ER is mainly composed of cisterns and is found in cells actively involved in **protein synthesis.**

Functions:

- It provides the structural base for synthesis of protein
- Provides channel for the transport of materials among cell organelles.

b. **Ribosomes:** These are particles of about 200Å in diameter and are composed of RNA and protein.

- Ribosomes are synthesized and assembled in ***nucleolus***.
- Ribosomes are site of ***protein synthesis.***

On the basis of their sedimentationribosomes are classified in to:-

(i) *80 S sizes:* In plants, they dissociate into **60 *S*** and **40 *S*** sub-units.

(ii) *70 S sizes:* They are made up of **50*S*** and **30*S*** sub-units (Found in mitochondria and plastid).

Structural differences between eukaryotic and prokaryotic cells

SN	Eukaryotic cells	Prokaryotic cells
1	All membrane-bound structures e.g. ER, Mitochondria, Golgi bodies, Lysosomes, Vacuoles etc. are present.	All membrane bound structures are absent.
2	True nucleus is present	Nucleus lacks nuclear membrane
3	Nucleolus is present in the nucleus	Nucleolus is absent
4	Both 70 *S* and 80 *S* ribosomes are present	**Only 70** S type of ribosomes are present
5	Cytoplasm has microtubules	Cytoplasm doesn't have any microtubules
6	E.g. **Cells of higher plants and animals, Protozoa, Fungi**	E.g. **Bacteria, Archaea, Actinomycetes, phytoplama, BGA**

c. **Golgi body:** Also called ***endomembrane system*** because it associated with ER & Lysosomes. Provides site of synthesis of lytic enzymes and supports cell wall formation.

d. **Lysosomes:** Produced from the Golgi complex and aresite of hydrolytic enzymes (mainly ***acid phosphatise***). Lysosomes common in animal cell but also present in plant cell. Main function of Lysosomes is autolysis; known as ***suicidal bags*** (digest cell itself).

e. **Peroxisome:** Mainly occur in photosynthesing cells of higher plants and concerned with *glycolate metabolism* or ***photorespiration*** of C_3 plants.

f. **Spherosomes:** These are site of hydrolytic enzymes and are found in plant cells only; they don't occur in animal cells. Contains up to 98 % lipid; their function may be in *lipid storage*.

g. **Glyoxysomes:** Concerned with ***glyoxylate metabolism*** and help in conversion of lipids of germinating seeds into carbohydrates.

h. **Vacuole:** plant cells have one or more vacuoles of variable size. It contains cell sap and large part of cytoplasm up to 90 % is occupied by a large vacuole. Vacuoles are surrounded by a unit membrane; the membrane is referred to as ***tonoplast.***

i. **Centriole:** these are confined to animal cells and aren't found in plant cells.

4. **Mitochondria:**

- These are site of **Krebs cycle** and **electron transport chain.**
- The Krebs Cycle (Citric Acid Cycle) and Electron Transport Chain (ETC) are the main energy-producing stages of aerobic cellular respiration, occurring in the mitochondria.
- Originated by fusion of pre-existing mitochondria. An average cell may have 200 to 800 mitochondria.
- Mitochondria are surrounded by two concentric unit membranes; named outer membrane and inner membrane.
- The respiratory complex contains three large enzyme complexes (i.e. NADH dehydrogenase complex, b-c1 complex and cytochromes oxidase complex) these enzyme complexes are embedded in the inner membrane.
- The infolds of inner membrane are known as ***crystae***
- The crystae greatly increase the area of inner membrane and have oxysomes known as electron transport particles (ETP).
- The space on the inside of the inner membrane is known as ***matrix***. It contains a variety of transport proteins that make it selectively permeable to molecules metabolized in the matrix. In addition to proteins it carries out the oxidation reactions of respiratory chain, i.e. the electron transport chain and ATP synthesis in the matrix. The respiratory chain carries out oxidative phosphorylation that generates most of the ATP in the animal cells, so it named as ***power house of cell.***

5. **Chloroplasts:**

- Actual **site of photosynthesis**.
- Chloroplast contains ***chlorophyll*** which imparts the characteristic green color to plants.

- The membrane free space enclosed by membranes of chloroplast is referred to as ***stroma***.
- Within the stroma are embedded about 40-80 ***grana***. Each granum contains 5-25 flat cisternae stacked on top of each other are called ***thylakoids***. It has quantosomewith 250-300 photosynthetic units.

In addition to chloroplast, plant cells contain two more type of plastids called as:

a. **Leucoplast:** these are color less plastid and function in the storage of starch, protein and fats.

b. **Chromoplasts:** contain pigments other than chlorophyll e.g., phycocyanin, phucoxanthin etc.

6. Nucleus:

- Nucleus contains ***chromosomes*** which control functions of cell by having almost all the genetic information of a cell.
- It produces ribosomes, t-RNA, m-RNA, and certain other RNAs.

Unit -2. Cell division

- The process of formation of new cells from the pre-existing cells is referred to as **cell division.**

Two types:

1. Mitosis:

- Cell division in which one cell gives rise to two cells of genetically identical to the cell which produced them e.g.Occurrence in somatic cells responsible for vegetative growth.

2. Meiosis:

- Cell division in which one cell with two set of chromosomes gives rise to four cells with one set of chromosomes each e.g. Occurrence in **gamete formation** and responsible for sexual reproduction.

1. Mitosis:

- It is the process by which a cell divides into two daughter cells, each having a nucleus with sets of chromosomes exactly identical to the parent cell.

Cell cycle:

- Divided in to **Interphase**, **Mitosis** and **Cytokinesis**.

A. Interphase:

✓ It is the **longest** stage of cell cycle.

✓ The longest and thinnest 'chromosomes' (chromatin fibers) are found during interphase.

- ✓ **Giant chromosomes**, specifically polytene chromosomes, are characteristic features of the interphase nuclei in certain cells, like insect salivary gland cells.
- ✓ Interphase is further divided into three sub stages.

a. G1 (first gap):
- Pre-DNA replication phase
- Protein & RNA synthesis takes place

b. **S (synthesis, DNA):**
- DNA synthesis takes place
- Histone proteins

c. G2 (second gap):
- Post-DNA replication phase
- RNA & protein particularly component of microtubules synthesis.

B. Mitotic phase:

a. **Prophase**:
- In mitotic cell division small sub units of prophase are absent.
- Appearance of definite thread-like structures in nuclei.
- Chromosomes become condensed.

b. **Metaphase**:
- Disappearance of nucleolus.
- Appearance of spindle apparatus.
- Chromosomes arranged in ***equatorial plate*** called as ***Metaphagic plate***
- Shortest and thickest state of chromosomes in this phase.
- Chromosomes are **clearly visible** so **chromosome number** can be counted easily during this phase.

c. **Anaphase**:
- Chromatids separate at centrosomes and move towards opposite sides called ***Anaphagic movement***
- ***Smallest*** form of mitosis.
- ***Smallest*** chromosomes in size are found in this phase.

d. **Telophase**:
- Nucleolus and nuclear membrane reappear
- It is **apposite** to prophase.

C. Cytokinesis:
- ✓ It is complete by the end of Telophase.

- ✓ At the equatorial plate, membranous elements of ER and products of Golgi bodies organize into the phragmoplast gives rise to cell plate and subsequently to middle lamella.

2. **Meiosis**:

- ✓ Meiotic cell division was first described by van Beneden in 1883, in the round worm.
- ✓ Meiotic division of one somatic (2n) cell gives rise to four haploid (n) daughter cells.
- ✓ The reduction in chromosomes number of the daughter cells is mainly due to a single S phase followed by two successive nuclear divisions.
- ✓ Hence, during meiosis chromosomes replicate only once, while the nuclei divide twice.
- ✓ **Meiosis** takes place during gamete/spore formation. Therefore it is confined to reproductive cells.

First meiotic division:

- ✓ Results in reduction of chromosomes numbers to just half of the mother cell.
- ✓ It is also referred to as ***reductional division***
- ✓ It consists of four different phases as:

A. **Prophase-I**:

- This stage is the longest in duration and divided into **five sub stages.**

a. **Leptotene**:

- Chromosome condensation so that they visible as fine threads formed a structure of ***Bouquet.***
- Chromomeres, the bead-like structures of condensed chromatin, are clearly visible during the early stages of meiosis, specifically in leptotene and zygotene phases of prophase I.
- RNA synthesis occurs.

b. **Zygotene**:

- The main event is ***synapsis;*** pairing between homologous chromosomes.
- ***Bivalent***: Haploid (n) pair of chromosomes

c. **Pachytene**:

- The main events that occur during Pachytene are as follows.

(i) ***Tetrad***: Each bivalent has four chromatids as strands, hence bivalents are said tobe in four strand or ***tetrad stage***.

(ii) ***Crossing over***: Interchange of parts between non sisters chromatids of chromosomes during meiotic prophase.

d. **Diplotene**: in this stage occurs chaismata.

- ***Chaismata***: attachment of homologous chromosomes to each other.

e. **Diakinensis**:

- It begins after complete termilization of chaismata.

Unit-3. Chromosome: structure and function

✓ Chromosomes were first described by Strausberger in 1875.

✓ The term chromosome was first used by Waldeyer in 1888.

✓ Chromosomes number can be counted with relative case only during mitotic metaphase.

Somatic chromosomes numbers of common plants and animals:

SN	common name	Botanical name	Chromosome No.
1	Plants		
	Barley	*Hordeum vulgare*	14
	Tomato	*Lycopersicon esculentum*	24
	Onion	*Allium cepa*	16
	Tobacco	*Nicotiana tabacum*	48
	Rice	*Oryza sativa*	24
	Bajra	*Pennisetum americanum*	14
	Pea	*Pisum sativum*	14
	Rye	*Secale cereal*	14
	Potato	*Solanum tuberosum*	48
	Jowar	*Sorghum bicolor*	20
	Maize	*Zea mays*	20
2	Animals		
	Honey bee	*Apis mellifera*	32
	Roundworm/ Horse nematode	*Ascaris megalocephala Univale/ Paras carisequorum*	2
	Fruit fly	*Drosophila melanogaster*	8
	Man	*Homo sapiens*	46
	House fly	*Musca domestica*	12
	Rat	*Rattus norvegicus*	42

Genetic materials:

1. **Chromatin**: It is composed of DNA, protein and RNA.
2. **Histones**: Constitute about 80 % of the total chromosomal proteins.
3. **Chromosomes**: Composed of nucleic acid (DNA or RNA)
4. **Nucleic acid**: Contains many nucleotides
5. **Nucleotides**: It consists of nucleoside and phosphoric acid.
6. **Nucleoside**: Contains sugar + nitrogenous base

7. **Nitrogenous base:** Two types Purine (A, G) and pyramidene (T, C or **Uracil** in RNA).

1. RNA (Ribose Nucleic Acid):

- It is found in cytoplasm as well as in nucleus.
- It is generally involved in protein synthesis and act as genetic material for all plants viruses e.g.TMV (contains **ss-RNA)** except cauliflower mosaic viruses which contain ds-DNA.
- **Heinz Fraenkel-Conrat and Bea Singer** in 1957 proved RNA is the genetic material in the Tobacco Mosaic Virus (TMV) by showing that RNA alone, not the protein coat, determined the characteristics and infectivity of new viruses.

Types of RNA:

a. **Messenger RNA** (m-RNA):
   - It carries the genetic information present in the DNA.
   - Forms about 5-10% of the total RNA present in Cell.

b. **Transfer RNA** (t-RNA):
   - Also known as soluble RNA (s-RNA)
   - It carries amino acids to the site of protein synthesis.
   - It forms 10-15 % of total RNA present in cell.

c. **Ribosomal RNA (r-RNA):**
   - Most stable type of RNA and is found associated with ribosome.
   - It forms about 80% of the total cell RNA.

2. DNA (De-oxyribose Nucleic Acid):

- It is a double stranded helix made up from many nucleotides.
- DNA is present in nucleus, mitochondria and chloroplast.
- It encodes genetic information; serve as the template for DNA replication and transcription.
- DNA is only component of the cells that is highly conserved.
- The **sugar-phosphate** backbone forms the structural framework of DNA, consisting of alternating **deoxyribose sugar** (a 5-carbon sugar) and **phosphate groups** linked by strong **phosphodiester bonds, creating** a continuous strand with a 5' to 3' directionality, providing stability and forming the sides of the DNA double helix with the nitrogenous bases attached to the sugars.
- **Hydrophobic forces** (base stacking) and **hydrogen bonds** (between complementary bases) are both crucial, weak interactions that stabilize the DNA double helix, working together.
- **Hydrogen bonds** provide specific pairing (A+T, G+C), while the **hydrophobic effect** forces the nonpolar bases inward, away from

water, causing them to stack and greatly contributing to the helix's overall strength and structure.

Unit-4. Chromosomal aberration

- Any change which alters the normal structure of a chromosome is known as structural chromosomal aberration.

Types of structural changes:

a. **Deletion (deficiency):**
- Loss of a segment of a chromosome.
- It plays an important role in species formation and releasing variability through mutations.
- Deletions are important cytological tools for mapping genes.

b. **Duplication (repeat):**
- A chromosome segment present in more than two copies in the same nucleus.
- It results in addition of one or more genes to a chromosome.
- Duplications are less harmful than detection.
- They don't reduce the viability of an individual
- Duplications lead to addition of some genes in population.

c. **Translocation:**
- One way or reciprocal transfer of segments between non homologous chromosomes is known as translocation.
- Translocation alters the chromosome size, chromosome number thus play important role in the formation of species
- They are useful tools in breeding programmes for transfer of **desirable characters** from one species to another.
- **Phenotype effect** of translocation in *human Down syndrome* (**Mongolism**) can arise in the progeny of an individual heterozygous for a translocation involving chromosomes number **21**.

d. **Inversion:**
- It refers to structural change in a chromosome in which a segment is oriented in a reverse order.
- Gene order is changed in the inverted segment of a chromosome.

Unit-5. Genome

- A basic or monoploid set of chromosomes of an individual is called **genome.**

Monoploid:
- Contains a single chromosome set and are characteristically sterile.
- A monoploid can be haploid but all haploid can't be monoploid.

Haploid: Single set of chromosomes (denotes by 'n')

Uses of haploids:

- ✓ Used for development of pure lines
- ✓ Disease resistance
- ✓ Development of inbred lines and
- ✓ Production of aneuploidy.

Double haploid: double haploid plants are developed from haploids by doubling the chromosome number of haploid plants by **Colchicine treatment**.

Polyploidy:

- Presence of more than two genomes in an individual is known as polyploidy.
- It is of two types:

1. **Auto polyploidy**:

- ✓ If all the genomes present in an individual are identical, it is called ***autopolyploid***.
- ✓ In many species, autopolyploids show an increase in vigour and size this phenomenon is known as' ***gigantism***.
- ✓ It includes auto triploids (3x), tetraploids (4x), penta (5x) and other polyploidy.

a. **Autotriploids:**

- Having three set of homologues chromosomes.
- Autotriploids are generally ***highly sterile*** due to defective gamete formation. But triploids of some species are **highly fertile** e.g. Spinach (*Spinacea oleracea*)
- Triploids are useful only in those plant species which propagate asexually like **banana**, **sugarcane**, and **apple**.

(i) **Banana:** Mostly cultivated varieties of banana are auto-triploids (3n) and are seedless.

(ii) **Water melons:** The **triploid** water melons are **seedless** and have the capacity of early maturity and disease resistance. These are produced by crossing between tetrapod and diploids

(iii) **Sugar beets:** triploid sugar beets produce larger roots and more sugar contents than diploids.

b. **Autotetraploids**:

- They have **four copies** of the same species.
- Tetraploids are usually very stable and fertile because pairing partners are available during meiosis.
- E.g. Rye, **grasses,** alfalfa, berseem (var. Pusa giant), groundnut and **potato** are well known example of autotetraploids.

2. **Allopolyploidy:**
   - Allopolyploids contain two or more distinct genomes, which are derived from different species.
   - Allopolyploids can be developed by interspecific crosses and fertility is restored by chromosome doubling with **colchicine** treatment.
   - It found in about **50** % of crop plants.
   - Polyploidy is mainly induced treatment with a chemical known as ***Colchicine***.
   - Examples of polyploids i.e. wheat; the common bread wheat (*T. aestivum*) is an ***allohexaploid*** and tobacco, brassica, cotton etc, are ***allotetraploid***.
   - **Triticale** (man-made cereal) is an example of inter-generic polyploidy and developed by crossing between wheat (*Triticum aestivum*) and Rye (*Secale cereal*) to derived combine benefits of high yield of wheat and disease, drought resistant of rye.
3. **Aneuploids**: a loss or gain of one or few chromosomes as compared to the somatic chromosome number of a species is known as *aneuploidy*.

Aneuploids are following types:-

a. **Monosomic:**
   - An individual lacking one chromosome from a diploid set is called ***monosomic*** and this condition is known as ***monosomy.***
   - Monosomic = $2n - 1$
   - Double monosomic = $2n - 1 - 1$

b. **Nullisomic:**
   - The absence of a chromosomes pair from a diploid set is termed as ***nullisomic.***
   - Nullisomics = $2n - 2$

e. **Trisomics:**
   - An individual having one extra chromosome in the diploid set of chromosomes known as ***trisomics***.
   - Trisomic = $2n + 1$

f. **Tetrasomics:**
   - An individual having one pair of extra chromosome in diploid complement is known as ***tetrasomics***.
   - Tetrasomics = $2n + 2$

Aneuploidy in man:

- Aneuploidy in humans is having an abnormal number of chromosomes (not 46), resulting from extra (trisomy) or missing (monosomy) chromosomes, common causes of miscarriage, infertility, and genetic conditions like Down Syndrome (Trisomy 21), Turner Syndrome (XO), and Klinefelter Syndrome (XXY).

Aneuploidy types and examples-

1. Trisomy (extra chromosome):
   - **Down syndrome/mangolian idiocy (47):** Trisomy 21 extra copy of chromosome 21.
   - **Edwards's syndrome (47):** Trisomy of 18.
   - **Patau's syndrome (47):** Trisomy 13.
   - **Klinefelter Syndrome (47):** XXY (extra X in males).
   - **Extreme klinefelter's syndrome :** XXXY(48)
   - **Meta female :** XXX (47)
   - **Retarded mental and secondary sex growth:** XXYY(48)
   - **Normal trisomic male (Jacobs syndrome): XYY (47)**
2. Monosomy (missing chromosome):
   - **Turner Syndrome (45):** XO (missing one X in females).

Ploidy levels of different plant cells involved in fertilization of angiosperms are -

SN	Plant cell	Ploidy level
1	Microspore /pollen mother cell (PMC)	Diploid (2n)
2	Microspore Pollen grains	Haploid (n)
3	Vegetative (somatic) cell	Diploid (2n)
4	Vegetative cell nucleus of pollen grain	Haploid (n)
5	Generative cell in a pollen grain is	Haploid (n)
6	Megaspore mother cell (MMC) is	Diploid (2n)
7	Functional megaspore	Haploid (n)
8	Mature Egg Cell	Haploid (n)
9	synergid cells and antipodal cells within the mature embryo sac	Haploid (n)
10	Secondary nucleus	Diploid (2n)
11	Zygote	Diploid (2n)
12	Endosperm	Triploidy (3n)

Unit-6. Classical genetics

1. **Alleles:** The various forms of a **gene** are called alleles.
2. **Gamete:** Gamete refers to a sexual unit.
3. **Contrasting characters**: The individual plant features with marked phenotypicdifference are known as **contrasting characters** or **traits** e.g. Red and white **flower colour**, yellow and green **seed coat colour.**
4. **Back cross:** The crossing of F_1 with one of its parents is called ***back cross***. Back cross play important role to development disease resistance in plants
5. **Test cross**: Crossing of F_1 with of its recessive parents is called ***test cross***.
6. **Top cross:** A cross between an inbred line and an open pollinated variety also known as ***inbred varieties cross***.

Mendalian laws:

1. **Law of Dominance**
   - When two different alleles for a trait are present (heterozygous), the dominant allele's trait is fully expressed, while the recessive trait remains hidden.
   - **Example**: A pea plant with one allele for purple flowers (dominant) and one for white flowers (recessive) will have purple flowers.

2. **Law of Segregation**
   - The two alleles of a gene present in the F_1 do not contaminate each other; they separate and pass into different gametes in their original form producing two different types of gametes in equal frequencies.
   - **Also called**: The 'law of purity of gametes' as they only carry one allele.

3. **Law of Independent Assortment:**
   - This law states that when two pairs of gene enter in F1 combination, both of them have their independent dominant effect means alleles for different genes (on different chromosomes) assort independently during gamete formation; the inheritance of one trait doesn't influence the inheritance of another.

Gene:
- Smallest and functionally part of genetic material.
- An alternative form of gene called **allele**.

Allele:
- When a gene has two alternative forms called alleles.

Multiple allele:
- The many allele of a single gene are called ***multiple alleles***
- In the case of multiple alleles, only the monohybrid ratio is observed in the F_2 generation.

- These control same character
- No crossing over in the *multiple* ***allelism***.

Example of multiple allele:

- Fur colour in rabbits
- Wing type in drosophila
- Eye colour in drosophila
- AB & O blood group in human

Pseudo alleles:

- Govern different expression of the same character.
- They exhibit cis-trans position effect.

Types of gene interaction:

Typical di-hybrid ratio for a single trait:	**9:3:3:1**
Duplicate gene action:	**15:1**
Complementary gene action:	**9:7**
Supplementary gene action:	9:3:4
Inhibitory gene action:	**13:3**
Masking gene action:	12:3:1
Polymeric gene action:	9:6:1
Additive gene action:	1:4:6:4:1

Antigen:

- An antigen is a molecule of immunogenic, which specifically interacts with an antibody

Antibody:

- An immunogen is a molecule, which when introduced into the system of an animal, induces the production of a class of proteins, known as antibodies.
- Antibodies belong to a class of proteins called ***immune-globulins.***

Unit-7

1. **Mutation:** Sudden heritable changes in the characters of organisms are called ***mutation.***
2. **Mutants:** The individuals showing these heritable changes are known as ***mutants.***
3. **Mutagens** Agents, both chemical and physical, increase the frequency of mutations several-fold are called ***mutagens.***

Examples commonly used mutagens:

SN	Class	Mutagens
1.	Physical mutagens	
	Ioning radiations	α-rays, β-rays, γ-rays (^{14}c, ^{60}co), & x-rays
	Non- ioning radiations	UV rays
2.	Chemical mutagens	
	Alkalating agents	Mustard gas, EMS, MMS, EES
	Base analogues	5-Bromo-uracil
	Acridine dyes	Acriflavin
	Deamination agents	Ethium bromide

Note:-The rate of spontaneous mutation is very low generally ranges $10^{-4} – 10^{-6}$ for eukaryotes and $10^{-5} – 10^{-7}$ for prokaryotes.

Linkage

- All the genes on a chromosome are said to be linked to one another and belong to the linkage group.
- The phenomenon of inheritance of linked genes in same linkage group is called ***linkage.***
- It opposites to **crossing over**.

Phases of linkage:

a. **Coupling**: Linkage between two or more either dominant (AB) or recessive (ab) allele is referred to as ***coupling.***

b. Repulsion: Linkage of dominant allele with that of recessive allele (Ab or aB) is known as ***repulsion.***

Detection of linkage:

- **Test cross** is the most common method of detecting the linkage.

Interference (Muller):

- Tendency of one crossover to reduce the chance of another crossover in its adjacent region.

Coincidence (Muller):

- It explained the strength or degree of interference.

Sex determination:

- It utilizes various genetically concepts to decide whether a particular individual will develop into male or female sex.
- Sex determining genes are present on both **x-chromosomes** as well as autosomes.

Sex-linked characters:

- The characters for which genes are located on sex chromosomes are known as ***sex–linked traits.***

- Such genes are called sex linked genes, and linkage of such genes is referred to as ***sex linkage***.

Main features:

a. **Location**: Sex –linked genes are located on X chromosomes only.

b. **Number**: In diploid organism, each homogametic sex (xx or zz) has two copies of linked alleles.

c. **Expression**: In homogametic sex, a recessive gene can express. In heterogametic sex, a recessive allele can express.

d. **Transmission**: Sex linked genes are transmitted from female to male and not from the male, because males receive their**x chromosome** from the female only. Sex linked genes are first inherited from male to female in the first generation and then back to male in the second generation.This type of inheritance in which a sex linked gene is inherited from grandfather to grandson through daughter is called ***cris-cross inheritance***.

Sex linked characters:

a. **In drosophila** (white eye, vermilion eye).

b. ***Hemophilia*** is the well-known example of sex linkage in human. In case of a marriage between hemophilic woman and normal man, the disease will be transmitted to 50% of the sons even if the gene is in heterozygous condition in the carrier.

c. **The *color blindness*** is another example of sex linkage in humans. Sons from the marriage between color blind man and normal woman will be normal, but daughters will carry such genes in heterozygous condition. Marriage of such carrier girl with color blind boy will yield children in which both male and female children will be color blind each in 50% cases.

Sex limited characters:

✓ The characters which are expressed in one sex only are referred to as ***sex limited characters***. E.g. breast development in woman and beard in man.

Unit-8 Protein synthesis-

Involve two main processes:

1. **Transcription :**

- The process of synthesis of messenger RNA from a DNA template is known as ***transcription.***
- It is first step in gene regulation and protein synthesis is **Transcription**
- It takes place during interphase and continues up to prophase of cell division.

- The enzyme involved in transcription is RNA polymerage.
- **DNA** ____(RNA polymerage)___ **m-RNA**
- In prokaryotes, transcription begins at a specific site within the DNA sequence called the **promoter site**
- This site serves as the binding location for the enzyme RNA polymerase, which unwinds the DNA and synthesizes a complementary RNA strand.
- **The sigma** (σ) factor in prokaryotes recognizes promoter sequences (sites) of DNA.

Reverse transcription (teminism):

- Formation of DNA from mRNA is known as ***reverse transcription.***
- Reverse transcriptase is RNA–dependent DNA polymerage which involve in process of reverse transcription.
- **m-RNA** ___(Reverse transcriptase)____ **DNA**

2. **Translation:**

- The process of protein synthesis from information in m-RNA is known as ***translation***.
- **DNA** ___(transcription)____**mRNA**__(translation)___**Protein**

Start Codon:

- In protein systhesis, the start codon initiates the translation process, while the stop codon signals its termination.
- Standard start codon: AUG
- Amino acid: Methionine
- RNA **type: t-RNA** present with anticodons

Stop Codon:

- It signals the end of protein synthesis
- Standard Stop Codons: UAG, UGA and UAA
- Amino acid: None
- **t-RNA** with anticodons are not present

Cytoplasmic inheritance:

- Most of phenotypic characters are controlled by genes present in the chromosomes but some characters expressed by factors present in the cytoplasm. These factors lying in the cytoplasm are called ***plasmagenes***.
- The transmission of characters controlled by plasma genes is called ***Cytoplasmic inheritance.***
- The first case of Cytoplasmic inheritance was reported by **Correns** in 1909 in four O" clock (*Mirabilis jalapa*) for leaf colour.

- It used for development of **Cytoplasmic male sterility** (CMS) lines in several crops like **maize, pearl millet, sorghum, cotton** etc.

Self incompatibility:

- Self incompatibility term was coined by A.B. Stout (1917) but self incompatibility first reported by Joseph Koelreuter.
- It is a genetic mechanism that prevents plants from self-pollinating ensuring cross-pollination and genetic diversity.
- It can be used in hybrid seed production by using self-incompatible plants as the female parent and maintaining parental lines.
- It is a powerful genetic defense in flowering plants (angiosperms) that prevents self-pollination by recognizing and rejecting its own pollen, ensuring genetic diversity through cross-pollination (outcrossing) and preventing harmful inbreeding.

Types of Self-incompatibility Systems:

1. **Gametophytic Self-Incompatibility** (GSI): self incompatibility is controlled by the genetic constitution of gamets (pollen's genotype). Found in Solanaceae (tomato, tobacco) and Papaveraceae (poppy).

- Gametophytic self-incompatibility was first discovered by East and Mangelsdorf in 1925 while studying the tobacco species ***Nicotiana sanderae.***

2. **Sporophytic Self-Incompatibility** (SSI): Self incompatibility is goverened by the genotype of pollen producing parent plant's (sporophyte), as seen in the mustard family (Brassicae) - cabbage, cauliflower, broccoli.

- Sporophytic self-incompatibility (SSI) was first reported in **1950** by **Hughes and Babcock** in ***Crepis foetida***, and independently by **Gerstel** in ***Parthenium argentatum*** (guayule).

Autogamy (self pollination):

- Autogamy, also known as self-fertilization, is a reproductive process where a single individual produces both male and female gamates, which then fuse to form a new individual.
- Various mechanism promoting autogamy are follows- Bisexuality, Homogamy, Cleistogamy, Chasmogamy, position of anther.

Allogamy (cross pollination)

- Allogamy, also known as cross pollination, is a reproductive process where a male gamete (pollen grain) from one plant fertilizes a female gamete (egg) of another plant of the same or a closely related species.

- Various mechanism promoting allogamy are follows- Dicliny, Dichogamy, Heterostyly, Herkogamy, Self-incompatibility, male sterility.

Segregation:

- In plant breeding, segregation refers to the process where different versions of a gene (alleles) separate during gamete formation, ensuring that each gamete (pollen or egg) carries only one allele for each gene.

Methods for handling segregating populations are follows-

1. Pedigree method
2. Bulk method
3. Backcross method

Male sterility source utilization in hybrid seed production:-

SN	Crop	Source
1	**Bajra**	Tift 23A
2	**Sorghum**	Combine kafir 60
3	**Wheat**	*Triticum timopheevii*- cytoplasmic male sterility (T-CMS) system
4	**Rice**	**Wild Abortive (WA),** Honglian (HL), and Boro II (BT)
5	**Maize**	**Texas** (CMST/GMS), gene like **Ms44**
6	**Cotton**	Gregg (associated with the *G. hirsutum* GMS line)
7	**Tobacco**	*Nicotiana suaveolens* (CMS).
8	**Brassica**	Ogura CMS)
9	**Sunflower**	PET1(CMS system)
10	**Chilli**	MS-12 GMS line

Unit -9. Plant breeding

Contributions of breeders:

1. **Thomas Fairchild (1717):**
   - He was the first person to scientifically develop an interspecific hybrid, **Dianthus** (*Caryophyllus barbatus*), known as **'Fairchild's Mule'**, by crossing Sweet William (*Dianthus barbatus*) and Carnation (*Dianthus caryophyllus*) during 1716-17.
2. **Rimpu (1890):**
   - He made the first successful **intergeneric cross** to create the man-made cereal, **Triticale** by crossing between bread wheat (*Triticum aestivum*) and rye (*Secale cereale*).

- This breeding was done to combine desirable traits from both parents: the high yield and grain quality of wheat with the hardiness and disease resistance of rye.
- **DT-46** is the variety of triticale

3. **Jones, D.F. (1917):**
   - He made **double cross** in maize.
   - It was cross the two single crosses (ABxCD) to produce the final commercial hybrid, which is more vigorous, stable, and produced higher yields than open-pollinated varieties.
4. **Vavilon, N.I.:**
   - He gave the concept of **'centres of origin'** and identified the world's major centers of origin for cultivated plants.
   - He identified **8 main centers** and **3 sub centers** of crop diversity.
   - He is known as the **'father of genebanks'** for his efforts to collect and conserve global plant diversity and establishment of gene bank.
5. **Nagaharu, U (1935):**
   - He gave ***U's triangle-*** a fundamental concept in plant genetics named after botanist U Nagaharu.
   - This triangle illustrates the evolutionary relationships and genomic origins of six major Brassica species (cabbage, rapeseed, mustard).

Harrington, J.B. (1937):

- He published a paper titled **'The mass-pedigree method in the hybridisation improvement of cereals'** in the Journal of the American Society of Agronomy.
- This paper introduced a modified plant breeding approach where single plant selection is delayed until later generations, particularly useful for selecting for resistance against abiotic stress.
- J.B. Harrington was a prominent figure who also coined the term '**agrologist**' for the agricultural science profession.

6. **Flor, H. H. (1956):**
   - He proposed the **gene-for-gene hypothesis** in 1956 based on his research of the pathogen-host relationship between flax (*Linum usitatissimum*) and flax rust (*Melampsora lini*).
7. **Vander Plank, J. E. (1963):**
   - Developed the concept of vertical and horizontal resistance.
   - **Vertical resistance:** This is **race-specific resistance**, meaning it is completely effective against some races of a pathogen but not against others. It is typically controlled by a single gene (monogenic) or a few genes (oligogenic).

- **Horizontal Resistance:** This is **non-race-specific** resistance, effective against all races of a pathogen. It is controlled by **many genes** (polygenic) and acts to reduce the rate of disease spread rather than providing complete immunity. It is generally more stable and durable over time than vertical resistance.
- **The Vertifolia effect**: describing the erosion of horizontal disease resistance when breeding for specific (vertical) resistance was introduced by plant pathologist J.E. Van der Plank in his 1963 named after the German potato variety '**Vertifolia**', which showed this trait for late blight.

8. **Donald, C.M. (1968): (RAS Exam-09)**
   - Introduced the **Crop Ideotype Concept** working with wheat.
   - **Crop Ideotype**: A theoretical model of a plant with specific, desirable characteristics (e.g., erect leaves, short stature, high harvest index) to improve yield and efficiency.

Plant Breeding methods:

1. **Pure line selection:**
   - This method was developed by **Johannsen** using the '**Princess**' variety of common beans (*Phaseolus vulgaris*)
   - It refers to the **homogenous progeny** of a self-pollinated homozygous plant.
   - Development of new variety through identification and isolation of single best plant progeny is known as ***pure line selection*** or individual plant selection.
   - This method is commonly used in **self-pollinated** species e.g. **Wheat, barley.** (*RAS Exam-09).*
2. **Mass selection:**
   - Refers to a method of crop improvement in which individual plants are selected on the basis of phenotype from a mixed population, their seed are bulked and used to grow the next generation.
   - This is one of the **oldest methods** of crop improvement.
   - Mass selection is **mostly used for cross pollinated crops** but can be applicable to **both self and cross pollinated** species.
   - E.g. Maize, pearl millet, mustard & often cross pollinate crops like cotton, sorghum etc.
3. **Clonal selection:**
   - They are homogenous within a population, but individual plant or cells can be heterozygous in origin.

- Progeny of a single plant obtained by asexual reproduction is known as ***clone***.
- Crops which are propagated asexually or by vegetative means are known **as asexually propagated** or **vegetative propagated** or clonal crops.

4. **Pedigree method** (first outlined by **H.H. Love, 1927)**
   - Pedigree refers to record of the **ancestry** of an individual selected plant.
   - Pedigree breeding is a method of genetic improvement of self-pollinated species in which superior genotypes are selected from segregating generations and proper record of the ancestry of selected plants are maintained in each generation.
   - It is widely used for the improvement of self-pollinated species for development of new varieties, while in cross pollinated species it is used for development of inbred lines.
   - The pedigree method in plant breeding offers breeders maximum opportunity to use their skill and judgment for selecting superior plants, especially for simply inherited traits, allowing them to meticulously track parent-progeny lineage and promote promising individuals while eliminating undesirable ones early, leading to rapid homozygosity and recovery of novel trait combinations.
5. **Bulk breeding methods:** (first proposed by **H. Nilsson-Ehle, 1908)**
   - It refers to selection procedures in which the segregating population of self-pollinated species is grown in bulk plot.
   - This method is also known as **mass or population method.**
   - This method is used for the **genetic improvement of self-pollinated** crop plants.
6. **Backcross method: (Harlan and Pope, 1922)**
   - Backcross refers to crossing of F_1 **with either of its parents.**
   - When the F_1 is crossed with homozygous recessive parent, it is known as **test cross.**
   - Back cross method is applicable in all three groups of crop plants, viz. **self-pollinated**, **cross pollinated** and **asexually propagated**.
   - Backcross method has been widely used for the development of **disease resistance varieties** in both self and cross pollinated species of wheat and cotton.
   - It has also been used for inter-specific gene transfer and development of multiline varieties in self-pollinated species.

7. **Recurrent selection:**
- It is a strategy involving repeated cycles of selecting superior plants, interbreeding them and then reselecting from the progeny to gradually improve a population's performance for a trait, enhancing desirable genes through constant genetic recombination and maintaining diversity for continued progress, effectively creating a better base for hybrid or synthetic varieties.
- It was originally developed as a method of breeding cross pollinated species, but now it is also used for self-pollinated species.

8. **Single Seed Descent (SSD) method:**
- This method is a modification of the bulk method, was initially proposed by **Goulden (1939)** and later refined by **Brim (1966).**
- It is a plant breeding technique used to rapidly advance inbreeding and achieve **homozygosity** in population.
- It involves selecting one seed randomly from each plant in a segregating population and bulking them to grow the next generation.

Breeding methods only for self pollinated plants are-
- Pure line selection
- Pedigree method
- Bulk method,
- Single seed descent

Unit-10. Seed technology

Seed:
- Botanically seed is a **fertilized ovule** consisting of intact embryo, stored food and seed coat which is viable and has got the capacity to germinate.
- **National (Indian) Seed Act passed in:1966**
- **National Seed Corporation in: 1963**

Classes of seed:
- Seeds are groped in to five main classes.

1. **Nucleus seed:**
- It is first stage or class of improved seed
- Such seeds are the initial seed of an improved variety and are available in limited quantity.
- It is produced directly by the originating plant breeder at research institute.
- It has highest genetic purity (about 100%).

2. **Breeder seed:**
- It is progeny of nucleus seed.

- It is produce under strict supervision of plant breeder at research institute.
- Breeder seed is used to **produce foundation** seed.
- Genetic and physical purity is almost same to nucleus seed.
- It has **golden yellow** colored tag.

3. **Foundation seed**:
   - It's progeny of breeder seed.
   - Known as **mother seed** and it is the source of registered and certified seed classes.
   - It is genetically 99.9% pure. However, physical purity of **98%** is permissible.
   - It is mainly produced by National Seeds Corporation (NSC).
   - It is produced on government farms, at experiment stations, agriculture universities or on cultivator's field under strict supervision of research scientists and expert from state seed certification agencies (**SSCA**).
   - **White tag** is denoted for foundation seed.
4. **Registered seed**:
   - It is produced from foundation seed or registered seed itself.
   - It has satisfactory genetic **identity** and **purity.**
   - In India it is generally **omitted** (avoided) and certified seed is produced directly from foundation seed.
   - It has **purple** or orange colored tag on bags.
5. **Certified seed**:
   - It is progeny of either registered seed or foundation seed or certified seed itself.
   - It has satisfactory genetic identity and purity.
   - It is approved and certified by the state seed certifying agency.
   - The bags are denoted **blue (azora blue)** colored tag.
   - It is available for general distribution to the farmers for commercial crop production.

Viability of seeds:

- **Seeds viability** is defined as the capacity of the seed to remain capable of germination for some specific period of time.

Seed viability tests:

a. **Potassium permanganate method**:
   - It's **qualitative** test to find out whether seeds are viable.

b. **Tetrazolium chloride test**:
   - Known as **biological** test

- Seeds are soaked in 0.5-2% solution of **2, 3, 5, triphenyl tetrazolium chloride**.
- The viable or living seeds take **bright red** coloration and the colour becomes more intense in the embryo.
- It is a **quickest** method of seed viability test.

c. **Grodex test**:
- Grodex is a seed germination indicator
- It's a brand name of **triphenyl tetrazolium bromide** in powder form.
- The

d. **GADA test: it** is a biochemical method of seed storability developed by P. Linko in 1961 to measure **glutamic acid decarboxylase activity (GADA)** in seeds, like **wheat and rice.**

Isolation distance:

- Isolation of seed crop is essential to avoid genetic & physical impurities.

Minimum isolation requirements of some crops:

Crops	Minimum isolation distance (in m)	
	Foundation seed	Certified seed
Self-pollinated crops like rice, wheat, barley, oat	**3**	**3**
Pearl millet hybrid	**1000**	200
Maize, sorghum	**400**	200
Cotton hybrid	**50**	**30**
Groundnut, Soyabean	**3**	3
Rapeseed & Mustard	400	200
Cucurbits	800/1000	400/500
Okra	400	200
Tomato	**50**	**25**
Onion	1000	1000

Real value of seed: -

- **Real Value =** $\frac{\text{Purity \%} \times \text{Germination \%}}{100}$
- It's **percentage of germinability** of **percentage of purity** of seed lot of a crop plant.
- It is expressed in **percentage** and also known as **utility percentage of seed.**

- Impurity percentage is called **Dockage**.

Genetic purity:

- Genetic purity of a seed means it is **'true to type',** essentially confirming the seed carries the exact genetic makeup of the variety, ensuring the resulting plant shows the breeder's desired traits (yield, disease resistance, quality) without contamination from other varieties or species.
- Genetic purity is required **100%** as far as an impurity by seeds of other varieties of same crop is concerned but in case of impurity by seeds of other crop species; it is permitted up to **0.1%** only in foundation seed and 0-0.5% in other seeds.
- **GOT (Grow Out Test)** – it is done to determine genetic purity on the basis of plants morphological characters.

Minimum genetic puriety of different categories of seeds:-

- Nucleus and breeder seed**: 100%**
- Foundation and certified seed : **99%**
- Composite/synthetic/multiline seed : **98%**
- Hybrids (e.g. maize): **95%**
- Hybrids of vegetables (i.e. tomato, brinjal, musk melon):**90%**
- Castor hybrid**: 85%**

Physical purity:

- Physical purity of a seed refers to its puriety from contaminants like dirt, weed seeds, other crop seeds, and inert matter (stones, chaff), ensuring the seed lot is composed primarily of the desired pure seeds.
- Physical impurities in a seed lot, such as dirt, stones, weed seeds, chaff, or other crop seeds, are collectively known as **dockage**.

Minimum physical puriety standards of different categories of seeds are:-

- Bhindi (okra) and cucurbits: **99%**
- Major field crops -maize, bajra, wheat, rice, legumes: **98%**
- Mustard, soyabean and millets (except bajra): **97%**
- Groudnunt, spinach and sugarbeet: **96%**
- Carrot, amaranth : **95%**

Germination percentage of seeds:

- Seed germination percentage is the measure of viable seeds in a lot that successfully sprout into seedlings, calculated by dividing the number of seeds that sprout by the total seeds sown and multiplying by 100.

Minimum Germination Standards of different crop seeds-

- **Maize (hybrids):** (90%)
- **Cereals:** Wheat (85%), Paddy (80%)
- **Millets:** Pearl Millet and Sorghum (75%).
- **Cotton :** 65%
- **Vegetables:** Onion (70%), Okra (65%), Cucurbits and Carrot (60%)

Rapid varietal identification test are-

- At field level for verifying seed purity by traditional field Grow-Out Tests (GOT).
- Other rapid methods primarily include chemical, biochemical, and molecular techniques.

1. **Chemical Tests:**
   - **Phenol Test:** Used for **Wheat**.
   - **Peroxidase Test:** Used for **Soybean**.
   - **KOH (Potassium Hydroxide) Test:** Used for **Rice** and **Sorghum**.
   - **NaOH (Sodium Hydroxide) Test: Wheat**
   - **Gibberellic Acid Test:** Used for **Cotton** and **Wheat**.
   - **Alkaloid Test:** Used for **Lupinus species** (Lupin)
2. **Biochemical Tests**
   - **Protein Electrophoresis (SDS-PAGE):** for **Maize**, **Wheat**, **Rice**, and **Barley**.
   - **Isoenzyme Analysis:** to identify varieties in crops like **Tomato**.
3. **Molecular (DNA-Based) Methods-**
   - These are the fastest and most precise techniques, identifying varieties directly through their DNA fingerprint.
   - **PCR-Based Markers (SSR/RAPD):** for **Rice**, **Peppers**, **Maize**.
   - **SNP (Single Nucleotide Polymorphism) Assays:** in **Wheat** and **Cotton**. Technology like VISTa (Variety Identification System of Triticum aestivum) allows for rapid barcoding and identification using a limited number of SNP markers.
4. **Modern Optical Techniques-**
   - **Hyperspectral Imaging (HSI):** A 2025-ready emerging technology used for **Pea**, **Rice**, **Oat**, and **Cotton** to classify seeds with over 95% accuracy using deep learning model.

Seed index:

- Weight of **100** seeds (seed index used in case of **bold seeds** like gram, maize).
- Seed index of important field crops are-

- Desi gram = 140-259g kabuli gram= 260-450
- Maize = 200-250g

Test weight:

- Weight of **1000** seeds (wt. of **small seeds** like rice, wheat etc.).
- Test weight of field crops are-

✓ Wheat = 40g *Phalaris minor* = 2g
✓ Rice = 25g Basmati rice = 21g
✓ Bajra = 5-7g Mustard = 3-5g

Seed quality index: It is the vigor of seed germination.

- Germination percentage in maize is: 90 % (highest in field crops)
- In cotton germination percent is: **65 %** (lowest among field crops)

Seed dormancy:

- Unability of viable seed to **germinate** even under optimal conditions provided for germination.

Breaking of dormancy: two methods

1. **Scarification**: Any treatment i.e. physical or chemical that weakness the seed coat, is known as scarification.
   - **KNO_3 (1-3 %):** **Strongest** dormancy breaker.
   - **Thiourea:** Mainly used in case of **potato**.
2. **Stratification**: Seeds are subjects to **temperature** treatment (**chilling temperature**).
   - For chilling treatment**:** Seeds should be kept 2-8 °C for 12-24 hours.

Seed plot technique of potato:

- Developed by **Pushkarnath** *et al*. in **1967**
- It is a technique of multiplication of **disease free** (Virus free) seed in Northern plains of **country**
- The viruses are transmitted by **aphids**.
- **Nucleus seed** of potato is produced during April to August when aphid population is low in hills. It is brought to plains and is stored in cold-storage for planting the seed crop in October.

Other categories of seeds:

1. **Orthodox seeds**:
   - Seeds remain viable for long time at low temperature and low moisture content. They don't lose their viability at low temperature e.g. Seeds of cereals and beans.
2. **Recalcitrant seed**:
   - Seeds don't remain viable for long time at low moisture & low temperature, because they show very drastic loss of viability with decrease in moisture content below 12-14%.

- E.g. Mango, coconut seeds

Seed Act:

- The seed Act was enacted by the parliament on 29.12.1966 and the seed rule were notified in 1968.
- This act came into force throughout the country on **2nd Oct. 1969**.
- Seed was declared as an **essential commodity** under the **Essential Commodity Act 1955**.

Seed production agencies:

a. **National Seed Corporation** (**NSC**): it was registered in **1963** having objectives of:

- To promote development of a seed industry in India.
- To produce and supply foundation seeds of various crops

b. **State Seed Corporation** (**SSC**): it concerned with;

- Production and supply of certified seed
- To reduce the work load of **NSC**.

Agmark:

- According to Agriculture Production Act 1937
- It's used **National mark** to express **quality/Purity** of agriculture and animal (dairy) **products**.
- Central laboratory of agmark is situated at: **Nagpur.**

Unit-11. Terminology

1. ***Genetics:*** The branch of biology concerned with the study of heredity and variation
2. ***Eugenics***: The application of principles of genetics for the improvement of humankind.
3. ***Gene pool:*** The sum total of all the alleles present in the breeding or reproductive members of a random mating population.
4. ***Gene frequency:*** The proportion of one allele relative to all the alleles of a gene in a Mendelian population.
5. ***Cytogenetics:*** Concerns with the study of various aspects of chromosomes and their effects on the development of characters of organisms
6. ***Genome***: A complete set of chromosomes of a diploid species
7. ***Euphenics:*** Deals with the control of hereditary diseases especially inborn errors of metabolism.
8. ***Epistasis:*** When one gene affects in any way the expression of another gene, the phenomenon is called ***epistasis.***
9. ***Genomics:*** It refers to the study of the structure and function of entire genome of an organism.

10. ***Biodiversity:*** Refers to the total variability present within and among species of all living organisms and their habits
11. ***Acclimatization:*** Refers to adaptation or adjustment of an introduced variety to the new environment.
12. ***A-line:*** *M*eans the male sterile parent used in crossing to develop a new hybrid seed. It is the male sterile line.
13. ***B-line:*** It is the fertile counter parent of the 'A' line. It is also called the maintainer line.
14. ***R-line:*** It is an inbred line which when crossed with a male sterile strain, the resultant hybrid is male fertile.
15. ***In vitro***: biochemical process taking place in a test tubes (in lab.)
16. ***In vivo***: biological process taking place in a living cell or organism.
17. ***Cris-cross inheritance*****:** Inheritance of sex linked genes from grandfather to grandson through daughter.
18. ***Gene cloning***: Isolating a gene and process of producing identical copies is known as ***gene cloning.***
19. ***Electroporation:*** A technique uses electric discharge to produce pores on cell membrane for intake of recombinant DNA.
20. ***Heterosis***: The superiority of F_1 in one or more character over its parents known as heterosis or hybrid vigour.
21. **Mid-parent heterosis (MPH)** is a specific measure showing how much better the hybrid performs compared to the average of its two parents.
22. ***Heterobeltosis*** *(better parent heterosis):* When F_1 is found superior to either parent.
23. ***Allogamy/cross pollination***: Pollen grains from flowers of one plant transferred to receptive stigmas of flower of another plants.
24. ***Self-pollination/autogamy*****:** Pollen grains from an anther fall on receptive stigma of the same flower.
25. ***Chasmogamy:*** Flower opens, but only after the pollination has taken place e.g. **Moong**
26. ***Cleistogamy*****:** Flower does not open at all, ensuring complete self-pollination e.g. **Wheat, cereals.**
27. ***Geitonogamy:*** Pollen grains from flowers of one plant fall on the receptive stigmas of other flower of the same plant e.g. **Maize**
28. ***Apogamy:*** it is the development of a sporophyte from a gametophyte without fertilization or the fusion of gamete. Eg. **Ferns** (Pteridophytes).
29. ***Monoecious:*** Male and female reproductive structures are found on the same plant but in separate flowers e.g. **Maize**

30. ***Dioecious:*** Male and female reproductive structures are found on separate plants e.g. **Papaya, date palm**
31. ***Xenia:*** Pollen's effect on the seed itself (e.g., seed coat color, size) most commonly seen in maize where pollen color determines kernel color. This term coined by W. O. Focke in 1881, it's a result of double fertilization in angiosperms.
32. **Metaxenia:** Pollen's effect on the fruit wall/pericarp (e.g., flesh quality, size, shape). **Date Palm** is one of the most classic and studied examples of metaxenia, where the pollen source significantly influences the fruit's size, shape, color, ripening time, and chemical composition (e.g., sugar and tannin content).
33. ***Pseudo-heterosis:*** Superiority of F_1 over its parents in vegetative growth, but not in yield and adaptation also called ***Luxuriance***.
34. ***Inbreeding:*** Mating between closely related individuals such as selfing and sub-mating.
35. ***Inbreeding depression***: Loss or decrease in vigor and fitness as result of inbreeding.
36. ***Single cross***: it is common in hybrid maize. Produced by crossing two genetically distinct, purebred inbred lines, resulting in a highly uniform first-generation (F1) plant with superior traits like high yield, hybrid vigor, highest heterosis and most uniform.
37. ***Three way cross hybrid*:** The hybrid progeny between a single cross and an inbred *viz.* ***(A x B) × C.***
38. ***Double cross hybrid:*** The hybrid progeny from a cross between two single crosses *e.g.* ***(A x B) × (C x D).***
39. ***Top Cross:*** A cross between an inbred line and an open pollinated variety. Also known as inbred varieties cross.
40. ***Synthetic varieties:*** A variety which is developed by inter-mating in all possible combinations a number of inbred lines with good general combining ability and mixing the seed of F_1 crosses in equal quantity is referred to as synthetic variety (first in maize).
41. ***Composite varieties:*** In cross pollinated crops, a variety developed by mixing the seed of various genotypes which are similar in maturity, height, seed size, seed color, etc.

9. PLANT PHYSIOLOGY

Important points:

- ✓ Photo-respiration is: ***Energy spending process.***
- ✓ Mass flow is affected by: ***Transpiration.***
- ✓ Conversion of fat to sugar occurs in: ***Glyoxysomes.***
- ✓ Photorespiration occurs in: ***Chloroplast.***
- ✓ Photosynthesis in an: ***Oxidation –reduction process.***
- ✓ Photo system II is absent in: ***C_4 plants.***
- ✓ Conversion of fat into carbohydrate is in: ***Glyoxylate cycle.***
- ✓ Precursor of IAA is: ***Tryptophan.***
- ✓ Father of plant physiology is: ***Stephen hales.***
- ✓ Cobalt as a constituent of: ***Vit. B_{12}***
- ✓ Law of tolerance introduced by: ***Shelford.***
- ✓ Glycolysis occurs in the part of cell: ***Cytoplasm***
- ✓ Krebs cycle & ETC occurs in: ***Mitochondria***
- ✓ Final product of Glycolysis is: ***Pyruvate***
- ✓ Light or hill reaction takes place in: ***Grana of chloroplast***
- ✓ Dark reaction or Calvin cycle takes place in: ***Stroma of chloroplast***
- ✓ Most abundant protein in the world: ***Rubisco***
- ✓ One molecule of glucose is produced: ***686 Kcal.***
- ✓ Highest water use efficiency order: ***CAM > C_4 > C_3***
- ✓ Kranz type leaf anatomy found in: ***C_4 plants***
- ✓ Calvin cycle & hatch- slack occurs in: ***Chloroplast***
- ✓ Photosynthetic rate highest in: ***C_4 plants***
- ✓ Glucose is a type of sugar: ***Monosaccharide***
- ✓ Sugar which is sweetest among all sugars: ***Fructose***
- ✓ Examples of disaccharides are: ***Maltose, lactose, sucrose***
- ✓ Non reducing sugar: ***Sucrose***
- ✓ Sugar found in germinating seeds: ***Maltose***
- ✓ Lactose (milk sugar) is a combination: ***Glucose + Galactose***
- ✓ Term protein was coined by: ***Moulder in 1840***
- ✓ The term enzyme was coined by: ***W. Kutins***
- ✓ Lock & key model was proposed by: ***Fisher***
- ✓ Term vitamin was proposed by: ***Funk***
- ✓ Bond that joins amino acids between the carboxyl group of amino acids is: ***Peptide bond***
- ✓ Kranz anatomy was shown by which group of plants: ***C_4 plants***

- ✓ Who used the term ***Florigen*** for a flowering stimulus hormone: ***M. Chailakyan***
- ✓ Mineral element which is involved in stomatal movement is: ***K***
- ✓ Light reaction of photosynthesis takes place in: ***Thailakoid***
- ✓ Breaking of seed dormancy by low tempreature treatment of moist seeds is termed as: ***Stratification***
- ✓ RNA synthesis takes place in: ***Ribosomes***
- ✓ This is essential factor for germination of seed: ***Air***
- ✓ Soaking up of water by dry substances due to hydrophilic colloids is: ***Imbibitions***
- ✓ Shrinkage of protoplasm due to outward flow of water in a hypertonic solution: ***Plasmolysis***
- ✓ The effect of global warming is somewhat good on C_4 plants because they ***efficient*** *user of* CO_2.
- ✓ **Emerson** & **Lewis** (1943): They observed that quantum yield declined sharply at wavelength greater than **600 nm** in red zone also called ***red drop***.
- ✓ The **red spectrum** (wavelength of **580** nm to **680** nm) of light which is the **most effective** for inducing flowering in both short day and long day plants.
- ✓ **Otto Warburg 1920:** found that photosynthesis in algae was inhibited by O_2. This occurs in all C_3 plants and this effect is known as ***Warburg effect***.
- ✓ The light intensity, at which the photosynthetic intake of CO_2 is equal to the respiratory output of CO_2, is called ***compensation point***.
- ✓ Therefore net photosynthesis is **zero** at thecompensation point.
- ✓ The **cold treatment** to a plant bud or seedling in order to fulfill a specific low temperature requirement (especially for cool loving plants) is known as: ***Vernalisation.***
- ✓ The term "***vernalisation***" was coined by: ***T.D. Lysenko.***
- ✓ Translocation of water and nutrients from roots to above ground parts of plants takes place through: ***Xylem.***
- ✓ Mineral deficiency in plants can be quickly corrected by: ***Foliar spray***
- ✓ In water molecule angle between H_2 and O_2 molecules is: $\boldsymbol{105^0}$
- ✓ The total amount of water present in the soil is called as: ***Hollard.***
- ✓ Mass flow/pressure flow theory for the movement of food was given by ***Munch.***
- ✓ Most abundant chemical present in the cell wall is: ***Lignin***
- ✓ Accumulation of hormone during stress is: ***Proline***

- ✓ Photorespiration occurs in C_3 plants in: ***Light only***
- ✓ Nutrient evolved in the biosynthesis of IAA is: ***Zn***

Types of protein:

Proteins	Found in
Collagen	Muscles
Keratin	Hair & nail
Fibroin	Silk
Elastin	Insect wings
Regulatory	Enzymes

Unit-1. Plant physiology

- Plant physiology is the study of vital or **functional** activities of plants.

1. **Diffusion**:

- Means to spread or to **disperse in all directions.**
- The movement of the molecules of gases, liquids and solids from the regions of **higher concentration to the regions of lower concentration** until a equilibrium is reached among molecules throughout the available space is known as ***diffusion*** e.g. ***Spreading of scent molecules in rooms.***
- Other examples of diffusion are:

States of diffusion:	Formed
• Gas into liquid:	***Foam.***
• Liquid into gas:	***Clouds*** (vapor).
• Solid into gas:	***Smoke.***

2. **Osmosis**:

- It is a special type diffusion of liquids though ***a semi permeable membrane.***
- Means diffusion of liquids from lower concentration of solution to higher concentration through ***semi-permeable membrane.*** Osmosis occurs only in liquids.
- Plasma/cell membrane is a semi-permeable membrane e.g. ***Swollen of Kismis in water.***
- In a pure solvent, the value of osmotic potential is zero.

a. **Endo osmosis**: The solvent of lower concentration enter inside the cell is called ***endo-osmosis.***

- Eg. Raisins (***Kismis***) swell in water due to **endosmosis.**

b. **Exo osmosis**: The out ward movement of solvent from high concentration medium is called ***exo-osmosis.***

- E.g. In pickles salt act as exo-osmosis on bacterial cells and killed microbes.

3. **Plasmolysis:**
   - Plasmolysis is the **shrinking** of protoplasm due to outward flow of water in a **hypertonic** solution.
   - If a plasmolysed cell is kept in pure water, the cell attains their original shape and size due toendosmosis. This phenomenon is called ***de-plasmolysis.***
   - E.g. Excessive use of **fertilizer** in the soil kill the plant due to plasmolysis.
4. **Diffusion pressure deficit** (DPD):
   - It is also called ***suction pressure***.
   - It is the force per unit area by which water enters into a cell.
   - DPD = OP – TP (WP)

Where,

- OP = Osmotic pressure
- TP = Turgor pressure
- WP = Wall pressure

- When water enter in a cell, due to swelling a (outward) pressure exerted by the cell solution on the cell wall is known as ***turgor pressure***.
- When water enters in cell, the cell wall exert an opposite (inward) pressure on cell membrane is called ***wall pressure.***
- Water potential is **always negative** and the maximum value is **zero.**

5. **Solutions:**
   a. **Hypotonic solution**: means **low concentrate** solution.
   b. **Hypertonic solution**: **Strong** or higher concentrate.
   c. **Isotonic:** Neither weak nor strong solution e.g. ***eye drops***
6. **Imbibitions:**
   - The adsorption of water by dry woody substances due to hydrophilic colloids is called ***imbibitions.***
   - E.g. **Absorption of water by dry woods.**

Unit-2. Water relation in plants

Absorption of water by plants:

- The absorption of water takes place in the terminal portion of roots i.e. ***root hairs.***
- Root hairs are more developed in xerophytes and absent in hydrophytes.

Types of absorption:

- **Active absorption**: Water is absorbed due to activities of roots and metabolic energy spends in non osmotic active absorption. Only a small portion of water is absorbed by active absorption.
- **Passive absorption**: Passive absorption is controlled by activities of areal parts i.e. shoots, leaves of plants due to pressure exerted by transpiration, hence energy is not required. Plant absorbs most of water by passive absorption.

Ascent of cell sap:

- The upward movement of water from stem base to tree top is called **'ascent of sap'.**

Mechanism of ascent of sap:

a. **Vital theories**:

(i) **Relay** or clambering pump theory: given by **Godlewski** 1884.

(ii) **Pulsation theory**: given by **Sir J. C. Bose** (1923).

b. **Root pressure theory:**

- It coined by **Stephan Hales** (1727).
- The hydrostatic pressure developed due to the accumulation of water absorbed by the roots called ***root pressure***.
- It measured by ***manometer***.

c. **Physical theories:**

(i) **Capillarity theory**: **Boehm 1809**: He stated that water rises in narrow tubes due to surface tension.

(ii) **Imbibitional theory**: According to **Sachs 1878**, imbibitions activity of cell walls of xylem is responsible for ascent of sap.

(iii) **Cohesion and adhesion theory/transpiration pull theory**:

- This theory was propounded by **Dixon & Jolly** 1894.
- This theory is most accepted theory.

Loss of water in plants:

a. Transpiration:

- The loss of water in the form the living aerial parts of the plant in the form of water vapour is termed as ***transpiration***.
- Transpiration is a **necessary evil.**
- It cooled down the leaf surface of plants.
- Nearly 99% of the water absorbed by the plants is lost in transpiration.
- It is a vital and unavoidable phenomenon of plants.
- Transpiration is regulated by stomata activities.
- The principle organ of transpiration is **leaf.**
- Again foliar transpiration is of two types' *viz*. **Stomatal** (80-90 %) and cuticular (up to 20 %).

(i) Stomata:

- These are **specialized epidermal cells** which are distributed all over leaf surface but in case of terrestrial plants, mainly on lower surface of leaves, therefore approx. 97% of transpiration takes place from the **lower surface** in such plants.
- Each stomata has two **kidney shaped guard cells**, inner wall of guard cell is thick and outer wall is thin.

Mechanism of stomata opening and closing:

- Opening and closing of stomata are due to its **turgidity and flaccidity** respectively. Means stomatal movement is governed by ***Turgor pressure.***
- When **T.P. of guard cells increases**, stomata are **opened** and when decreases, stomata are closed.
- Rate of transpiration is determined by ***Photometer***.

b. Guttation:

- Term is given by ***Burgerstein***
- The exudation of water with salt and minerals through **hydathode** is called ***Guttation***.
- **Hydathode:** a structure present at tips of veins of leaves.
- It is caused due to ***root pressure***
- Guttation normally occurs at **night**.

Unit-3. Photosynthesis

1. Photosynthetic pigments:

a. **Chlorophylls pigments:**

- Chlorophylls occur **in the grana plastid** and are associated with the **thylakoids membrane.**
- The centre of the **porphyrine ring** of chlorophyll A & B contains a single atom of **magnesium** (Mg).

Chlorophyll pigments are two types:

(i) **Chlorophyll'a':**

- Empirical formula is **$C_{55}H_{72}O_5N_4Mg$.**
- Occurrence is universal in green plants.

(ii) **Chlorophyll 'b':**

- Empirical formula of Chlorophyll 'b' is **$C_{55}H_{70}O_6N_4Mg$.**
- It present in higher plants and in green algae but absent in red algae.

b. **Carotenoids pigment**:

- Carotene and xanthophylls are together called '***Carotenoids***'.
- It belongs to large group of compounds called ***terpenoids.***
- These are fat soluble **yellow pigments.**
- Carotenoids are located in **chloroplast** and **chromoplast**

(i) **Carotene**:

- **Orange-yellow** coloured having empirical formula **$C_{40}H_{56}$.**
- It is abundant in roots of carrot hence the name **carotene**
- Its most common form is **β-carotene.**
- β -carotene is the precursor of Vitamin A.
- It's quickly oxidized in air hence the rapid change of colour takes place in the scalped carrot.

(ii) **Xanthophylls**:

- It's **more abundant than carotenes** having yellow to brown colour.
- Empirical formula is $C_{40}H_{56}O_2$.
- It is also called ***carotenol.***
- The principal yellow pigment of maize is ***zeaxanthin***.

c. **Anthocyanin:**

- It is a **purple** pigment
- Soluble in water hence it occurs in **solution form** in the water of the cells.

- It **doesn't take** part in photosynthesis
- It is present in **sugar beet, apple**

2. Photosynthesis/CO_2 assimilation/food production:

Definition:

- The formation of carbohydrates from utilizing CO_2 and water by illuminated green cells and O_2 liberated as by product is called ***photosynthesis.***

$$6CO_2 + 6H_2O \xrightarrow{\text{(Light + Chlorophyll)}} C_6H_{12}O_6 + 6O_2$$

- Reaction of the light phase is light sensitive hence called ***photochemical reaction***.
- **Source of oxygen** during photosynthesis is ***H_2O***
- The conversion of light energy in to chemical energy i.e. ATP is called ***'Photophosphorylation'***.
- The breakdown of water molecule H_2O into hydrogen and oxygen by light energy is called as ***Photolysis*** of water.

It has two phases' viz. **light phase** and **dark phase**:-

a. **Light reactions:**

- The site of light reaction is **grana** and dependent on pigment.
- $$\textbf{Light} + \mathbf{H_2O} \xrightarrow{\text{(Chloroplast)}} \mathbf{O_2}\ \textbf{released}$$

Emerson effect:

- Green plant exposed to monochromatic light i.e. wavelength **greater than 680 nm** in **red zone** this prolonged duration of red zone decreased the rate of photosynthesis (i.e. decrease in quantum yield) this effect is called **red drop**.
- The quantum yield produced by the mixed light was greater than the total yield got from the two beams of light used separately. This enhancement effect is known as ***'Emerson effect'***.

b. **Dark reaction:**

- Dark phase of photosynthesis i.e. **CO_2 fixation**:
- The reactions of the dark phase are temperature sensitive and **don't requires light.**
- This is purely chemical reaction and called ***Blackman reactions*** on the name of F.F. Blackman who first demonstrated its existence.

(i) **Calvin cycle/C_3-cycle:**

- This is predominantly found in **wheat**, **rice**, barley, **pulses, soyabean, cotton, sunflower** etc. and such plants are called **C_3 plants** because in the Calvin cycle, **first stable product is** (PGA) **C_3 compound** (having three number of carbon in a compound).

Blackman's: **law of minimum or low of limiting factor:**

- CO_2____(Light, ATP + $NADH_2$) ___Reduction in CH_2O
- If light intensity is increased by twice but CO_2-concentration is constant, there will be no increase in photosynthetic rate.
- It means CO_2 acts as a limiting factor because its concentration is available in minimum quantity.

(ii) **Hatch & slack cycle/C_4 cycle:**

- It is found in **sugarcane**, **maize**, **sorghum**, **bajra** etc. such plants are called **C_4 plants.**
- C_4 plants have **"kranz anatomy"** in leaves.
- 1st stable product is Oxalo-acetic acid (**OAA**) or Malic acid, both is having 4 carbons compound hence the name C_4-cycle.
- PEP carboxylase has high affinity for CO_2 and hence C_4 plants are able to absorb CO_2 strongly from a much lower CO_2 concentration than the C_3 plants.
- Thereby resulting higher rate of photosynthesis.
- Most of **weeds are C_4 plants.**
- C_4 plants lack **photo respiration**, hence photosynthetic rate is higher.

(iii) **Crassulacean acid metabolism (CAM):**

- All CAM plants have **succulent habit.**
- Such plants have also the **capability of fixing the CO_2** lost in respiration.
- Such plants behave like C_4 plants during the night and as C_3 plants during the day.
- These plants have slowest photosynthetic rates.
- In CAM plants open the stomata **during the night** and closed at day time.
- **E.g. Onion**, **Garlic**, **Sisal**, **Lilli** and plant of family **crassulaceae** (eg. cactus, pine apple).

Difference between C_3, C_4 and CAM plants:

S.N	Particulars	C_3 Plants	C_4 Plants	CAM Plants
1	Site of the cycle i.e. leaf structure	Mesophyll	Mesophyll and bundle sheath cell i.e. **kranz anatomy**	Mesophyll and vacuole
2	1st stable compound	**PGA (3-carbon)**	**OAA** or Mallic acid (4-C)	Both
3	CO_2 acceptor	**RuBP**	**PEP**	PEP
4	Dark reaction through	**Calvin cycle**	**Hatch & Slack** cycle and Calvin cycle	-
5	Compensation point (CO_2 concentration)	50 ppm	1-5 ppm	-
6	Photosynthetic rate	Medium	**High**	Slow
7	Biomass production	Medium	**High**	Low
8	Light saturation point	Low	**High**	Medium
9	Photorespiration	**High**	Low	Low
10	Water use efficiency	**Low**	**medium**	**High**
11	Transpiration rate	**High**	Medium	Low

Unit-4. Respiration

1. Photorespiration/glycolate pathway:

- ✓ Krtokov (1963) introduced the term photorespiration to refer to the release of CO_2 in respiration in presence of light.
- ✓ In photorespiration high conc. O_2 induced by high light, checks the production of two molecules of PGA and forms glycolate instead of one molecule of PGA.
- ✓ **H_2O_2** is formed.
- ✓ Known as **C_2 cycle** (2-carbon compound; glycolate)
- ✓ Don't formed **ATP** (energy) hence;
- ✓ This is **energy consuming**/harmful process.
- ✓ It occurs between ***chloroplast***, ***peroxysomes*** and ***mitochondria.***

2. Respiration:

- The energy is released by the oxidation of various photosynthetic products especially carbohydrates by the process called ***respiration.***

$$C_6H_{12}O_6 \longrightarrow CO_2 + H_20 + 686 \text{ kcal}$$

Mechanism of respiration:

- ✓ First and common stage in respiration is called **glycolysis or EMP pathway**.
- ✓ One molecule of glucosegives: two molecules of pyruvic acid.
- ✓ Hence, end product of glycolysis is pyruvic acid.
- ✓ It occurs in cytoplasm under anaerobic condition.
- ✓ **Acetyl Co –A**: link molecule between glycolysis and Krebs cycle.
- ✓ 2 molecules of **pyruvic acid** and further will be **oxidized** through link and Krebs cycle
- ✓ Two molecule of $NADH_2$: further will be used for electron transport system to release $\mathbf{H_2O}$
- ✓ Net production: **2 ATP in glycolysis**. But two molecules of $NADH_2$ will be oxidized aerobically to yield 6 molecules of ATP.
- ✓ Thus total molecule of ATP through glycolysis in presence of O_2 will be **eight** instead of **two**.
- ✓ Both **ETS** & **Krebs cycle** (citric acid/TCA cycle) takes place in **mitochondria** under **aerobic** condition.
- ✓ Energy released during respiration:
- ✓ During the break down of one molecule of glucose, **Net 36 ATP** molecules are formed.
- ✓ **ATP** (Adenosine-Tri-Phosphate) was introduced by: **Karl Lohmann**
- ✓ **Father of ATP Cycle** is considered **: Fritz Lipmann**

3. Respiratory quotient:

- **R.Q.** is the ratio of the volume of CO_2 released to the volume of O_2 absorbed in the respiration.
- It is also called **respiratory ratio**
- R.Q. = Volume of released CO_2/volume of absorbed O_2
- It is measured by **Ganong's respirometer**

R.Q. values of some products:

- Fat: 0.7
- Protein: 0.8-0.9
- Carbohydrates: **1.0**
- Succulents: 0
- Organic acid: >1 or 1.33

Unit-5. Photoperiodism

- **Response of plant to day light period** specially **flowering behaviour** is called photoperiodism.
- The term photoperiodism was coined by **Garner and Allard** (1920).

They classified plants into three categories:-

a. **Short day plants**:
- Such plants require the day length **less than 12 hours** for flower initiation.
- E.g. **tobacco, soybean, rice**, generally **kharif** crops and many tropical plants.

b. **Long day plants**:
- Such plants require **more than 14 hours** day light for flower initiation.
- E.g. **wheat, barley, sugar beet, castor**, generally **Rabi** crops

c. **Day neutral plants**:
- Such plants are **unaffected by the day length** e.g. **maize, tomato**, buckwheat, **sunflower, cotton, cucumber** etc.

d. **Intermediate short day plant**:
- These plants flowers best between 12 to 14 hours of day light e.g. sugarcane

10. AGRICULTURE EXTENSION

Extension: Derived from following two **Latin words.**

Ex = Out

Tensio = Stretching.

- Means ***stretched out to people*** in the rural areas.
- Extension term was originated in **England in 1866.**
- Extension education term was first used in **1873 by Cambrige university.**

Defination:-

- The **National Commission on Agriculture (1976)** refers to extension as an out of school education and services for the member of the farm family and other directly or indirectly engaged in farm production to enable them to adopt improved practices in production, management, conservation and marketing.
- The fundamental objective of extension is to develop the rural people economically, socially and culturally through education.

Priciples of extension:

1. **Principles of cultural awarness**
2. **Grassroots principle**
3. **Principles of indigenous knowledge**
4. **Principle of interest and needs**
5. **Principle of learning by doing**
6. **Principle of participation**
7. **Family principle**
8. **Principle of leadership**
9. **Principle of adaptability**
10. **Principle of satisfaction**
11. **Principle of evaluation**
12. **Principle of empowerment**

Levels of extension:

1. **Extension education**—main aim to create extension professionals. Agencies i.e. agricultural universities, research institutes, apex training centres educate, train and develop professionals for teaching and do research in extension and develope professionals for extension service.
2. **Extension service:-** The primary responsibility of extension service is with the State goverment. The department of agriculture and associated branches of state goverment carry out extension work with the farmers and rural people over the entire state.

Functions of extension-

- The main function of extension is to bring about desirable change in human behaviour by means of education. Changes may be brought about in rural people's knowledge, skill, attitute, understanding, goals, action and confidence.

Education:

- Production of **desirable changes** in human behavior through gaining desirable **knowledge**, **attitude** and mental/**manual skill** is called **education**.

Types of education:

1. **Formal education:**
   - It is highly institutionalized, chronologically graded and hierarchically strucatured education system e.g. school, college education.
   - Learners are homogenous and confined to the premises of educational institute with **top down (vertical) flow of information.**
2. **Non formal education:**
   - Extension education is a **non formal** education.
   - Learners are heterogenous and helps peoples to face actual life situations outside the educational institution.
   - **Horizontal flow of information** between the teacher and learner.
3. **Informal education-**
   - It is life long process and happens in everday life.
   - Self learning is the characteristic feature of informal learning.

Extension education process:

- According to Dr. J.P. Leagans (1967) 'Extension Education' process involves five steps (**SOTER**) –

1. **Situation-**
2. **Objective-**
3. **Teaching-**
4. **Evaluation-**
5. **Reconsideration-**

Teaching learning process -

- Cone of experience given by **Edgar dale** as modified by Sheal (1989). It is also known as the **learning pyramid**.

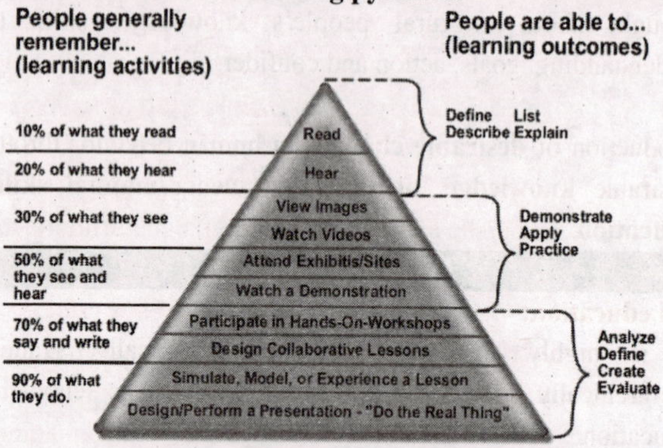

Learning pyramid

Steps in Extension Teaching- remember tricks *'AIDCAS'*

- According to Wilson, MC and Gallup, G. (1955), the six steps in effective extension teaching are (**AIDCAS**).-

1. **Attension-**
2. **Interest-**
3. **Desire-**
4. **Conviction-**
5. **Action-**
6. **Satisfaction-**

Adoption and diffusion of inovations: remember tricks *'AIETA'*

- Adoption is a – Mental process
  - **Adoption**- is a decision to make full use of an innovation as the best course of action available. Or adoption refers to the stage in which an individual select a technology for use.
  - **Diffusion**- is the process by which an innovation is communicated through certain channels over time among the members of a social system.
  - According to Hamberg five stages of adoption process (**AIETA**)-

1. **Awarness-**
2. **Intrest-**
3. **Evaluation-**
4. **Trial-**
5. **Adoption-**

- The categories of adopters are innovators, early adopters, early majority, late majority and laggards.

Categories of (new innovation) adopters:

1. **Innovators**: First people to adopt a new idea. These are venturesome, adventurous people and oriented to take risk. These are generally literate and have good financial resorces, but very little in number (**2.5%** of total population).
2. **Early adopter**: these are respectable and more integrated part of community. They have greatest opinion ledership. The people consider them as **'the man to check with'** before using a new idea. These share about **13.5** % of general population.
3. **Early majority**: these are considered deliberate by the social system. These peoples are called ***"neighbours and friends" from whom most farmers seek information***. (Their proportion is 34% of total population).
4. **Late majority**: these are considered skeptical and adopt new idea just after the average member of a social syste. (Their proportion is 34% of total population).
5. **Laggards:** these are considred traditional, and these are last to adopt an innovation. These possess almost no opinion leadership and are considered as **near isolates**. (Their proportion is 16% of total population).

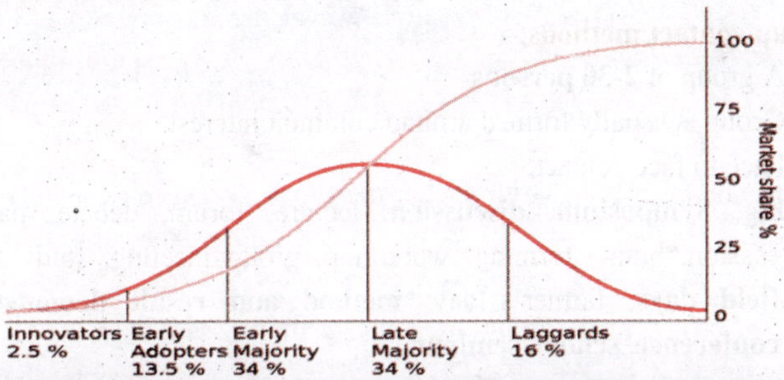

Communication:

- Term derieved from Latin word '***communis***' means common.
- Communication is the process by which messages are transferred from a source to receiver (Rogers and Shoemaker, 1971).

Models of communication-

1. **Aristole model of communication**- it consists of 3 elements.
   - Speaker-Speech-Audience (SSA)
   - Aristole also identified 3 elements i.e. ethos, pathos and logos.

2. **Legan's model of communication** (1963): 6 elements (CMCTAR).
   - Communicator-Message-Channel-Treatment-Audience-Response.
3. **Berlos's model of communication** (1960): it has 6 steps (SMECDR)-
   - Source-Message-Encoder-Chanel-Decoder-Reciever
   - It is also known as berlo **SMCR model of communication** because it has 4 main elements i.e. Source-Message-Channel-Receiver.
4. **Shanon-Weaver (1949)** - 5 elements (STSRD)-
   - Source-Transmitter-Signal-Reciever-Destination
5. **Schramn** (1961) - 5 elements (SESDD)-
   - Source-Encoder-Signal-Decoder-Destination
6. **Rogers and Shoemaker** (1971)- 5 elements (SMCRE)
   - Source-Message-Channel-Reciever-Effects
7. **Westley- Maclean's Model**: (SECDR)-
   - Sender-Encoding-Channel-Decoding-Receiver.

Classification of communication:

- According to use and nature of contact:

1. **Individual contact method**:
   - ✓ Contact is **face to face** or **person to person.**
   - ✓ It is **very effective** in teaching, quantum of feedback available is very high but slow communication
   - ✓ E.g. **Farm and home visit**, **personal contact**, **telephone call**, office call, minikit trial, personal letters and **result demonstration.**
2. **Group contact methods**:
   - ✓ A group of **2-30 persons**
   - ✓ Group is usually formed around common interest.
   - ✓ Face to face contact.
   - ✓ E.g. **Symposium**, **discussion**, lecture, forum, debate, panel,buzz session, brain storming, workdhop, group meeting, study tour and **field days**, farmer's day, **method and result demonstrations**, **conference**, **school, seminar**.
3. **Mass contact**:
   - ✓ **More than 30** persons
   - ✓ It is **more useful** for making people **aware** of the **new agricultural technology quickly**.
   - ✓ Eg. publications like **Bulletins**, news letter, leaflet/booklet/folder, **newspaper**, journal, magazine and circular letter, mass meeting, campaign, **exhibition** (ACF Exam-2021), **television**, **radio**, **fairs**, **posters, cinema**, drama & songs.

Instructional aids:

1. **Audio-Aids-**
   - **Devices** through which we can hear the message are called audio aids. E.g. telephone, radio, tape recorder, Public Adress System.
2. **Visual aids-** through which we can only be seen not hear the message.
   (a) **Non-projected**-e.g. photographs, diagram, map, chart, poster, models, specimen, flannel graphs, flash cards, chalk board.
   (b) **Projected** – slide, motion film, filmstrips, over head projector etc.,
3. **Audio-Visual aids-**
   (a) **Non-projected**- Drama, Puppet show,
   (b) **Projected**- Cinema, Video, and T.V.

Psychology-

- Psychology word derieved from two greek words-
- **Psyche** = Soul
- **Logos** = Science
- It is the science of human behavior and experience.

Intelligence quotient (I.Q.)-

- I.Q. is the ratio between mental age and the actual or chronological age and expressed in percentage.
- About two-thirds of the population scores between I.Q. 85 to 115.

Classification of I.Q.-

Terman Merril test		Shaker Udal classification	
I.Q.	**Category**	**I.Q.**	**Category**
Below 70	Mental defective	0-25	Idiots
70-75	Borderline or feeble minded	25-50	Imbeciles
75-90	Dull and backward	50-75	Morons or feeble minded
90-95	Low average	75-90	Borderline or dull
95-105	Average	90-110	Average
105-125	Superior	110-125	Superior
125-140	Very superior	125-140	Very superior
Above 140	Genius	140 above	Genius

Rural sociology-

- The term Sociology was coined by Auguste Compte from two latin-greek words.
  - Latin word ***Socious*** means ***campanion*** and
  - Greek word ***Logos*** means ***speech***
  - Sociology is the science of human relationship it studies the attitudes and behavior of rural people.
  - **Auguste Compte** reffered as father of sociology.
  - The rural society is self-sufficient.
  - The unit of production in rural society is family.

Social control:

- Social control is the pattern of influence the society exerts on individuals and groups to maintain order and establish rules in the society.

1. **Folkways:** - these are the informal rules of behavior based on habit and tradition. These are approved forms of behavior for specific situations.eg. Good manners, greeting others.
2. ***Mores***- these are socially accepted ways of behavior, which do involve moral standards. E.g. honest, monogamy, Saluting the national flag.
3. ***Norms***- are the established behavior pattern for the members of a social system. Norms are the general rules that govern or regulate social action.
4. ***Laws***- laws act as controls which formally define rights and obligation in society. e.g. enforcement of laws for prohibiting child labour, dowry

Family-

- Family is a group defined by a sex relationship sufficiently precise and enduring to provide for the procreation and upbringing of children.
- **Family** is the basic unit of: ***Society***

Types of family:-

1. **Nuclear family**- it also known as primary family or conjugal family include husband, wife and their unmarried children and share a high degree of emotional bonding.
2. **Joint family**- also known as consanguineal family, living together more than one primary family, on the basis of blood relationship and common residence.
3. **Polygyny family-** where a man marries more than one wife.
4. **Polyandry family-** where a woman is married to more than one husband.
5. **Patri-archal family-** male is considred the head and hold all power.
6. **Matri-archal family-**where head will be the female member.

7. **Patri-lineal family**- where determination of descent and property inheritance take place along the male line.
8. **Matri-lineal family**- where determination of descent and property inheritance take place along the female line.
9. **Patri-local family**- where wife goes to live the family of her husband.
10. **Matri-local family**- where the married couple and their childrens are live with the wife's family.

Social stratification:

1. **Caste-**

- A caste is social category whose members are assigned a permanent status within a given social hierarchy and whose contacts are restricted accordingly (Lundberg). E.g. Brahmins, kshatriyas etc.
- Caste is most rigid and determined only by birth.

2. **Social class-**

- The term social class means a numbers of individuals in the same society whose economical status is similar e.g. rich, poor.

Social Group:-

- Group is a unit of two or more people in reciprocal communication and interaction with each other.

Types of group-

1. **Primary group**- refers to those personel relations that are direct, face to face, relatively permanent, and ultimate. Generally small in size, mostly less than 20 to 30 persons are members. E.g. Family, group of close friends
2. **Secondary groups**- relatively large in size composed of indirect and goal oriented relationship, which are often temporary. E.g. students in a classroom, office employees, polital party, clubs.
3. **Formal group**- it is organization based on its hierarchical structure, with designated tasks related to its function e.g. panchayat, registered societies, committees, task forces etc.,
4. **Informal group**- these are not fomally organized, and rules and procedures are not rigidly enforced. E.g. play group, friendship group
5. **Voluntary group**- is a group of individuals of common interest, who enter into an agreement, usually as volunteers, to form an organization to accomplish a purpose.e.g. Co-operative society, youth club.

Culture-

- It is learned behavior, which has been organized into patterns and is shared and transmitted among the members of society.
- Culture is a way of life, mode of thinking, acting and feeling.

Elements of culture-

1. **Beliefs-** beliefs are the acceptance of a proposition as true is an important means of social control.
2. **Values-** values regarded as important ratings which people attach to things, conditions and circumstances.
3. **Mores-** mores are those forms of action which the group regards as eseential and right. Mores are the common ways of acting which are definitely regarded as right and proper than the folk-was and leads to a degree of severty of punishment, if violated.
4. **Taboos-** taboos are type of mores for which the society does not give consent. Taboos refer to negative action. E.g. marriage within the family among Hindus is a taboo, eating beef in Hindu religion. These are the unwritten laws of the society.
5. **Customs-** customs are socially prescribed forms of behavior and refined mode of falk-ways. It is the mode of behavior, accepted procedure of eating, meeting folks, etc.

Leaders-

- Leader is a person who is clearly distinguished from other individuals in power, status, vision and such traits of character as intelligence, integrity, courage, wisdom and judgment (Chitambar, 1973).

Types of leadership-

1. **Autocratic leader**- operates as if he cannot trust people.
2. **Democratic leader**- shares with the group members decision-making and the planning of activities. The participation of all is encouraged and makes decision in consultation with his followers.
3. **Free-rein leader**- gives complete freedom to his followers to establish their own goals and policies. He does not lead and avoid power.
4. **Laissez-faire leader**- believes that if you leave workers alone the work will be done. He seems to have no confidence in himself.
5. **Paternalistic leadership**- he serves as the head of the family and treats his followers like his family members. He assumes a paternal or fatherly role to help, guide and protects the followers, provides them with good working conditions, welfare facilities and services.
6. **Professional leaders**- he received specific specialized training in the field in which he works full-time as an occupation, and paid for his work e.g. gram sevak, extension officer.
7. **Opinion leaders**: - these are peoples with expertise or influence in a specific area. These are also called ***key communicators***.

8. **Lay-leaders-** also called ***volunteer leaders*** these may or may not have received special training, not paid for his work, and generally works part-time with local group organizations, e.g. gram sahayak, youth club president.
9. **Formal leader**- who holds recognized positions or authority is called formal leaders e.g. Sarpanch, chairman of cooperative society, religious leader.

Important projects and associated persons:

SN	Project	Associate person/year
1	Gurgaon project (HR)	F.L. Brayne, 1920
2	Shantiniketan	Initially started in 8 villages of W. Bengal, by M.D. Tagore
3	Sriniketan	RavindraNathTagore, 1921
4	Rural reconstruction scheme	Daniel Hamilton, 1903
5	Marthandam project (TN)	Spencer Hatch, 1928
6	Sewagram (Wardha)	Sh. Mahatma Gandhi, 1921
7	Etawah Pilot project (UP)	Albert Mayor, 1948
8	Nilokheri Experiment	S.K. Dey, 1948
9	**Majdoor Manjil** (it is concept of Nilokheri project rehabilitation to 7000 immigrants from Pakistan)	S.K. Dey 1947
10	**Grow more food campaign**	K.M. Munshi, 1947-48
11	Firka Development (Madras)	T. Prakashan 1946
12	Bhoodan movement	Acharya Vinobha Bhave, 1951
13	Community Development	2 Oct. 1952
14	National Extension Service	2 Oct. 1953
15	Indian Village Service	T. Mosher, B.N. Gupta, 1945
16	Young Farmer Association	P. Deshmukh
17	Father of Extension	Leagnes
18	National Demonstration Scheme	Kalwar and Subramanian, 1964
19	All India coordinated project on National demonstration on major food crops	1964
20	Model village project (WB) given by	Daniel Hamilton 1903
21	IADP (Intensive Agricultural District Programme) also known as ***package programme.***	July, 1960
22	ICDP (Intensive Cattle Development Project)	1963-64
23	IAAP (Intensive Ag. Area Programme)	March, 1964
24	High yielding variety (HYV) Programme	1966-67
25	Green revaluation	1966-67

26	DPAP (Drought Prone Area Programme)	1973-74
27	Division of agricultural extension started	1971
28	Creation of DARE (department of agriculture research and education	1973
29	T& V (training and visit) system	1974, by Daniel Benor
30	CADP	1974
31	Desert Development Programme (DDP)	1977-78
32	IRDP (Integrated Rural Development Programme)	1978
33	Indira Awaas Yojana (IAY)	1985
34	Initiation of Institutional –Village-Linkage-Programme (IVLP)	1995
35	Establishment of National Gene Bank at New Delhi	1996
36	IRDP, TRYSEM, DWACRA SITRA etc merged into	SGSY, 1999
37	Lab to land program	1, June 1979
38	Nationalization of bank	1969
39	Extension is a	Two way flow of message
40	Concept of demonstration	Dr. Seaman A. Knapp (1885)
41	Father of extension in India	K. N. Singh
42	MANAGE (National Institute of Ag. Extension & Management	Hyderabad , 1986
	Agriulture Technology Information Centre (ATIC). Under NATP project it serves as 'Single Window System'.	1999
43	Agricultural Technology Management Agency (ATMA). It is a component of ICAR's NATP project.	2004
44	National Rural Employment Guarantee Act (NAREGA) enacted	Sept., 2005 and came in to force from 2 Feb., 2006
45	National Rural Employment Guarantee Act (NAREGA) is being replaced by:	Viksit Bharat Guarantee for Rozgar and Ajeevika Mission (Gramin) (VB-G RAM G) Bill, 2025.

Method demonstration:

- It shows how to carry out an entirely new practice or an old practice in better way.
- It is based on the principle of '***learning by doing***'. It is a skill training.
- It is most influence type of demonstration.

Result demonstration:

- A demonstration which shows the value or worth of the new practice is: ***Result Demonstration***
- It is work on the principle '***seeing is believing***'.
- It is an ideal way to present to farmers a comparision between traditionall and new practices but limitation is to take longer time.

Participatry Rural Appraisal (PRA)-

- It is a family of approaches and methods to enable local people to share enhance and analyse their knowledge of life and condtions (Chambers, 2004).

Farmers field school (FFS)-

- It is a group-based adult learning approach that teaches farmers to experiment and solve problems independently.
- It is specially used in IPM techniques of various fild crops and also known as 'schools without walls'.

Remote Sensing-

- Remote sensing is the process of acquiring information about an object or phenomenon without making actual physical contact with it.

Artificial Intelligence-

- It is the ability of a computer or computer-enabled robotic system to process information and produce outcomes.
- It is analogous to the brain and resembles the thought process of human in learning, decision making and solving the problem.

Ethnocentrism-

- It is tendency of the people to thinking one's own group's ways are superior to others or judging other groups as inferior to one's own.

Perception-

- Sensory experience of the world, it is through this experience that we gain information about environment around us.

Communication fidelity:-

- Fidelity is the faithful performance of communication process by all its elements: communicator, message, channel and receiver (Berlo, 1960).

Empathy:

- Ability on the part of one person to understand the other person's internal frame of mind and reference, and accept the same.

Frustration-

- It is a condition where in the goal directed behavior of an individual is blocked and feeling of anger or disappointment because you cannot get what you wants.

Emotions-

- An emotion is a feeling such as happiness, love, fear, anger, grief, affection, pleaser etc., which can be caused by the situation.

Motivation:

- A process of initiating a conscious and purposeful action or goal direted and needs satisfying behaviour is called as: ***Motivation***

✓ *Intrinsic motivation*- it generates from within one's own self.

✓ *Extrinsic motivation*- it generates from out side environment.

Important facts:

✓ The role of extension education in India is performed by: ***State Agriculture University***

✓ The person acting as a connecting link between higher department official & farming community, known as: ***Village extension worker***

✓ Expression of the ends towards which our efforts are directed are called: ***Objectives.***

✓ Agriculture extension worker is: ***A messenger men***

✓ Over–adoption may due to: ***Insufficient knowledge***

✓ Farm management is an: ***Intra farm science***

✓ The most effective approach for personal contact is: ***Telephone calls***

✓ Training and visit (T & V) system was first introdced in the state of: ***Rajasthan***

✓ The two important principles of extension education are: ***Participation and leadership***

✓ ***Sociometry-*** it is quantitative method for measuring social relationship.

✓ The unique, integrated and organized system of all behavior of a person is: ***Personality.***

✓ **Extravert personality**- these are socially adaptable (adjust to any situation) and interested in people, they go to the extent of scarifying themselves for others.

✓ **Introvert personality**- they are socially shy and remain interested in their own feeling and reactions.

✓ The group of people who have lived together long enough, sharing common values and general interest, to be considered as a social unit is known as: ***Society.***

✓ ***Andragogy:*** refers to the art or science of teaching adults.

✓ ***Pedagogy:*** refer to methods and practices used in teaching, especially of children.

- ✓ Large group when divided into smaller units for short period for discussion is called – ***Buzz session***
- ✓ ***Brain storming***- Free introduction of ideas without any limitations to feasibility is found in.
- ✓ ***Folder-*** it is a single piece of paper folded, the once or twice, and when opened, the essential information presented in a sequence.
- ✓ ***Leaflet***- is a single printed sheet of paper of small size containing preliminary information relating to a topic.
- ✓ ***Pamphlet-*** it contains complete information about the topic in a size of 2-12 pages.
- ✓ ***Bulletin-*** it is printed, bound booklet with 12-20 pages, containing comprehensive information about a topic.
- ✓ ***Booklet-*** it contains a relatively large amount of content material, and the number of pages in it ranges between 20 to 50.
- ✓ Systematic display of models, specimens, charts, photographs, pictures, posters, information etc. in a sequence around a theme to create awareness and interest in the community is called: ***exhibition.***
- ✓ ABC (attractive, Brief and clear) are the characteristics of: ***poster.***

11. AGRICULTURE ECONOMICS

The word 'Economics' is derived from the **Greek word** ***'Oikonomia'*** which means:-

- Oikos = Household
- Nomos = Management

Three main definitions of economics are follows—

- Adam Smith's **Wealth definition**, Marshall's **Welfare definition** and Robbin's **Scarcity definition.**

1. Economics is the Science of Wealth- ***Adam Smith***
2. Economics is the study of human behavior in daily business of life- ***Marshall.***
3. Economic as the science which studies human behavior as a relationship between ends and scarce means which have alternative uses- ***Robbin.***

✓ Micro-economics is also known as ***'price theory'*** while macro-economics is known as ***'income theory'*** **or** ***employement theory.***

✓ Book "***Wealth of Nations'*** had wrritten by **Adam Smith**.

✓ Book "***Science of Choice'*** had written by **Robbins.**

Theories in economics:

- Theory of **profit**: **A. Walker**
- Modern theory of interest: **Hicks-Hansen**
- Population theory: **Malthus**
- Theory of inflation: **A.P. Lerner**
- Theory of absolute advantage: **Adam Smith**

Factors of production- four main factors-

1. **Land-** its earnings or rewards as rent.
2. **Labour**- reward as wage, it is an active factor of production.
3. **Capital-** reward interest, it is a passive factor.
4. **Enterpreneurship**- reward as profit.

Farm efficiency measures:

- Net capital ratio = **Total assets/Total liabilities**
- Working capital = $\dfrac{\text{Working assets + Current assets}}{\text{Intermediary liabilities + current liabilities}}$
- Current ratio = Current assests / Current liabilities
- Debt Equity Ratio= Differed Liabilities/Net Worth
- Production efficiency = Yield per acre/Yield of locality × 100
- Operating cost ratio = Total operation cost/Total profit
- Rate of capital turn over= Gross income/Total farm income× 100.

Difference between:-

Production economics	Farm management
Macro approach	Micro approach
Inter farm science	Intra farm science
Decisions are made for aggregate resources of nation	Decisions are taken for individual farm

Variable cost	Fixed cost
Known as cost of variable inputs	Known as **sunk** cost
It is always in cash	It may be cash and non-cash both
It vary with level of output	It doesn't change with level of out put

Goods-

- Anything that satisfies human wants or needs are called as goods.

Utility:

- Any commodity or service which can satisfy a human want is known as utility.
- Utility means the power to satisfy a human want.

Types of utility:

1. **Form utility**- When you change a form of something into a thing which satisfies your want. Processing of potato into potato chips, which fetches superior price than raw potato because of processing.
2. **Place utility**-When you can transport goods from one place to another e.g. transportation of apples from Himchal Pradesh to other states of country to get better price.
3. **Time utility**- Keeping or storing goods at the times of surplus production over a period of time so that they can be used later when scarcity. E.g. cold storage of potato in main season, where from supply of these potato in lean time of production.
4. **Possession utility**- It is obtained due to possession or transfer of ownership of commodity. E.g. Agriculture land converted to commericial land would increase the utility for the same piece of land.

Demand –

- Desire and ability to purchase goods at various prices during a specific time.

Types of demand:

- **Price Demand:** Quantity demanded at different prices.
- **Income Demand:** Quantity demanded at different income levels.

The Law of Demand

- The **Law of Demand** is a fundamental principle in economics that states, all being equal (ceteris paribus), there is an **inverse relationship** between the price of a good and the quantity demanded.

This means-

- As the price of a product increases, the quantity demanded tends to decrease.
- As the price of a product decreases, the quantity demanded tends to increase.

Factors Influencing Demand:-

1. **Consumer Income:** For normal goods, demand increases as income rises. For inferior goods (e.g., generic brands), demand decreases as income rises.
2. **Tastes and Preferences:** Changes in fashion, trends, habits, and advertising can significantly shift consumer preferences for products.
3. **Prices of Related Goods:**
   (i) **Substitutes:** If the price of a substitute good (like coffee to tea) raises, the demand for the original good will likely increase.
   (ii) **Complements:** If the price of a complementary good (like petrol to cars) rises, the demand for the original good may decrease because they are used together.
4. **Consumer Expectations:** Beliefs about future prices or income levels can influence current purchasing decisions. For example, expecting a future price increase might raise current demand.
5. **Population and Market Size:** A larger number of consumers in the market generally lead to a higher overall demand for goods and services.

Agriculrure Credit-

- It is refers to one of several credit vehicles used to finance agricultural activity.

3 R's of credit-

1. Returns
2. Repayment capacity and
3. Risk bearing ability.

5 C's of credit-

1. Character
2. Capacity
3. Capital
4. Collateral and
5. Condition.

7 P's of credit-

1. **Principle of productive purpose**
2. **Principle of personality**
3. **Principle of productivity.**
4. **Principle of phased disbursement**
5. **Principle of proper utilization**
6. **Principle of payment**
7. **Principle of protection.**

Types of farmer (according to land holding):

- **Five major categories of land holding are:**

▪ Marginal farmer	**Less than 1 ha.**
▪ Small farmer	1-2 ha
▪ Semi- medium farmer	2-4 ha
▪ Medium farmer	4-10 ha
▪ Large farmer	10 ha or above

Finance according to period:

▪ **Loan**	**Term (years)**
▪ Short term/ crop loan	**1 to 1.5**
▪ Medium term	1.5 to 5
▪ Long term	5 to 30 years

Types and system of Farming-

Types of farming

a. Diversified farming
b. Specialized farming
c. Mixed farming
d. Dry farming
e. Ranching

System of farming

a. Individual farming
b. Capitalistic farming
c. State farming
d. Corporate farming
e. Joint farming
f. peasant farming
g. Collective farming
h. Coopeartive farming.

Types of coopeartive farming-

(i) Cooperative better farming
(ii) Cooperative joint farming
(iv) Cooperative tenant farming
(v) Cooperative collective farming

Diversified farming:

- A farm on which the income from a single product is less than 50 % of total

Specialised farming:

- A farm on which the income from a single enterprise is more than or equal to 50%.

Marketable surplus:

- Portion of total produce in stock which the farmer is willing to sell
- Total production - total requirement
- Marketable surplus – marketed surplus only under **ideal** condition.

Marketed surplus:

- Quantity of the produces **actual** sale is the market at particular time for sale.

Benefit: cost ratio in agriculture:-

- **B: C ratio =** ***Gross income /Total cost*** **of cultivation.**
- B/C ratio greater than one indicate that a project is beneficial.

Iso-product curve/ iso- quant:

- It is curve of **two input** combinations which have same output.

Indifference curve:

- It is a curve which shows combinations of **two commodities** which derive **equal** satisfaction to the consumer.

Agricultural credit:

- **Short term loan**: For the purpose of purchasing seeds, fertilizers and for meeting out familyrequirements.
- **Medium term**: To purchase agricultural equipments, animals, and land improvement equipments e.g. Tractor.
- **Long term**: To purchase land, farm buildings and expensive agricultural equipments and for repayment of old loan.

Types of assets:

- Fixed assets: Difficult to convert in cash e.g. **land, buildings**
- Working assets: **Farm machinery, equipments**
- Current assets: More liquid e.g. **cash, seed, fertilizer.**

Stage of production -

1. **Stage-I :-**
   - It is called ***irrational zone*** or the stage of increasing return to the factor of production.
   - In this stage, MP increases at first then decreases.
   - AP increases through out this stage i.e. upto point A.
   - TP increases sharply during this stage.
   - In this stage elasticity is more than one.
2. **Stage –II**
   - It is known as ***rational zone***.
   - This stage is also known as ***stage of diminishing returns***.
   - In this stage, AP starts to diminish (decrease) while MP continues to decrease although MP is still positive but less than AP.
   - Elasticity is zero to one.
   - Here, TP increases at a decreasing rate until it reachs the maximum point B.
3. **Stage-III**
   - This is also called ***stage of negative return.***
   - In this stage, TP starts decline, MP become negative and fall below the x-axis.
   - AP being positive but decreasing rate.
   - When TP decreases, MP is negative.
   - Elasticity less than one.

Figure 1

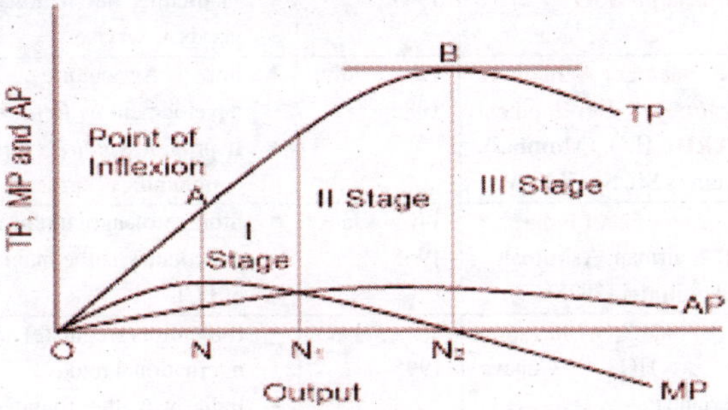

Fig-1 Three stages of production.

Important institutions related to agriculture economics:

SN	Name of Organization	Set up on	Objective
1	Reserve Bank of India (RBI) (presently governer is **Sh. Sanjay Malhotra, since 11.12.2024)**	**1935**	▪ To maintain financial stability in India. ▪ Its banker's bank ▪ To govern the issue of bank notes ▪ To run the currency and credit systems of the nation to its benefit.
2	**State Bank of India (SBI) (Chairman Sh. Challa Sreenivasulu Setty since 28.08.2024)**	**1955**	▪ It is an Indian multinational, public sector banking and financial service statutory body with H.Q. Mumbai (MH). ▪ It originated from bank of Culcutta, which established on 2 June, 1806. ▪ It is largest Indian bank.
3	Agricultural marketing (**AGMARK**)	1937	▪ This mark indicates the purity and quality of Ag. Products. ▪ This label is given under Central Agricultural Produce (Grading & Marking) Act, 1937.
4	**ECOMARK**	1991	▪ This mark is given to ensure that the product is environment friendly.
5	Bureau of Indian Standards (**BIS**)	1, April, 1987	▪ This marks provides third party guarantee to the buyer (that the goods are made in accordance with the provenance of Indian standards) ▪ It was earlier named as **ISI** (Indian Standards Institutions) established on 1947
6	International Organization for Standardization (**ISO**)	25, Feb., 1947	▪ To promote a universal standard for facilitating international exchange of goods & services.
7	National Bank For Agriculture and Rural Development (**NABARD**), H.Q. (**Mumbai**), **chairman** is **Sh. Shaji K .V.**	12, July, 1982	▪ For Agriculture and Rural development by ***Refinance***. ▪ It provides *indirect* ***finance*** to rural communities (farmers).
8	Food Corporation of India (**FCI**) chairman: **Ashutosh Agnihotri (2025)**	14, Jan., 1965	▪ Storage of agriculture commodity in the country is the main responsibility of FCI.
9	World Trade Organization (**WTO**), HQ- Geneva (Switzerland)	1, Jan., 1995	▪ It monitors (regulate) and encourages international trade ▪ India was the founder member of WTO. ▪ It was earlier named as General Agreement on Tariffs & Trade (GATT), set up on, Jan., 1948 ▪ Founded after Uruguay round (8th)

10	National Institute of Agriculture Marketing (NIAM), Jaipur	8, Aug., 1988	▪ To provide specified trainings, education, consultancy and research in agriculture marketing at national level. ▪ It renamed as Choudhary Charan Singh National Institute of Ag. Marketing. (CCS-NIAM).
11	Tribal Co-operative Marketing Federation (**TRIFED**)	1987	▪ To develop the system of marketing of forest product in tribal communities of country.
12	Agricultural & Processed Food Products Export Development Authority (**APEDA**), New Delhi	Feb., 1986	▪ To enhance export of processed agricultural products.
13	National Agriculture Co-operative Marketing Federation (**NAFED**), New Delhi	1958	▪ It is apex body of Co-operative marketing for agriculture related products in India.
14	Directorate of Marketing & Inspection (**DMI**), H.Q. (Faridabad, HR & Nagpur, MH)	1935	▪ Main responsibilities are Quality control of agriculture commodities.
15	Commission on Agricultural Cost & Price (CACP)	Jan., 1965	▪ It recommends the **Minimum Support Price (MSP), Statutory Support Price (SSP)** and **Procurement Price (PP).**

1. **Minimum Support Price (MSP):** announced by govt. of India on the recommendation of CACP before sowing of crops in two seasons i.e. Rabi & Kharif. The Indian government currently sets Minimum Support Prices (MSP) for 22 mandated crops, comprising 14 Kharif crops (like paddy, maize, cotton), 6 Rabi crops (like wheat, gram, mustard), and 2 commercial crops (jute, copra).
2. **Fair & Remunerative Prices (FRP)** – it is announcing for sugarcane.
3. **Procurement Prices (PP):** are announced by govt. of India on the recommendation of CACP before harvest of the crops.

Important points:

- ✓ Cooperative movement was started in India: ***1904***
- ✓ Father of economics: ***Adam smith***
- ✓ Father of co-operative movement in India was: ***F. Nicholson***
- ✓ In farm production capital is the: ***Passive factor***
- ✓ Labor in farm management is an: ***Active factor***
- ✓ The agricultural produce Act was passed in: ***1937***
- ✓ **AGMARK** established in the year: ***1937***
- ✓ Planning commission established in: ***1950, March***
- ✓ **VAT** (value added tax) is an: ***Indirect tax***
- ✓ Agriculture year: ***1stJune to 31st May***
- ✓ Financial year: ***1stApril to 31stMarch***
- ✓ Demand for luxurious good is: ***More elastic***
- ✓ Demand for necessary commodities is: ***Inelastic***
- ✓ One of the important fixed cost on farm is: ***Land rent***
- ✓ The demand of agricultural products in general is: ***Inelastic***
- ✓ The price elasticity of demand for '**foodgrain**' is: ***Relatively inelastic***
- ✓ Indian economy is concerned with: ***Mixed type***
- ✓ The period of 12th five year plan in India is: ***2012-2017***
- ✓ Risk is minimized in: ***Diversified farming***
- ✓ Short term credit facility is given for the purpose of: ***Crop production***
- ✓ Finance is provided by land development bank: ***Long term loan***
- ✓ Retail markets are considered to be: ***Perfect market***
- ✓ When marginal utility is zero, total utility is: ***Maximum***
- ✓ Principle of Equi-marginal returns is applied when: ***Resources are limited***
- ✓ At present, the farming system of India has become: ***Market oriented***
- ✓ **The U" shape** of cost curves could be best explained by the law of: ***Variable properties***
- ✓ **NABARD** established in 12, July, 1982 with the recommendation of: ***Shivaraman committee or CRAFICARD committee.***
- ✓ CRAFICARD (Committee to Review Arrangements for Institutional Credit in Agricultural and Rural Development in 1979 under the chairmanship of: ***Shri B. Sivaraman.***
- ✓ **RRBs** (Regional Rural Banks) established in 1975 with the recommendation of**:** ***Narsimham committee***.

- ✓ The ownership sharing patterns of RRBs are government of India (50%), nationalized banks (35%) and state government (15%).
- ✓ The International Bank for Reconstruction and Development (IBRD) was established in 1945. It is also known as World Bank.
- ✓ A market structure with a large number of firmes selling differentiated product is called: ***Monopolistic competition***
- ✓ Most important principle if farm management is done is: ***Law of diminishing return***
- ✓ The cost of production of the crop can be minimized by using economic principle: ***Principle of least cost combination***
- ✓ The apex body for institution finance for agriculture in India is: ***NABARD***
- ✓ The national centre for agricultural economics and policy research (NIAP) is located at: ***New delhi***
- ✓ According to law of diminishing return the economically relevant zone is: ***Zone-ll (Rational Region)***
- ✓ Principle of Equi-marginal return is applied when: ***The prices of resource is high***
- ✓ Law of dimising returns are operates when: ***One input are variable***
- ✓ Indian economy is a type of economics: ***Mixed economics***
- ✓ In economics, anything that can satisfy a human want is called as: ***Goods***
- ✓ In economics cash is a: ***Working capital***
- ✓ All possible combination of two inputs capable of producing same level of output refer as: ***Isoquant***
- ✓ All possible combination of two inputs which can be purchased with a given amount of fund refer as: ***Isocost line***
- ✓ The time gap between investment and return is: ***Gestation period***
- ✓ Risk bearing ability depends on: ***Net worth***
- ✓ When total utility is maximum then marginal utility is: ***Zero***
- ✓ Stage I of production is also called the stage of: ***Increasing return***
- ✓ In the case of fruits and vegetables the marketed surplus is nearly: ***80 %***
- ✓ The income elasticity of demand for inferior goods is: ***Negative***
- ✓ The income elasticity of demand for giffen goods is: ***Negative***
- ✓ In the monopoly market there is: ***One seller of the produce***
- ✓ When there is a only one buyer of the product, the market is termed as: ***Monopsony market***
- ✓ ***Oligopsony market-*** only few buyers are present in market.

- ✓ ***Oligopoly market-*** only few sellers are available in market.
- ✓ There is a large number of buyers and seller and they have perfect knowledge of demand, supply and prices termed as: ***Perfect market***
- ✓ A market in which business is done in accordance with the rules and regulations is called: ***Regulated market.***
- ✓ The cost involved in moving the product from the point of production to point of consumption is called: ***Marketing cost***
- ✓ Balance sheet of farmer's exhibit the: ***Financial position of a farmer.***
- ✓ Finance management at farm level is called as: ***Micro finance management***
- ✓ As per the RBI norms, 18 per cent of the Net Bank Credit should be deployed in agriculture
- ✓ In India marketed surplus is greater than the marketable surplus
- ✓ ***Land*** rent is an example of: ***Fixed cost***
- ✓ Sales tax is an example of: ***Indirect tax***
- ✓ Through cooperative both members and non-members are benefited through: ***Cooperative***
- ✓ The most common method exists in regulated market is: ***Open auction system***
- ✓ The technique which protect the traders from extreme falls in price is known as: ***Hedging***
- ✓ Storage function creates: ***Time utility***
- ✓ When one enterprise neither adds nor hinders the production of another enterprise, the enterprise relationship is known as: ***Supplementary***
- ✓ The percentage expenditure on luxurious commodities increases as income increases
- ✓ The market in which permanent or durable commodities are traded: ***Secular market***
- ✓ Crop farming and milk production is an example of: ***Complementary enterprise***
- ✓ In WTO, non-actionable subsidies like those on agricultural research, plant protection chemicals and subsidies for ecologically fragile zone are counted under: ***Green box***
- ✓ Limited permission on subsidies like those on electricity, irrigation, fertilizers for the purpose of agricultural production are counted under: ***Blue box***
- ✓ Agriculture produce (Grading and Marketing) Act was passed in the year: ***1937***

- ✓ Co-operative credit societies act was passed in: ***1904***
- ✓ The analytical concept-cum criterion developed recently to compare the overall development of countries is: ***Human Development Index (H.D.I).***
- ✓ The Human Development Index quoted in World Development Report is a quantitative measure of a country's: ***Standard of living, educational opportunities and Longevity of citizen.*** **(RAS-pre 2013)**

What is GST?

- **The goods and services tax (GST)** act were passed in the lok sabha on 29th march, 2017, and came into effect from 1st July, **2017.**
- Goods & Services Tax Law in India is a **comprehensive, multi-stage, destination-based tax** that will be levied on every **value addition**.
- In simple words, GST is an indirect tax levied on the supply of goods and services. GST Law has replaced many indirect tax laws that previously existed in India.

Multi-stage: There are multiple change-of-hands an item goes through along its supply chain: from manufacture to final sale to consumers.

- **CGST:** Collected by the Central Government on an intra-state sale (Eg: Within Rajasthan)
- **SGST:** Collected by the State Government on an intra-state sale (Eg: Within Rajasthan)
- **IGST:** Collected by the Central Government for inter-state sale (Eg: Rajasthan to Gujrat)

12. STATISTICS

- ✓ Father of vital statistics was: ***Captain John Grant.***
- ✓ The purpose of randomization in field experiment is to: ***Control variance***
- ✓ Unit less measure of dispersion is: ***C.V.***
- ✓ Coefficient of Variation is computed by: ***(S.D./Mean)× 100***
- ✓ Critical Difference (CD) is calculated by: ***√(2 MSE/r) × t***
- ✓ Expression for degrees of fredom in RBD is: ***(n-1) (r-1)***
- ✓ The range of probability lies between: ***0 to 1***
- ✓ Minimum degree of freedom for error in ANOVA is: ***12***
- ✓ T' test is applicable when number of treatments is: ***2***
- ✓ **Goodness** of test is used for distribution of: ***Chi square test***
- ✓ The first step of summarizing the data is: ***Classification***
- ✓ Probability of impossible event is always: ***Zero***
- ✓ To find the average height of plants we should use: ***Arithmetic mean.***
- ✓ When sample size is small and S.D. is unknown use: ***T' test***
- ✓ For testing the significance of correlation coefficient use: ***T' test.***
- ✓ Latin square design is suitable for comparing: ***5 to 12 treatments***
- ✓ In the field experiments the commonly used design is: ***RBD***
- ✓ The regression coefficient lies between: ***-∞ to +∞***
- ✓ The correlation coefficient lies between: ***-1 to +1***
- ✓ The student **t"** test was discovered by: ***W.S. Gosset***
- ✓ The value of X^2 (chi square) always lies between: ***0 to ∞***
- ✓ Most commonly used measure of central tendency is: ***Arithmetic mean***
- ✓ Measure of dispersion which is considered as best: ***Standard deviation***
- ✓ Which is not a absolute measure of dispersion: ***Coefficient of variation.***
- ✓ The value of standard deviation may vary between: ***0 to ∞***
- ✓ Correlation is used to analysis the behavior of: ***Two or more variable.***
- ✓ The RBD is available for a wide range of treatments: ***2 to 24***
- ✓ If observed frequency is equal to expected one than: ***The value of x^2 static is zero***
- ✓ To getting the higher precision is required: ***Larger samples***
- ✓ F-test can be used for testing the significance of: ***Several differences***
- ✓ The minimum sample size for using x^2 test should be: ***50***
- ✓ To find the average size of shoes sold in the market we should use: ***Mode***
- ✓ To present the whole data by only single value we used: ***Measure of central tendency***

- ✓ During the experimentation, the some lost information can be get by: ***Missing plot technique***
- ✓ Two study effect of two or more factors requiring different plot sizes: ***Split Plot Design***
- ✓ In industries for quality control, the most important measure of dispersion: ***Range***
- ✓ If all the variate values are negative the standard deviation will be: ***Positive***
- ✓ Under the assumption for analysis of variance, the parent population should be: ***Normal***
- ✓ The analysis of variance technique is used for: ***Comparing the means of more than two populations***
- ✓ The design where number of replication is equal to number of treatments: ***Latin square design (LSD)***
- ✓ For testing the independence of two attributes, the test used is: x^2 ***test***
- ✓ Correlation of continuity in 2 x 2 contingency table should be used when expected frequency of any cell is: ***Less than 5***
- ✓ When two variable moves in the same direction correlation is said to be: ***Positive***
- ✓ When two variables moves in the opposite direction correlation is said to be: ***Negative***
- ✓ In the case of perfect negative correlation the coefficient of correlation will be: ***-1***
- ✓ Probability of any event is a number lying between: ***0 to 1,*** but not practically ***zero***
- ✓ In describing the amount of variation in a population & measure often used is: ***Coefficient of variation***
- ✓ If the fertility variation in the field is in two directions at right angles, which of the experimental deign is suitable: ***LSD***
- ✓ Arithmetic mean is most commonly used because: ***It is based on all observations***
- ✓ To testing the homogeneity of correlation coefficient used: ***Chi square test***
- ✓ To reduce the experimental error with heterogeneous material are used: ***More replication & use of local control***
- ✓ The lack of homogeneity in the size of distribution is called: ***Dispersion***
- ✓ Such measure of dispersion which is affected least by fluctuation of sampling: ***Standard deviation***

- ✓ If the kurtosis of a distribution is less than 3, then it is called: ***Platykurtic curve***
- ✓ The number of independent constraints in the set of observation is called: ***Degree of freedom***
- ✓ The test whether theory fits well in practical can be judged by: ***Chi square test***
- ✓ For testing a single proportion of the population test is used when the sample size is large: ***Z- test***
- ✓ The correlation between price and demand of a commodity is: ***Negative***

Statistics:

Definition: By statistics we means aggregates of facts affected to a marked extent by multiplicity of causes, numerically expressed, enumerated or estimated according to reasonable standard of accuracy collected in a systematic manner for a predetermined purpose and placed in relation to each other.

Variable:

- A characteristic/phenomenon which may take on **different values** i.e. **rainfall records**, birth, **death rates** etc.

Variables are **two types:**

1. **Continuous variable**: Quantities which can take **any numerical** value within certain range e.g. Plant height, yield etc.
2. **Discrete/discontinuous variable**: can take only **integer** values.

Variate:

- A single **observation** or measurement.

Central tendency:

- A value of the Variate which thoroughly representative of the **series** or the **distribution** as a whole is known as **central tendency**.

Measurement of central tendency:

1. **Arithmetic mean**:
   - Most **commonly** used measure of central tendency, only measure of **practical** importance.
   - It is regarded as the **best** of all the averages.
   - It is rigidly defined and based on all the observations.
2. **Median**:
   - The value of the **middle** most when items are arranged in either ascending or descending order of their magnitude
   - E.g. 3, 6, 9, 5, 4, 7, 8 put in ascending order = 3, 4, 5, **6**, 7, 8, 9.
   - Median is = **6**

- Median is used for **qualitative** data such as **intelligence**, **ability**, **honesty** etc.

3. Mode:

- Value of the Variate which occurs most frequently or whose **frequency** is maximum
- E.g. 4, 7, 6, 5, 4, 6, 4
- Mode = **4**
- It used for model size of shoes, size of readymade garments and in business/meteorological fore casting.

4. **Harmonic mean**:

- It is the reciprocal of the arithmatic mean
- Used to find **average speed**, distance and rate.

Empirical relationship between mean, mode and median are:

a. In a **symmetrical** distribution: Mean = Mode = Median.

b. In a **skewed** distribution/asymmetrical distribution: Mean - Mode = 3(Mean-Median).

Measures of dispersion:

- Variation of the variable about the central value

Two types of measures of dispersion:

A. Absolute measures:

1. **Range**:

- It is **simplest** measure of dispersion.
- Range = Largest value – Smallest value.

Properties of range:

- Coefficient of range = L - S/L + S
- Quartile range = $Q_3 - Q_1$
- Coefficient of quartile range = $Q_3 - Q_1/Q_3 + Q_1$

2. **Mean deviation:**

- M. D. = $\frac{\sum(X_i - X)}{N}$

3. **Standard deviation**/root mean square deviation:

- It is defined as the **square root** of the **arithmetic mean**.
- Given by **Karl Pearson**, 1823 denoted by Greek word ***sigma*** (σ).
- S. D. = $\sqrt{}$ **variance**
- It is most commonly used measure of dispersion

4. **Variance**:

- Term **variance** proposed by **R.A. Fisher**.
- **Square root** of **standard deviation** known as ***variance***

B. Relative measure of dispersion:

1. **Coefficient of variation** (C.V.):
   - Given by **Karl Pearson.**
   - Most commonly used measure of relative variation.
   - **C.V. = S.D./Mean**
   - If standard deviation is expressed as **%** of mean is known as **C.V.**
   - If value of C.V. is **greater** it means more **variability** or less homogeneity, if values of C.V. **less** means vice versa.
2. **Standard error of mean**:
   - **SEM** defined as the **S.D.** of the **sampling distribution of means**.
   - It gives idea about **variability** of given data or sample.

Moment:

- Force × distance.
- The moment in statistics are used to describe the various characteristics of frequency distribution like central tendency (μ 1), variance (μ 2), skewness (μ 3) and kurtosis (μ 4).
- In a symmetrical distribution all odd moments i.e. μ1, μ3, μ5 would always be **zero** because positive deviation & negative deviation of symmetry are exactly balance or **equal**.

Difference between raw moment and central moment:

Raw moment or arbitrary mean	**Central moment** or central mean
It is denoted by **μ'r**	It is denoted by **μr**
Calculated about any point	Calculation about **mean** only & take more time
1st row moment about point A = 0 is equal to **mean**	1st central moment **always** zero
Raw moment will be changed of origin	Central moment aren't changed of origin

Note:

- Usually calculation of central moment takes the more time due to reason that mean may not be in whole number therefore calculation of the central moment first we calculate raw moment and with help of the formulae we calculate central moment.

Raw moment:

- The **r** th moment about arbimatary mean "**A**" is defined as:

$$\mu'r = 1/N \sum F_i (X_1 - A)^r$$

Central moment:

- The **r**th moment about mean (x) is defined as:

$\mu_r = 1/N \sum Fi (X_1 - X)^r$

1. Value of μ_0=
    - $1/N \sum F_i (X_1 - X)^0 = \mathbf{1}$
    - The 0th central moment is always **one**.
2. Value of **1st** central moment:
    - $\mu_1 = 1/N \sum F_i (X_1 - X)^1 = \mathbf{0}$
    - The first central moment is **always zero**,
    - Whereas" raw moment is **not zero**. It is **equal** to mean.
3. Value of 2nd central moment:
    - $\mu_2 = 1/N \sum F_i (X_1 - X)^2 = \boldsymbol{\sigma^2}$ (variance)
    - 2nd central moment gives an idea about measures of **dispersion.**
4. Value of 3rd central moment:
    - $\mu_3 = 1/N \sum F_i (X_1 - X)^3$ = measure the **skewness**.
    - μ_3 gives an idea either curve is symmetrical or not.
5. Value of 4th central moment:
    - $\mu_4 = 1/N \sum F_i (X_1 - X)^4$ = kurtosis.

Skewness:

- There are two type of frequency distribution:

1. Symmetric frequency distribution:
    - Mean = Median = Mode.
2. Asymmetrical frequency distribution:
    - Mean ≠ Median≠ Mode.
    - Quartiles are not equi distant from median.
    - $Q_3 - Md \neq Md - Q_1$
    - Curve is more inclined in one side than other.
    - The sum of positive deviation from median is not equal to the sum of negative deviation from the median.

Definition:

- Skewness may be defined as departure from symmetry or lack of symmetry of frequency distribution is known as **skewness**.
- It is denoted as $\boldsymbol{\beta_1}$.

Skewness is two types:

1. **Positive** skewed frequency distribution:
    - A distribution or skewness is said to be positive if the frequency curve has a longer tail on the right hand side as compare to left hand side.
    - Positive skewness = Mean > Median > Mode.

2. **Negative skewness:**

- The frequency curve is more inclined to the left hand side as compare to right hand side.
- Negative skewness = Mode > Median **or** **M**ean < Median < Mode.

Measures of skewness:

a. **Prof. Karl Pearson** coefficient of skewness:

- $S_K = \frac{\text{Mean} - \text{Mode}}{\text{SD}}$

b. Based on moments:

- Skewness (β_1) = μ_3^2 / μ_2^3
- y_1 or r = $\pm\sqrt{\beta_1}$
- kurtosis (β_2) = μ_4/μ_2^2
- y_2 = β- 3

For symmetric distribution:

- β_1 = 0
- β_2 = 3
- Skewness = 0

Kurtosis:

- Refers to degree of **flatness** or **peakness** of the frequency curve.
- It is denoted as $\boldsymbol{\beta_2}$.

Three types:

- Leptokurtic (narrow peak/base): $\beta_2 > 3$
- Mesokurtic (normal curve): $\beta_2 = 3$
- Platykurtic (flat peak/broad base): $\beta_2 < 3$

Distribution: Theoretical distributions are two types.

a. **Discrete** distribution:

- X can take only integer values like 0, 1, 2, 3, ------
- Eg. **Binomial distribution**, **Poisson distribution**

b. **Continuous** distribution:

- X can take **all possible** values in a given range.
- E.g. **Normal distribution**.

1. Binomial distribution:

- Given by **James Bernoulli** in 1700 A.D.
- Bernoulli trial: a trial is known as Bernoulli trial if it has only two out comes namely **success** and **failure** with probability **p** and **q** respectively.
- p + q = 1

Definition:

- A random variable "x" said to have binomial distribution if it assumes only non-negative values and its probability mass function is given by:
- $P(x) = \mathbf{(n/x)\ P^X.\ q^{n-x}}$
- X = 0, 1, 2, -----
- n = number of trial
- x = number of success of n trial
- p = probability of success
- q = probability of failure or **(1-p)**
- p + q = **1** (B.D. represent by **1**)
- **n p** are the parameters of B.D.

Properties of B.D.:

- Arithmetic mean (μ_1) of B.D = np
- Variance (μ_2) of B.D. = npq
- Skewness (μ_3) = npq (q-p)
- Kurtosis (μ_4) = npq (1 + 3 (n-2) p q
- Standard Deviation = √npq
- In case of B.D = **Mean > variance.** ***(JRF-07)***

2. Poisson distribution:

- Discovered by **S.D. Poisson**

Definition:

- It is a limiting case of B.D such that
- N tends to ∞
- p tends to zero
- and np = m

Therefore a variable **x**" is said to follow a Poisson distribution if it assumes only **negative** values and its probability mass function is given as:

- $$\mathbf{P(x) = \frac{e^{-\lambda}.\ \lambda^x}{X}}$$

Here,

- ✓ λ is called the parameter of P.D
- ✓ λ is always greater than **zero**.
- ✓ X = number of success
- ✓ e = 2.718

Examples of P.D:

- Number of deaths from a disease
- Number of defective materials

- Number of wrong calls receives in a month.

Properties of P.D:

- Mean (μ_1): λ
- Variance (μ_2): λ
- μ_3: λ
- μ_4: $3\lambda^2 + \lambda$
- **In case of P.D: Mean = Variance.** ***(JRF-09)***
- P.D is useful in theory of games, waiting time, problems of business.
- P.D tends to normal distribution when λ tends to ∞

Coefficient of P.D:

- $\beta_1 = 1/\lambda$
- $\beta_2 = 3+1/\lambda$

3. Normal distribution:

- Discovered by **de-Moivre** in 1733 also called *Gaussian distribution.*

Definition:

- A random variable **x'** is said to have a normal distribution with parameters **mean** and **variance** and its probability density function is defined as:

$$P(x) = \frac{1}{\sigma\sqrt{2\pi}} e^{-1/2(x-\mu/\sigma)^2}$$

Properties of normal distribution:

- The curve of normal distribution is bell shaped
- It is symmetric about the mean
- Mean = Mode = Median (are equal).
- Quartiles are equi distance from median: $Q_3\text{-}Q_2 = Q_2\text{-}Q_1$
- Coefficient of skewness for normal distribution is: *Zero*
- Coefficient of kurtosis for normal distribution is: *3*
- Range of normal distribution is from: $-\infty$ ***to*** $+\infty$
- But practically range equal to 6σ
- Mean deviation about mean: $4/5\sigma$
- Quartile deviation: $2/3\sigma$

Importance of normal distribution:

- It has got a large application in statistics quality control and important area of statistics used in industries for setting control limit.

Correlation:

- It studies the **relation** or association between two variables.
- Two independent variables are **uncorrelated**.
- The measurement of correlation is called the **correlation co-efficient** (r) or correlation index, which summarizes in one figure the direction & degree of correlation.
- **Range of correlation** are varies between **+1** to **-1.**

Types of correlation:

a. **Positive/negative**:

- If both the variables are varying in the **same direction** i.e. one is increasing the other also increasing and vice versa than it is known as **positive correlation.**
- If both the variables are varying in **opposite direction** i.e. as one variable increasing the other is decreasing and vice versa than it is known as the **negative correlation.**

b. **Simple, partial and multiple correlations**:

- **Simple** correlation: when only **two variables** are studied.
- **Partial** correlation: **More than** two variables are studied but consider only two to be influencing each other, the effect of other influencing variable being kept constant.
- **Multiple** correlations: **three** or more variable are studied simultaneously.

c. **Linear** and **non linear** correlation:

- If the amount of change in one variable **tends** to bear a constant ratio to the amount of change in the other variable is known as ***linear correlation***.
- If the amount of change in variable **doesn't** bear a constant ratio to the amount of **change** in other variable is known as ***non linear*** *correlation*.
- In the **most of** the practical situations we find a **non linear** relationship between variables.
- Absence of any **relationship** between the variable the value of correlation coefficient will be **zero**.

Regression:

- Term regression first used by **Sir Francis Galton** in 1877.
- The regression analysis is a mathematical average relationship between **two** or **more** variable. Relation may be linear or non linear.
- Original line of regression = $y = a + bx$
- Range of regression is varies between: $+\infty$ **to** $-\infty$

Transformation of data:

- When data do not follow the normal distribution the variate x_1 be transformed to some new variate say f (x_i) in such a way that f(x) is normally and independently distributed with mean (N) & Variance (σ^2)

Types of transformation:

a. **Square root transformation:**
- Used when countable data have **small** values between **0-10** & when data follow Poisson distribution.
- It is also used for percentage data when all values are **> 80** %.
- E.g. Transformation = $\sqrt{x}$

b. **Logarithmic transformation:**
- Used when variation in the countable data is **large** and values are also large.
- E.g. Transformation = **log x** or log x+1

c. **Angular or aresine transformation:**
- Used for data following binomial distribution and when data are in percentage based on count values.
- The percentage are transformed the degree.
- E.g. **Sin = p/100**

Hypothesis:

- Any **statement** about the population parameter.

Test of hypothesis:

a. **Null hypothesis:**
- A statement about parameters which to be rejected after testing is known as null hypothesis.
- Given by **R.A. Fisher**
- It is denoted by $\mathbf{H_o}$, It is the hypothesis of **no difference.**

b. **Alternative hypothesis:**
- This is the hypothesis contrary to the null hypothesis.
- Denoted by $\mathbf{H_1}$.

Types of error:

a. **Type 1st error:**
- It denoted by **Alfa (α).**
- Rejecting $\mathbf{H_o}$ when it is true or accepting $\mathbf{H_1}$ when it is false.

b. **Type 2nd error:**
- Denoted by **Beta** (β)
- **Accepting** Ho when it is false, **rejecting** H_1 when it is true.
- It is more **severe** than type I error.

Level of significance:

- The maximum probability of committing type 1 error is known as level of significance denoted by **Alfa**
- Generally we take **5% (field ex.)** or **1%** level of significance.

Degree of freedom (D.F.):

- It is difference in total number of items (n) & linear constraints.
- **D.F. =** N- Total number of constraints.
- D.F. = N-K (total no. of constraints)

Test of significance:

a. Chi-square (x^2) test:

- First used by **Karl Pearson** (1900)
- It is used for **enumerated** or non parametric/qualitative data in **Mendelian** or population genetics.
- Range: o to ∞
- $x^2 = \sum(O_i - e_i)^2/e_i$

Conditions for applying x^2 test:

- Total sample size ≥ **50**
- Expected frequency of each and every class should not be **less than 5.**
- The sample observation should be independent.

Application of Chi-square:

- To test the goodness of fit
- For testing the hypothesis about the ratios for various classes
- Testing the independence of two attributes
- For detection of linkage
- To study gene frequencies in population genetics
- Test of significance of difference between two mean:

b. Z test/ student z test:

- Given by **R.A. Fisher.**
- Used when sample size is **large > 30.**

c. Student t- test:

- Given by **WS Gosset** 1905
- Used when the sample size is **small (< 30).**

Basic principle of field experimentation:

a. **Replication:**

- **Repetition** of treatment under investigation is known as **replication** so that each treatment experiences every type of environment in the field.
- It increases precision, to estimate experimental error.

b. **Randomization**:

- **Allocation** of treatment to the various experimental units by a random process
- It is **minimizes** the random variation.

c. **Local control**/error control:

- The process of **dividing** the experimental material into different groups called **blocks** to make them greater **homogenous** is known as local control/error control.
- It **reduce** the experimental error hence make the test of significance more **powerful**.

Treatment:

- The objects of comparison which an experiment has to try out in the field or elsewhere for assessing their values are known as **treatment**
- **E.g. Varieties**, **manures**, **cultivation practices**, methods of **seed treatment**, **insecticides** etc. are treatments used in case of **field experiment**.

Experimental unit:

- The group of **material** to which a treatment is applied in a single trial of treatment.
- E.g. A plot of land, a **patient** in hospital etc.

Experimental error:

- The **variation** in the data due to uncontrolled factor is known as experimental error.

ANOVA:

- **ANOVA** = Analysis of **variance**
- The concept by **R.A. Fisher**
- It is a statistical technique of partitioning the total variation into component variations & computing them by **F test**.

Experimental design:

- A local **structure** of an experiment that helps in obtaining results with some precision.

a. CRD (completely randomized design):

- It is regarded as **one way** classification and **no way** control or elimination
- Applied when the experimental material is **limited** and **homogenous**, such as the soil in **Pot** experiments.
- This provides the **maximum** number of degree of freedom for error, for a given number of treatments as compared to other design.
- Local control **not** used in CRD.

- Error degree of freedom in CRD: **N-n2**

b. RBD (randomized block design):

- **Two way** classification and **one way** control
- This design appropriate when the fertility gradient in the field is in **one direction** only
- It is used up to **20 treatments** without appreciable loss of efficiency.
- In RBD the most number of blocks must be **equal** to the number of replications fixed for ceach treatment.
- The number of plots in each block should be equal to the number of treatments.
- It provides more accurate results than CRD due to **formation** of homogenous blocks.
- It is **most commonly** used design in field experiment.
- This design **utilizes** all the three basic principles of field experimentation.
- Error degree of freedom in RBD: **(n-1)** x **(r-1).**

c. LSD (Latin square design):

- Three way classification two way control of error
- The term **square** has been used in LSD because the number of **rows** and **columns** are **equal**.
- This design used when fertility gradient is in **two directions**.
- Suitable for **5-8** treatments or from **5-12 at the most**
- Because the minimum error degree of freedom should be **12**
- **Shouldn't** be used for **less than 5** treatments
- Number of **replications** = Number of **treatment**.
- Number of **rows** = Number of **column** = Number of **treatments.**
- **Randomization** of treatments is done in such a way that each treatments occurs once and only once in **each row** and each **column**.
- Error degree of freedom in LSD: **(n-1) × (n-2)**

d. SPD (split plot design):

- A lay out in which **one set** of treatments is **assigned** to **large plots** called **main plots** in to which a replicate is divided and **another set** of treatments to **sub divisions** of the main plots called **sub plots** is termed as a **split plot design.**

This design used under following condition:

(i) When factors of the different nature are to be tested in same experiment that is level of one factor require larger area as compared to other factor e.g. **Depth of ploughing** and **nitrogen levels, date of**

sowing and **varieties,** varieties and nitrogen level, irrigation level and varieties.

(ii) When all factors aren't important

(iii) When the levels of one factor produce larger differences as compared with the levels of other factor.

(iv) When we want to study one factor with **higher precision** as compared to other factor and take it in sub plots.

e. SrPD (strip plot design):

- This design is used when both the factors **require** relatively **large area**.
- In this case **sub plot** treatment is applied **in strips** instead of randomizing in each unit.
- In this design number of error is **equal** to the number of plot size.

Confidence limit:

- The **range** within which the true **populations mean** lies is called confidence limit or **fiducially limit.**

Biometry:

- The application of statistical concepts and procedures to study of **biological problems.**

13. ANIMAL PRODUCTION

- ✓ Total livestock population in India (2009-10): ***529.7 Million***
- ✓ Share of world livestock **population** in India: ***15 %***
- ✓ **Cattle** population in India: > ***205 millions***
- ✓ Buffalo population: ***84 Millions***
- ✓ Cattle disease that is communicated to man is: ***Anthrax***
- ✓ Cow's milk is **light yellow** in color due to: ***Carotene***
- ✓ The milk is white in color due to presence of: ***Casein***
- ✓ The carbohydrate or sugar constituent of the milk: ***Lactose***
- ✓ Milk is deficient in element: ***Iron*** **(BHU Exam-2015)**
- ✓ Milk production is maximum in: ***Uttar Pradesh***
- ✓ The state which is second in milk production is: ***Rajasthan***
- ✓ Fat globule in the milk are in the form of: ***Emulsion***
- ✓ Protein present in the milk in the form of: ***Colloidal***
- ✓ Lactose and Minerals are present in milk in the form of: ***Solution***
- ✓ Milk is the poor source of: ***Iron and Vitamin-C***
- ✓ India **ranks first** in production of: ***Carpet wool***
- ✓ Best crop for silage making is: ***Maize***
- ✓ Temperature of artificial vagina: ***42 °C***
- ✓ First clone of adult (**sheep**) animal: ***Dolly***
- ✓ In India wool production is maximum in: ***Rajasthan (35%)***
- ✓ Disbudding in calf of newborn is done at the age of: ***4-10 days***
- ✓ **Swelling** due to the **collection** of gases in **rumen** of the animal is: ***Bloat***
- ✓ Protein content in colostrum of cow is 17.8% & in buffalo colostrum is: ***21.4%***
- ✓ Generally ruminant have ***four chambers*** stomach e.g.Cattle, sheep, goat, deer except **camel** that has ***three chambered rumen***.
- ✓ Milk is rich source of **Calcium**, **Phosphorus** and excellent source of **vitamin A**.
- ✓ India's position first in world **milk production**, total **buffaloes**, **wool production** and **total cattle production.**
- ✓ The world's first in vitro fertilized buffalo calf is: **Pratham**
- ✓ The most commonly used cry protectant for freezing of semen: ***Glycerol***
- ✓ ***Collagen*** is the main proteins of skin/connective tissues
- ✓ Light receptor pigment in animals is: ***Melanin***
- ✓ Cannibalism is seen in poultry due to: ***Sodium chloride deficiency***
- ✓ Cracked feet in poultry is caused by: ***Deficiency of biotin***

- ✓ Degnala and Alkali diseases are caused due to toxicity of: ***Selenium***
- ✓ Crazy chick disease is caused by deficiency of: ***Vitamin-E***
- ✓ Breed of poultry is excellent for broiler purpose: ***Plymouth Rock***
- ✓ A pregnant cow doesn't come in heat due to presence of: ***Corpus luteum***
- ✓ Animal protein factor is known as: ***Cobalamine***
- ✓ Fat content in double toned milk is: ***1.5 %***
- ✓ Hay can be stored if moisture per cent below: ***16 %***
- ✓ Legal standard for solid not fat (SNF) in cow milk is: ***8.5 %***
- ✓ Let down of milk in cow is due to: ***Secretion of Oxytoxine***
- ✓ Milk lipid is: ***Lecithin***
- ✓ Scientific castration is done by using an instrument known as: ***Burdizzo castrator***
- ✓ Only essential fatty acid of poultry is: ***Linolic acid***
- ✓ Glucose + Glucose = ***Maltose***
- ✓ Glucose + Fructose = ***Sucrose***
- ✓ Glucose + Glactose = ***Lactose*** *(Milk sugar)*
- ✓ Sulphur containing vitamins are: ***Thiamine and Biotin***
- ✓ Sulphur containing amino acids are: ***Cystine, Cysteine and Methionine***
- ✓ The best way to prevent fatal septicemia in young calves is to make sure that they get: ***Colostrum***
- ✓ The biggest compartment of ruminant stomach is: ***Ruman***
- ✓ The boiling point of milk is: ***100.17 ^{0}C***
- ✓ The main storage polysaccharide in animal cells is: ***Glycogen***
- ✓ ***Casein*** constitute about 80 % of the total protein present in milk
- ✓ The most variable constituent of milk is: ***Fat***
- ✓ The process of dehorning is accomplished by applying: ***Castic potash***
- ✓ The poultry reared for meat production is known as: ***Broiler***
- ✓ The protein which can't be digested: ***Elastin***
- ✓ Which is used for treatment of oxalate poisoning: **$CaCo_3$**
- ✓ Freshly drawn milk has a pH value of: ***6.5-6.7***
- ✓ The disease in which animals aborts or gives birth to a dead or weak calf is known as: ***Brucellosis***
- ✓ The deficiency of manganese in poultry causes: ***Perosis***
- ✓ ***Parakeratosis*** is caused by the deficiency of: ***Zinc***
- ✓ Oxidation of fat is called: ***Beta oxidation***
- ✓ Protein which is soluble in water and heat coaguable is: ***Albumin***

Pasteurization:

- It is done to kill all the **pathogenic** e.g. ***Clostridium botulinum*** and **most of** the **non-pathogenic** bacteria.

Methods:

- **Low Temperature Long Time** (LTLT): Exposing of milk at **61-63^{0}C** for **30** minutes.
- **High Temperature Short Time** (HTST): About **71-76.6^{0}C** for **15** seconds.
- **Ultra HighTemperature** (UHT): **148** ^{0}C **1- 5** seconds or **149** ^{0}C for **0. 5** sec. Followed by rapid cooling by **10** ^{0}C or below.

Common terms in livestock production:

Details	Cattle	Buffalo	Sheep	Goat	Pig	Poultry
Group	**Herd**	Herd	**Flock**	Flock/trip	Drove, herd	**Flock**
Adult male	Bull	Buffalo bull	Ram	Buck	Boar	Cock
Adult female	Cow	Buffalo cow	**Ewe** or Dam	**Doe**	Sow	Hen
Young male	Bull calf	Buffalo Calf	Male lamb	Male kid	Boarling	Cockrel
Young female	Heifer calf	Heifer calf	Female lamb	Female kid	Female piglet	**Pullet**
New born	Calf	Buffalo calf	Lamb	Kid	Piglet	Chick
Meat	**Beef,** calf-**veal**	**Buffin**	**Mutton**	**Chevon**	**Pork**	**Chicken**
Castrated male	Bullock Or steer	Bullock buffalow	**Wedder**	Wether	Hog, stage	**Capon**
Mating	serving	Serving	**Tupping**	Serving	**Coupling**	Serving
Parturition	Calving	Calving	Lambing	Kidding	Farrowing	Hatching
Sound production	Bellowing	Bellowing	Bleating	Bleat	-	-

Body temperature, respiration and pulse rate of domesticated animals:

Live stock	Body temperature		Respiration rate/Min.	Pulse rate/Min.
	(^{0}C)	(^{0}F)		
Cow	**38.5**	**101.5**	12-16	45-55
Buffalo	37.5	100	12-16	45-55

Goat	39.1	102.5	12-20	70-80
Sheep	39	102	12-20	70-80
Camel	37.5	99.5	-	-
Pig	39.2	102.6	-	-
Hen	**41.7**	**105-107**	15-30	300

Reproductive cycle of domestic animals:

Animals	Onset of Puberty (females) months	Length of estrous cycle	Length of heat period	Gestation period (in days)
Cattle	24-30	21 days	18 hours	282
Buffaloes	28-30	21 days	24 hours	316
Goat (Doe)	8-10	20 days	2-3 days	150
Sheep (Ewe)	9-14	15-18 (16.5) days	1-3 days	148
Sow (pig)	8-10	21 days	1-5 days	114
Camel	36-48	10 days	3-5 days	391

Composition of milk (% by weight) of different species:

Species	Water	Fat	Protein	Lactose
Cow	86.61	4.25	3.58	4.96
Buffalo	82.76	**7.38**	3.80	4.9
Goat	87.00	3.5	3.1	4.6
Sheep	79.50	5.3	**5.5**	4.6
Camel	-	4.9	3.7	5.1
Human	-	4.5	1.1	**6.8**

Nutrient:	**Bacterial fermentation**
▪ Carbohydrate	***Streptococcus lactis***
▪ Protein	***Bacillus subtilis***
▪ Fat	***Pseudomonas fragi***
▪ Conversion of milk into curd	***Lactobacillus thermophillus***

Classification of cattle breeds in India:

SN	Breed name	Native of	Features
A	**Dairy breed**		
1	**Sahiwal**	Montgomery (Pakistan)	**Highest and sweetest milk producing** cow breed in India. Due to loose skin known as ***'Lola'*** or ***multani***.
2	**Gir**	Kathiawar (Gujarat)	In Rajasthan found Ajmer & Bhilwara districts. Ears are long and pendulated and fore head is long and bony.
3	Sindhi or red sindhi	Karachi (Pakistan)	Red brown in color and intelligent facial expression.
B	**Draught breed**		
1	Malvi	Malwa (MP)	In Rajasthan found in Jhalawar district and in M.P. found in Mansour and Bhopal districts.
2	Nagori	Nagaur and Jodhpur (Raj.)	Superior among drought breeds. Color is white or grey
3	Ammritmahal	Mysore (Karnataka)	Narrow face long horns, grey colored body.
C	**Dual purpose** breeds		
1	Hariana	Hissar/Rohtak (Haryana)	Best dual purpose breed Horns are short & stumpy
2	Kankrej	Kutchh (Gujarat)	Highest body weight in Indian cows, powerful body, famous for ***'swaichal'***.
3	**Rathi**	Bikaner (Rajasthan)	Pure breed is found in Lunkaransar of Bikaner & Ganganagar districts of Raj.
4	**Tharparkar**	Sindh (Pakistan)	In Rajasthan found Jaisalmer & Jodhpur dist. Also known as ***white sindhi.***
5	Ongole	Guntur (Andhra Pradesh/KR)	Pure white in colour
6	Mewati	Kosi (Mathura)	Alwar and Bharatpur districts of Rajasthan

Exotic cattle breed:

- **Jersey**: Native of **English Chanel** (highest fat % in exotic breeds)
- **Holestein Friesian:**Native to **Holland,** Heaviest exotic breed, highest milk producing cow in the world (5000-6000 litre/year)
- **Frieswal** (Holstein x Sahiwal): High milk yielding breeds (14 lit/day)
- **Brawn Swiss:** Native of **Switzerland**

Hybrid developed form **NDRI, Karnal:**

- **Karan Fries:**(Tharparkar (cow)x H. Friesian)
- **Karan Swiss:**(Sahiwal (cow) x Brown Swiss)

Buffalo breeds:

SN	Name of breed	Native of place	Features
1	**Murrah**	Punjab, Delhi, Rohtak	Best breed of buffalo in world for milk production, av. fat 7%. ***Jet black*** colored body coat, short tightly curled horns. Skin is soft, smooth with scanty hairs.
2	**Bhadawari**	Agra (UP)	Highest Fat % (**13%), *Copper colored*** body coat, wedges shaped body.
3	Zaffrabadi	Kathiawar (Gujarat)	Heavy and loose body known as mini elephant, very prominent fore head and heavy horns
4	Nili Ravi	(Montgomery & Firozpur) Pakistan	White eyes known as '***Wall eyes'***
5	Mehsana	Baroda (Gujarat)	Developed from Murrah & Surti.
6	Surati	Baroda (Gujarat)	***Sickle shape horn*** is characteristic feature
7	Godavari	Andhra Pradesh	-
8	Toda	TN	Native of tribes of Nilgiri (TN), large sized animals

Common diseases of cattle:

- **Mastitis**: Inflammation of udder
- **FMD**: Viral disease
- **Riderpest**: Viral, known as cattle **plague**
- **Anthrax**: Bacterial disease, it is called **Spleenic fever**
- **Cow pox**: Viral disease.

Sheep breeds:

SN	Region	Breed name	Native	Remarks
1	Himalayan	Gaddi, Rampur-bushier, Bhakarwal	Jammu & Kashmir	**Fine quality** wool *Loei* is obtained from Bhakarwal
2	Western	Lohi	Pakistan	**Largest** Sheep
		Bikaneri/Pugal	Rajasthan	Best **Carpet** wool
		Marwari	PB & RJ	Black face
		Jaisalmeri	Rajasthan	Desert breed
		Malpura	Tonk, Rajasthan	Suitable for meat
		Chokla	**Shekhawati** region	Known as **'Merino'** of Rajasthan
3	Southern	Deccani	MH, KN, AP	**Good for mutton**, fit for weaving ***kambal***
		Nellore	Andhra Pradesh	**Tallest** breed
		Mandya	Karnataka	**White** in colour
4	Eastern	Shahabadi	Muzaffarnagar **(UP)**	Best wool producing breed in India.
		Hisardale	Developed at **Hissar**	Cross of **Bikaneri x Merino**

Exotic breeds of sheep:

Merino:

- Native of Spain
- **Fine wool** breed
- About **80%** of wool that is produced in the world comes from Merino and its crosses.

Rambouillet: (RPSC, AO-09).

- Native of **Spain**
- Dual purpose breeds.

Diseases of sheep:

- **Anthrax**: caused by Bacteria (**contangious** to man).
- **Stiff lamb disease**: caused by **Vitamin E** deficiency.

Goat Breeds of India:

✓ **Pashmina** used in winter clothes is obtained from the animal: **Goats**.

✓ **Morocco**: fine **leather** is prepared from **goat** skin.

SN	Name of breed	Native place	Features
1	**Jamunapari**	It is originated from city of barbera (somalia) in India found in Etawa (UP)	Dual purposes breed. Pendulated curved long ears; **'roman nose'** or parrot mouth appearance and thick long hairs on hind legs are major characters. **(RAS- 2013)**
2	**Barbari**	Somali (Africa)	**Dairy breed,** suitable for stall feeding (Known as city goat). Usually give birth **two kids (twins)** in one parturition
3	**Beetul**	Gurdaspur (PB)	Resembles as **'Jamunapari'**. Good dairy type breed, presence of **beard** in male
4	**Black Bengal**	Western Bengal	**Excellent for meat (RPSC, AO-09)** **Skin color is black, meat is tender**
5	**Marwari**	Marwar (Raj)	Disease resistant breed, good for meat, color is jet black
6	**Sirohi**	Sirohi (Raj)	Dual purpose breed, ears are flattened, drooping & leaf like
7	**Jhakrana**	Behror of Alwar (Raj)	Dairy breed large sized, beetle like in appearance.

Exotic breeds of dairy goat:

Angora:

- Native of Turkey
- Its hair used as wool known as ***Mohair***

Alpine:

- Europe (milk breed)

Anglo-Nubian:

- It is cross between Jamunapari and Nubian (England).
- Known as **'Jersey cow'**
- It is most dual purpose exotic breed of goat

Saanen:

- Native of Switzerland; presence of **beard** in **both** the **sexes** is the **characteristic** feature.

- This breed is also known as **Queen of milk**.

Toggenberg:

- **Chocolate** colored
- **Hornless** breed, skin is soft
- Native of **Switzerland**.
- Milk breed

Swine:

- Salted smoked meat of Pig is known as **Bacon**.

Exotic Breeds of swine:

- **Lindrace**: **white** in colour: native of **Denmark.**
- **Duroc**: **red** in colour: native of **USA.**
- **Yorkshire**: **white** in colour: native of **U.K.**
- **Chestor white:** white colored, U.S.A breeds
- **Hereford:** new breed of swine from U.S.A

Viral diseases of cattle:

• Swine fever	Pox
• FMD	Plague
• Blue tongue	New castle (Ranikhet)
• Leucosis	Rinderpest
• Marek's	Bronchitis

Fungal diseases:

• Bronchomycosis	Ringworm
• Brooder pneumonia	Aflatoxicosis

Protozoa diseases:

• Coccidiosis	Anaplasmosis
• Theileriasis	Trypanosomiasis (surra)

Bacterial diseases:

• Anthrax (*Bacilus anthrasis*)	Black quarter
• Brucellosis	Mastitis
• H.S. (*P. maltocida*)	Pneumonia
• Tuberculosis	Cholera
• Typhoid	

Poultry:

Poultry breeds:

SN	Name	Native	Features
1	Rod Island Red	Rhode island (England)	Dual purpose breed, produce 100 eggs/year, rectangular body
2	White Leghorn	Italy	Highest eggs production breed in the World (Av. 220 eggs/year)
3	Red Cornish	England	Famous in America for meat
4	Plymouth Rock	America	Produce best quality meat
5	Aseel	UP	Famous for **fighting sports**
6	Kadaknath	-	**Black** coloured meat

- A medium sized **egg** contains **6.0** g **protein, 6.5 g$CaCO_3$, 14g Vitamin.**
- **Andhra Pradesh** has largest number of poultry and is the largest producer of eggs in India.
- Colour of egg yolk is due to: **Xanthophyll.**
- Usually internal quality of eggs can be determined by: **Candling.**
- In our tropical climate necessary to construct poultry house **East to West** in order to **avoid** sun ray directly falling into the shed and heating up of the shed.

Important disease of poultry:

- **Bacterial** diseases: Fowl Cholera, **Diarrhea**
- **Viral** diseases: **Ranikhet,** Marek's disease
- **Protozoal** disease: **Coccidiosis**

Nutritional diseases:

- **Nutritional roup**: Vitamin **A** deficiency
- **Rickets:** Vitamin **D** deficiency
- **Curled toe paralysis**: Vitamin B_2 deficiency

Fish production:

- ✓ India is the **second** largest fish producer from **inland** sources after **China.**
- ✓ Maximum fish production is in **West Bengal.**
- ✓ Maximum **marine** fish production is in **Gujarat.**
- ✓ Largest contributor in agricultural **export**: **Marine products.**

Hay

- It is grass, legumes or other herbaceous plants that have been cut, dried and stored for use as animal fodder particularly for grazing animals.
- Lucern, berseem etc are used for hay making. (leguminaceae and thin stem crops suitable for hay making)
- Crops which have crude protein are suitable for hay making
- Storage moisture for Hay 15-18% (mostly < 16%)

Silage

- It is green fodder kept in air-tight silo, which get fermented and high moisture (60-65%) stored fodder which can be fed to cattle.
- Maize (most suitable), sorghum and oat are used for making silage
- Crops should have good percentage of sugar on their stem and appropriate moisture suitable for silage.

Terminology:

1. **Alveolus**: A small structure in mammary glands in which milk is manufactured by female.
2. **Anorexia**: Lack of appetite.
3. **Biological value**: Percentage utilization protein within animal body.
4. **Crude fiber**: Composed of cellulose, hemi cellulose, lignin etc. it is high in forage and other poly saccharides and low in grain.
5. **Culling**: The separation of diseased animals from healthy animals.
6. **Casting**: Throwing the animals on the ground
7. **Docking**: Removal of tail in sheep.
8. **Flushing**: Feeding extra concentrates to pregnant sheep is known as ***flushing.***
9. **Dehorning**: Suppression of emerging horn in goat at an age of 10 days.
10. **Mohair**: The hair of Angora breed of goats, soft white wool in great demand for clothing.
11. **Pashmina**: The undercoat of Kashmiri goats, soft wool in great demand for clothing.
12. **Pica**: An appetite for materials not usually considered to be food such as is observed in phosphorus deficient animals; a deprived appetite.
13. **Puberty**: The period of life at which the reproductive organ first become functional. This characterized by estrus and ovulation in the female and semen production in the male.
14. **Surra**: Common disease in camel caused by protozoa.

14. AGRICULTURAL ENGINEERING

- ✓ Father of agricultural engineering in India – **Professor Mason Vaugh**
- ✓ 1960-61– a tractor manufacturing started in india by **first manufacturer m/s Eicher good earth**
- ✓ Agricultural Institute, Allahabad (1942-43)
- ✓ At present agricultural machinery population is estimated at about 150 million which includes about 3 million tractor (**2.5 lakh tractor/year, 10000 power tiller /year)**
- ✓ Average man can develop maxium power of about **0.1hp (74.6 watts)** for doing farm work in a day
- ✓ Power developed by an average **pair bullocks** is about **1hp (746 wants)** for usual farm work in a day
- ✓ Power developed by a camel is about **1hp** for usual farm work in a day
- ✓ The averge force a **draft animal** can exert is nearly **one –tenth of its body weight**
- ✓ First bachelor's degree course in agricultural engineering in India – **Allhabad**
- ✓ Indian society of agricultural engineers (ISAE)– **1960 at IIT Kharagpur** and presently its headquarter is at New Delhi
- ✓ ISAE publications: 1) Journal Of Agaricultural Engineering (JAE), 2) Agricultural Engineering Today (ATE)
- ✓ Of the total energy produced in India the share of **thermal power** is about **56%, hydro electrical** Power (36%), nuclear power (16%) and rest comes from diesel and gas based power.
- ✓ The most popular tractor is found in **31-40 hp segment** , which accounts for 60% of the total sales in the country
- ✓ Thermal efficiency for diesel engine = **32 to 38%**
- ✓ Thermal efficiency for petrol engine= **25 to 32%**
- ✓ Unit power available for crop production (India) is about **1.35 kw/ha** but for desirable level of agricultural production power requirement could about **2.15 kw/ha.**

Biogas

- ❖ **Biomass**: mixture of **methane (45-70%)** and **carbon dioxide (30-35%)**
- ❖ **Buffaloes** : 15 kg dung/day, **bullocks or cows** : 10 kg dung /day and **calves** : 5kg dung/day
- ❖ Biogas calorific value = **4500 kcal/m^3**
- ❖ Suitable condition for biogas production= **pH (7-8), temp. (35^{0}C)**
- ❖ Cattle **dung : water** ratio for biogas slurry : **4:5or 1:1**

Wind energy

The speed required for operation of wind mill= **10 to 15 km/hr**

Two types of wind mill (**horizontal axis** and **vertical axis** rotor)

i. **Horizontal axis rotor** –axis of rotion is parallel to the direction of wind

i . Multi blade

ii. Propeller – most commonly used

iii. Sail type

Solar energy

- Temperature of sun = 5777^0k
- Radition range from sun :
  - **0.4 um** – maximum available range,
  - **<0.4 um** – ultraviolet radiation (8% of radiation),
  - **0.4-0.7 um** -visible radiation (46%)
  - **>0.7** um infrared radiation (46%)

Solar constant: solar radiation received per second by surface of unit area held normal to the direction of sun rays at mean earth–sun distance. Its value **1350** **w/m^2** or **1.94 cal/s/m^2.**

Radiation measurement-

- ✓ **Pyrheliometer** : beam radiation (direct radiation)
- ✓ **Pyranometer** : total radiation (global radiation)- accurate
- ✓ **Solarimeter** : total radiation (global radiation)- inaccurate
- ✓ **Pyrometer**: very high temperature measurement

Collection of solar radiation (three ways):

1. **By flat plate collector**: temperature range **40 centigrade** to **100 centigrade**
2. **Focussing or concentrating colletore** (>100 °C)
3. **Photovoltaic cell (solar cell):** directly convert **solar energy in to electricity** made of Silicon separated by thin barrier with conversion efficiency is about **10%.**

Salt content in water:

- **Portable water** contained **<550 ppm** of salt
- **Sea water** contained **30000-40000 ppm** of salt
- **Ground water** contained**<2000-3000 ppm** of salt

- **Solar still** –device converts saline water in to portable water.
- **Green house** is structure made of polythene/rain forced fibre to provide controlled condition for crop production.

15. MICRO BIOLOGY

- ✓ **Alexander Fleming (1929):** Discovered the world's first antibiotic '**Penicillin**' obtained from ***Penicillium notatum***, a mold (fungus).
- ✓ **Selman A. Waksman**: coined the term '**antibiotic** and discovered **streptomycin** (first treatment for tuberculosis in 1943) for this work he got '**Nobel Prize**' in 1952, that's why he called the '**Father of Antibiotics**'.
- ✓ **Anton van Leeuwenhoek:** invented the simple microscope in 1674, he was the first to report the existence of bacteria seen through his microscopes. He called them little '**animalcules**' and discovered microbial world with his microscope. He called **'Father of Microbiology'**
- ✓ **Louis Pasteur:** he discovered the **Germ Theory of Disease**, proving microorganisms cause **illness**, and invented **pasteurization** (heating liquids to kill microbes). He also developed the first vaccines for rabies and anthrax. He gave the principles of vaccination, microbial fermentation and pasteurization that's why he known as founder of modern microbiology and bacteriology.
- ✓ **M. W. Beijerinck (1898)**: he named 'virus' (infectious poison agent) and identified viruses as unique, filterable agents causing tobacco mosaic disease and demonstrated their ability to replicate within living cells, coining the term contagium vivum fluidum that's why he is considered the **"Father of Virology.**
- ✓ **J.O. Brefeld (1872-74)**: Developed **pure culture techniques** for culturing and isolation of micro-organisms (fungi).
- ✓ **Edward Jenner**: Developed the world's first vaccine by inoculating a boy with cowpox matter, protecting him from smallpox. This established the foundation for modern immunology, leading to the term 'vaccination' and proposed the Principle of Immunization. He is called '**Father of Immunology'.**
- ✓ **D. Iwanowsky (1892)**: Discovered viruses by showing that the infectious agent causing tobacco mosaic disease (TMV) was a '**filterable agent**'.
- ✓ ***Ernst Haeckel:*** proposed the kingdom **Protista** for single-celled organisms (including bacteria, fungi, protozoa) to fix limitations in two-kingdom systems. He coined certain terms like **Ecology**, **Phylogeny**, **Ontogeny**, and **Metazoa.**
- ✓ **M.J. Berkeley** (1803-1889): proved the fungus (*Phytophthora infestans*) cause of the **Great Irish Potato Blight** around 1845, supported the germ theory of disease and identifying numerous other plant-pathogenic fungi, establishing mycology as a vital field.

- ✓ **Hesse** : First introduced agar-agar as a solidifying agent in culture media
- ✓ **Petri:** Designed and developed glass dishes known today as **petridishes.**
- ✓ **John Tyndall (1820-83):** he disproving spontaneous generation by showing germs are in dust, developing a dust-free chamber to prove air doesn't create life, and discovering heat-resistant bacterial forms (endospores), leading to **Tyndallization,** a sterilization method using intermittent heating to kill them.
- ✓ **Joseph Lister** (1827–1912) was a British surgeon whose pioneering work on antiseptic surgery by using carbolic acid (phenol) to kill germs, drastically reducing surgical infections and mortality, effectively becoming the father of **modern surgery** and laying groundwork for modern aseptic techniques.
- ✓ **V. Behring** and **S. Kitasato** (1890): established **serum therapy**, showing that blood serum from immunized animals could neutralize toxins and cure diseases like **tetanus** and **diphtheria**, for this earned Behring the first Nobel Prize in Medicine (1901). They identified **'antitoxins'** (antibodies) and proved immunity could be transferred, laying foundations for vaccines and treatments, with Kitasato also pioneering pure bacterial cultures and plague research.
- ✓ **S. Winogradsky** (1856–1953): The father of **soil microbiology** OR **founder of microbial ecology,** revolutionized the field by discovering chemosynthesis (energy from inorganic compounds, not light) and elucidating biogeochemical cycles, particularly the **nitrogen cycle**, by isolating nitrifying bacteria and showing they convert ammonia to nitrite and nitrate.
- ✓ **Charles Chamberland** (1884), He developed the **Chamberland filter**, a porous **porcelain filter** for removing bacteria and the **Chamberland autoclave**, crucial for sterilization, both vital tools for Pasteur's work and for establishing germ theory, leading to vaccine development and water sanitation, and even hinting at the existence of viruses.
- ✓ **Elie Metchnikoff** (1845–1916): discovered **phagocytosis** (cells engulfing pathogens), establishing cellular/**innate immunity**, pioneering the use of probiotics (beneficial gut bacteria like **Lactobacillus**) for health, and founding the field of **gerontology**, linking aging to microbes and proposing ways to extend life. Awarded the 1908 Nobel Prize for his work on phagocytosis and regarded as the **'father of innate immunity'.**
- ✓ **Frederick Twort and Felix d'Herelle** (1915-1917): discovered **bacteriophages** (viruses that infect bacteria) around, with Twort observing their effects and **d'Herelle** coined the term **phage therapy**,

using phages to treat bacterial infections like dysentery, pioneering the use of viruses as medicine and laying groundwork for modern phage research and antibiotic alternatives.

✓ **Robert Koch** (1843–1910): he regarded as the **founder of modern bacteriology**. His work provided the definitive scientific proof for the **germ theory of disease** by demonstrating that specific microorganisms are responsible for specific infectious diseases.

Key Contributions to Microbiology-

1. **Discovery of Pathogens:** Koch identified and isolated the causative agents of major diseases like **Anthrax (*Bacillus anthracis, 1876*), Cholera (*Vibrio cholera, in 1883*) and Tuberculosis (*Mycobacterium tuberculosis in 1882*).** He isolated the ***tubercle bacillus***, a feat for which he received the **Nobel Prize in Physiology or Medicine in 1905**.
2. **Koch's Postulates:** He formulated a set of four rigorous criteria to establish a causal relationship between a microbe and a disease:
   (i) The microbe must be present in every case of the disease but absent from healthy individuals.
   (ii) The microbe must be isolated from the diseased host and grown in **pure culture**.
   (iii) The disease must be reproduced when the cultured microbe is inoculated into a healthy, susceptible host.
   (iv) The same microbe must be re-isolated from the experimentally infected host.
3. **Laboratory Techniques:** Koch revolutionized how bacteria are studied by introducing essential methods:
   (i) **Solid Culture Media:** to grow distinct bacterial colonies, a method known as **'cloning'** bacteria.
   (ii) **Staining Methods:** He introduced the use of aniline dyes (like methylene blue) to make transparent bacteria visible under a microscope.
   (iii) **Microscopy Improvements:** He was the first to use an **oil immersion lens** and an Abbe condenser to enhance resolution, and he pioneered **microphotography** for documentation.

✓ **Whittaker 5 kingdom classification-**

1. **Monera** – Prokaryotes
2. **Protista** – Unicellular eukaryotes
3. **Plantae (Photo syn.)** – Multicellular plants and higher algae.
4. **Fungi (adsorption)** – Multinucleate higher fungi.
5. **Animalia** (ingestion)

❖ **BERGY's manual of systematic bacteriology – is the standard for taxonomy-**

1. Eukaryotic microbes: Protozoa, fungi
2. Prokaryotic microbes : Bacteria, actinomycetes, BGA

❖ **Size (approx.):**

1. Bacteria – 0.5-3.0 microns
2. Fungi – 1.5-10 micronss
3. Protozoa – 2-200 microns
4. Viruses – 0.02- 0.4 microns or mostly 20-400 nm
5. MLO –0.1-0.3 micron
6. Algae – 0.1 micrometer (BGA) to 10 feet (higher algae)

Antibiotics:-

- Penicillin – *Penicillium* sp – G + ve; cellwall synthesis
- Tetracycline - *S. aureofacines* – G + ve, G-ve; Protein synthesis
- Bacitracin – *Bacillus subtilis* – G + ve, cell wall synthesis
- Chloramphenicol – *S. venezuelae* G + ve; G-ve, Protein synthesis
- Cycloheximide – *S. venezuelae,* protein synthesis, Eukaryote
- Nystain – *S. nouresii,* Eukaryote membrane
- Erythromycin – *S. nouresii* – fungi – Prokaryote
- Erythromyci – *S. erythreus* G + ve, G-ve, Protein synthesis
- Neomycin – S. *fradiae* G + ve, G-ve, protein synthesis

Microscopy:

The title of 'father of microscopy' or 'father of microbiology' belongs to **Antonie van Leeuwenhoek.**

4. **Dark field microscope:**
   - Applications- For gross morphology in the living specimen

2. **Phase contrast microscope:**
   - Applications – For revealing cellular structures in living cells

3. **Bright field:**
   - Used for studying gross morphology of yeasts, molds, a:gae etc.

4. **U V microscope:**
   - Application – For differentiating cellular components

5. **Fluorescence microscopy:**
   - Used for detecting specific types of antigens using an antibody tagged with fluorescent dye

6. **Electron microscope:**
   - Uses electromagnetic lenses and an electron beam.

Importat points:

- Bacteria not responsible for N-fixation- ***E.coli***
- Co-factor needed for nitrogen fixation – **Cobalt (Co)**
- Elements involved in nitrogen fixation – **Molybdenum**
- Leguminous crop not fixing nitrogen –**Rajma (*Phaseolus vulgaris*)**
- Nif gene is associated with – *Rhizobium bacteriod*
- Nitrogen fixation in rice field occurs due to presence of association of Anabaena (BGA) and azolla (water fern)
- Nitrogen fixing enzyme – **Nitrogenase** (First discovered in 1960 from *Clostridium pasteurianum*).
- Nitrogen fixing gene – **Nif genes**
- Non legumes N fixing plants – ***Alnus, Casuarina, Myrica, Frankia sp.***
- Bacteria involve in phosphorus cycle – ***Bacillus, Pseudomonas, Micrococcus, Flavobacterium, Aspergillus, Penicillium, Fusarium***
- Red pigment in the root nodules is known as **Leg haemoglobin**
- Stem and root nodules are present in – **Sesbania, *Azorhizobium caulonicans***
- Bacteria involve in sulphur cycle–Thiobacillus, Arthrobacter, *Desulfovibrio desulfuricans.*

16. BIOCHEMISTRY

- Term biochemistry derived from Greek word 'bio'-meaning 'life' and chemical 'relating to chemistry.
- The name biochemistry was coined by a german chemist **Carl Neuberg** in 1903. He is known as father of **modern biochemistry**.
- Biochemistry is the study of chemical substances and processes within and relating to living organisms.

CARBOHYDRATES-

- Carbohydrate is a biomolecule consisting of carbon, hydrogen and oxygen atoms.
- These are polyhydoxy aldehydes or polyhydroxy ketones the substances which yield these derivatives on hydrolysis.

Functions of carbohydrates-

- These are provided energy and stored energy to our body.
- These are structural constituents, build macromolecules.
- Spare protein and fat for other uses.

Classification of carbohydrates- 3 major classes

- Monosaccharides
- Oligosaccharides
- Polysaccharides

5. Monosaccharides-

- These are simplest sugar or carbohydrates, since they cannot be hydrolyzed into smaller units.
- These are aldehydes or ketons possessing two or more hydroxyl groups.

(a) Aldoses-

- Aldotrioses e.g. glycerose
- Aldotetroses e.g. erythrose
- Aldopentoses e.g. ribose
- Aldohexoses e.g. glucose (dextrose), galactose, mannose
- Aldoheptose e.g. glucoheptose (D-gulo-aldoheptose)

(b) Ketoses-

- Ketotrioses e.g. di-hydroxy-acetone
- Ketotetroses e.g. erythrulose
- Ketopentoses e.g. ribulose
- Ketohexoses e.g. fructose
- Ketoheptose e.g. sedoheptulose

6. **Oligosaccharide-**

- These are polymers of monosaccharides (simple sugars) containing two to ten residues accumulate in vacuole.

(a) **Disaccharides:** These sugars are formed by two monosaccharide units.

(i) **Reducing disaccharides**: maltose (glucose+glucose), lactose (glucose + galactose), and cellobiose.

(ii) **Non- reducing disaccharides**: sucrose (glucose + fructose).

(b) **Tri-saccharides**: These sugars are formed by three monosaccharide chains e.g. raffinose (glucose + fructose + galactose), melezitose.

(c) **Tetrasaccharides**: These sugars are formed by four molecules of monosaccharides e.g. stachyose (1 molecule glucose + 1 fructose+2 galactose)

3. **Polysaccharides:**

- These are long chain polymeric carbohydrates composed of many units of monosaccharides.
- These served as structural components of the cells, also serve as nutrients and stored form of energy.

Classification of polysaccharides on the basis of function-

(a) **Storage** e.g. starch (in plants), glycogen (in animals)

(b) **Structural** e.g. cellulose, hemicellulose, pectin.

(c) **Others e.g.** pentosan, hexosans, gum, agar etc.,

NUCLEIC ACID

- Nuclic acids are polymers of nucleotides joined by phosphodiester bonds formed between their sugar and phosphate groups.

Functions of nucleic acids:-

- Transmission of hereditary characters from parents to offsprings.
- Synthesis of proteins

Components of nucleic acids:-

4. Sugar e.g. ribose or deoxyribose
5. Phosphate group (phosphoric acids)
6. Nitrogenous bases e.g. adenine, guanine (**purines**) and thymine, cytosine, uracil (**pyrimidines**).

Nucleoside: (nitrogenous base + sugar)-

- They consist of a purine (adenine, guanine) or pyrimidines (thymine, cytosine and uracil) and a ribose or a deoxyribose sugar joined by beta glycosidic linkage.
- Pyrimidine has a single nitrogen containing ring and purine each has two nitrogen containing rings.

Nucleotide: (nucleoside + phosphoric acid)-

- A nucleotide contains a nitrogenous base (one of four adenine, thymine, guanine and cytosine) + a molecule of sugar + one phosphoric acid.

Types of nucleic acid:-

1. Ribonucleic acid (RNA)
2. Deoxyribo nucleic acid (DNA)

Ribonucleic acid (RNA):-

- It is present in all living cells.
- It is a molecule that plays an important role in the coding, decoding and expression of genes.
- It is generally single stranded and has ribose sugar while DNA is double stranded molecule which has deoxyribose sugar.
- It works as protein synthesis via translation and it carries genetic information that is translated by ribosomes.
- It mostly occurs in cytoplasm.

Types of RNA-

(i) Transfer RNA (t-RNA)
(ii) Messenger RNA (m-RNA)
(iii) Ribosomal RNA (r-RNA)

Deoxyribonucleic acid (DNA)-

- It is hereditary material in almost all organisms.
- It is essential for the synthesis of protein.
- It is composed of pentose sugar (deoxyribose), phosphoric acid and nitrogenous base.
- It is present in nucleus and organells like chloroplast and mitochondria.
- Cell organelles like ribosomes, peroxysomes, lysosomes and vacuoles don't have DNA.

AMINO ACIDS-

- Amino acids are organic compounds that contain both amino and carboxylic acid groups.
- Each amino acid connects to another by a covalent bond, known as ***peptide bond***.
- Methionine and cysteine are the two sulphur containing amino acids in mammals.
- Amino acids are molecules that form proteins.
- Generally 20 amino acids required by our body among these 10 are essential amino acids and remaining 10 are non essential.

Essential amino acids:-

- These amino acids can not be synthesized in body therefore they externally supplied with food.
- These are ten in number i.e. tryptophan, valine, methionine, isoleucine, leucine, lysine, phenylalanine, arginine and threonine (remember tricks '***TVMILLPATH***').

Non essential amino acids:-

- These amino acids can be synthesized by our body and are not taken up as food supplements.
- These include alanine, asparagines, aspartic acids, cysteine, glutamic acid, glutamine, glycine, proline, serine and tyrosine.

Functions of amino acids-

- Formation of protein
- Build muscles
- Grow and repaire body tissue
- Formation of enzymes, hormones and antibodies.

PROTEINS

- Proteins are complex, organic compounds composed of many amino acids linked together through peptide bonds.
- Proteins are required for the structure, function and regulation of body's tissues and organs.

Classification of proteins-

Based on composition-

1. **Simple protein: -**
   - On hydrolysis they yield only the amino acids and some carbohydrate compounds e.g. albumins, globulins, glutins, albuminoids, histones and protamines.
2. **Conjucated proteins:** amino acid + prosthetic group
   - These are simple protein combined with some non-protein material in the body e.g. nucleoproteins, glycoproteins, phosphoproteins, haemoglobins, chromoproteins, lipoproteins and lecithoprotein.
3. **Derived protein:-**
   - These are derived from simple or conjugated proteins e.g. denatured proteins and peptides.

LIPIDS

- **Lipids** are defined as esters of glycerol and fatty acids or as the triglycerides of fatty acids.

- Lipids include fats, waxes, sterols, fat soluble vitamins, monoglycerides, di-glycerides and phospholipids.
- **Phospholipids**- fatty acids + glycerol + phosphoric acids + nitrogenous base e.g. lecithin.
- **Glycolipids**- glycerol + fatty acids + carbohydrates
- **Lipids** (triglycerides) = glycerol + fatty acids.

The functions of lipids:-

- Storing energy
- Signally
- Acting as structural components of cell memberanes

Fatty acids-

- Fatty acids are the building blocks of the fat in our bodies.
- Fatty acid molecules are usually joined together in groups of three, forming a molecule called a ***triglyceride***.

Essential fatty acids:

- They can not be synthesized by the body and must be supplied from other source.
- These are four in number (Remember tricks ***'OLLA'***)
- *Oleic acid, Linoleic acid, Linolenic acid, Arechidonic acid.*

Non essential fatty acids-

- They can be synthesized by our body e.g. stearic acid, palmatic acid and arachidic acid.

Saturated fatty acids:-

- Fatty acid chains have all single bond, chain is saturated e.g. lauric acid (12C), Myristic acid (14C), Palmatic acid (16C), Butyric acid (4C), capric acid (10C), Stearic acid (18C), Arachidic acid (20C).

Unsaturated fatty acid:

- These are a component of the phospholipids in cell membranes and help maintain membrane fluidity.
- These have atleast one double bond in the fatty acid chain.
- Examples- myristoleic acid, palmitoic acid, oleic acid, linolenic acid and linoleic acid
- Source- avocado oil, olive oil, peanut butter, sunflower oil, nuts and seeds, such as almonds, peanuts, cashews and sesame seeds.

Vitamins and enzymes:

- The term **vitamin** was introduced by: **Funk** (1912)
- **Vitamins** are accessory food factors **play vital** role in growth, nutrition and metabolic process.
- **Fat soluble** vitamin: **A, D, E & K.**

- **Water soluble** vitamin: **B & C.**
- **Folic acid** is essential for maturation of **RBC** (Red blood corpuscles).
- The largest concentrations of **minerals** are found in **bones** and **teeth.**
- **Calcium** is essential component of all living cells, helps in **body building**.
- The term enzyme was coined by: **W. Kuhne** (1898)
- **Enzymes** are those **organic substances** (proteins) which are capable of **catalyzing** chemical reactions in living system.

Deficiency, their effects and major source of vitamins

Vitamin	Effect/deficiency	Source
A (Retinol)	***Night blindness***, growth failure, ***Xerophthalmia***	Liver is rich source, milk, carrot, papaya, mango etc.
B_1 (Thiamine)	***Beri-beri*** (disturbance in digestive system)	Rice polishing, leafy veg. fish, egg, meat
B_2 (Riboflavin)	***Photophobia***, inflamed lips, cracks on skin/***ariboflavinosis***	Milk powder, fish, pulses
B_3 (Niacin or Nicotinamide)	***Pellagra***/black tongue	
B_5 (pentothonic acid)	Necessary to making blood cells	
B_6 (Pyridoxine)	Degeneration of nerves	Yeast, wheat germ, legumes
B_{12} (Cobalomine)	***Pernicious anemia***	Animal food (liver, meat, fish, egg)
C (Ascorbic acid)	***Scurvy*** (soft tissues around the Joints become swollen and painful)	Aonla, guava, citrus fruit
D (Calciferol)	***Rickets*** (bone deformation), ***Pigeon chest*** in children & ***Osteomalacia*** in adults	Sun light, fish & liver oil
E (Tocopherol)	***Sterility*** (reproductive failure)	Outer layer of grains
H (Biotin)	*Dermatitis*	
K (Phylloquinone)	***Delayed blood clotting***	Cauliflower, liver, soybean
Folic acid	***Anaemia***	Palak, karonda, green leafy veg.

Growth substances and their function:		
SN	Hormone	Functions
1	Auxin	Named by F.W. Went, It has apical bud dominance, inhibit root elongation and promote cell division in ***tissue culture***
	a. IAA	Prevent premature fruit drop, it is **natural** auxin hormone
	b. IBA	Mainly used for **Root formation**
	c. NAA	Fruit thinner, **prevent fruit drop**
	d. 2,4-D	Fruit setting hormone (<20 ppm), and also used as a **herbicide** (>20ppm)
	e. Dicamba	It is active herbicide
2	Cytokinin	promote **cell division**
3	Gibberellins	**Stimulate Cell elongation**, increase fruit size, seed dormancy breaker
4	Abscisic acid (ABA)	Dormancy inducer, stomata closer, also called as **anti-gibberellins**.
5	Ethylene	**Ripening of fruits**, isodiametric growth of stem & roots Available in two forms:- (i) **Ethrel**: - it is liquid form of ethylene and used for banana fruit ripening. (ii) **Ethephone**: - it is gaseous form of ethylene and used for early ripen of sugarcane, wheat, coffee, tobacco and rice.
6	Clormequat (CCC)	Lodging preventer, also known as **cycocel**
7	Mallic hydrazide	Growth retardant, prevent onion sprouting in storage, sold under trade name ***Sproutstop***

Minerals	Deficiency symptoms
Calcium:	***Rickets & milk fever**(in animals)*
Phosphorous:	***Rickets/Osteomalacia & Pica*** *(in animals)*
Iron:	***Anaemia***
Zinc:	***Parakeratosis***
Manganese:	***Paresis***
Iodine:	***Goiter***
Cobalt:	***Coast disease***
Minerals	**Excesses (toxicity)**
Mo:	***Molybdenosis***
Fluorine	***Fluorosis***
Selenium:	***Degnala/alkali disease*** *(in animals)*

Elements:	Disease
Ca & Mg oxalate	*Kidney stone*
Lead	*Mental retardation*
Fluoride	*Fluorosis*
Arsenic	*Poisoning*
Iron	*Constipation or Dysentery*

Common medicinal plants and their uses:

Common Name	Botanical name	Part used	Chemical Content	Uses
Belladona	***Atropa belladonna***	Leaves	**Atropine,** scopolamine	Belladonna is highly toxic in its natural form **known as deadly nightshade.** Atropine is a crucial antidote for poisoning by certain insecticides that overstimulate the nervous system.
Ashwagandha	***Withania somnifera***	Roots	Withanine	General medicine for stress relief, improving sleep, boosting cognitive function, and enhancing physical stamina by lowering cortisol, but also for inflammation, blood sugar, fertility (male), and supporting conditions like arthritis
Sarafgandha	***Rauvolfia serpentine***	Roots	**Serpentine Reserpine**	Treat hypertension High blood pressure
Safed musli	***Chlorophytum spp.***	Roots	**Saponin** (sapogenins)	Diabeties, Known as **2nd Shilajeet**, because it has Aphrodisiac Properties for both male and females.
Opium poppy	***Papaver sominiferum***	Fruit (latex)	Morphine Codine Narcotine	Painkiller, god of sleep, cure leukemia Drug **'Heroin'** prepared from morhine
Isabgol	***Plantago ovata***	Husk, Seed	Mucilage	Laxative, used against irritation in

				gastrointestinal tract
Guggal	***Commiphora wightii***	Gum, Resin	Guggulipids	Treat obesity, arthritis Reduce cholesterol
Shatavari	***Asparagus racemosus***	Roots	**Saponins** (Shatavarin)	Useful in female Reproductive Health, stomach ulcers and diarrhea. Male Health Adaptogenic and Immune Support,

Other Medicinal and Aromatic plants and their uses:

Common Name	Botanical name	Part used	Chemical Content	Uses
Lemongrass (Two best varieties are – **Pragati & Sugandhi)**	*Cymbopogon flexuosus*	Stalks, leaves & roots	Oil is rich in **Citral** (isomer of **Geranial**)	It has a strong lemon aroma and flavor, used in perfumes, foods, and cosmetics. It has anti-inflammatory, antibacterial/ antiviral properties.Some species produce **Citronella** oil which is used for insect (mosquitoes) repellent.
Vetiver (khus)	*Vetiveria zizanioides*	**Roots**	**Vetiverone** & vetiverols	Its fragrant roots used in perfumes, cosmetics, food and making ***'khus sarbat'***. Grass is used in cooler pads for cooling. This plant also serves roles in soil/water conservation (Vetiver Grass System - VGS) and **phytoremediation**.
Holy Basil/Tulsi	*Ocimum sanctum*	**Leaves, seeds, roots**	**Eugenol**-an essential oil	**Adaptogenic** and anti-stress, **antimicrobial**, **anti-inflammatory**, and **antioxidant** properties. Grown in homes for its spiritual significance, often placed in the east or northeast for prosperity and it purifies the air, brings positivity.
Lavender	*Lavandula angustifolia*	Flowers (buds),	**Linalool** and Linalyl	**Aromatherapy:** Essential oils from flowers

		leaves, stem	acetate	and leaves are used to reduce anxiety, stress and restlessness. **Culinary:** Flowers and leaves add flavor to baked goods, drinks, soups, and salads. **Home Fragrance:** Dried flowers in sachets or potpourri. **Bath & Body:** used in oils, soaps, or baths.
Mint	*Mentha sps.*	Leaves	**Menthol**	**Aromatherapy and Flavoring:** used in flavorings for toothpaste; chewing gum, candies and beverages due to its refreshing taste and smell **Medicinal Uses**
Sandalwood	*Santalum album*	Heart-wood, roots	**Santalol**	Relaxation and Stress Relief, Sleep Aid, Meditation and Focus, Perfumes and Cosmetics, Incense and Room Fresheners
Rosemary	*Salvia rosmarinus*	Leaves, flowering tops	**Carnosic acid (carnosol), rosmarinic acid.**	Anti-inflammatory, Pain Relief, Digestive Health, Hair Growth, Antioxidant, Stress and Mood Regulation.
Gudhal/ Hibiscus	*Hibiscus rosa-sinensis*	Flowers and leaves	**Anthocyanins** (sambubioside) and **phenolic acids**	*Hibiscus sabdariffa* (Roselle) are used to make a popular, tangy **Cardiovascular Health:** Hibiscus herbal tea (Karkade) is known for its ability to help manage **high blood pressure** and cholesterol levels.
Gurmar/ Madhunashini	*Gymnema sylvestre*	Leaves	**gurmarin** (gymnemic acids)	**Diabetes management** It is known as **'sugar destroyer'** medicine for managing blood sugar. **Suppressing sweet cravings**. Roots are remedy for **snakebites.**